THE FRESHWATER ECOSYSTEMS OF SURINAME

MONOGRAPHIAE BIOLOGICAE

VOLUME 70

Series Editors

H.J. Dumont and M.J.A. Werger

The titles published in this series are listed at the end of this volume.

The Freshwater Ecosystems of Suriname

Edited by

PAUL E. OUBOTER

SPRINGER-SCIENCE+BUSINESS MEDIA, B.V.

Library of Congress Cataloging-in-Publication Data

The Freshwater ecosystems of Suriname / edited by Paul E. Ouboter.
p. cm. -- (Monographiae biologicae ; v. 70)
Includes indexes.
ISBN 978-94-010-4918-4 ISBN 978-94-011-2070-8 (eBook)
DOI 10.1007/978-94-011-2070-8
1. Freshwater ecology--Surinam. 2. Freshwater biology--Surinam.
3. Limnology--Surinam. I. Ouboter, Paul E. II. Series.
QH125.6.F73 1993
574.5'2632'09883--dc20 93-5180

ISBN 978-94-010-4918-4

printed on acid-free paper

Contents

Swamp forest destroyed by a peat fire, Cusewijne area (Photo P.E. Ouboter).

Preface

This book on the freshwater ecosystems of Suriname is first and foremost a compilation of the knowledge on the subject. It is, however, also a celebration. A celebration of the birthday of the largest scientific institute of Suriname. On the first of November 1968 the University of Suriname was born with the inception of the faculty of Law; afterwards a faculty of Medicine, a faculty of Social and Economic Sciences, a faculty of Natural Resources, and finally a faculty of Engineering were established. On 1 November 1993 the University of Suriname will celebrate its 25th anniversary. There will be a comprehensive jubilee programme, yet the Board of the University expressed the desire to have a book published showing the results of scientific research done in Suriname.

During the 25 years of its existence, the University of Suriname has played an important role in educating technicians and scientists, in conducting research, and providing services to the Surinamese community. Notwithstanding the limited resources, scientists in Suriname did research, and published articles in local and international journals. In order to single out their work and our university, we initiated the publication of a book on a subject which is important for Suriname, our geographical region, and the rest of the world.

Compared to other fields of biology, relatively much research has been done on aquatic ecosystems in Suriname. Part of the knowledge on this subject has been published in scattered publications, part remains hidden in the minds of individual researchers or in internal reports. The aim of this book is to provide a survey of this knowledge in a form which is easily accessible to all scientists interested.

We realize that this publication is far from being complete. Undoubtedly, some very important topics on the Surinamese freshwater ecosystems will be missing. We therefore hope that this publication will also be an incentive for scientists to continue research in this field, and to publish as a follow-up, a supplement to this publication. Hopefully, we will be able to fill some of the remaining gaps in knowledge in this field in the coming years.

We would like to thank the chairman of the Publishing Commission, Ine Tsai-Meu-Chong, for her support of this project. Wilma Shewnaik typed up and corrected many of the manuscripts. Mia Quik-Stregels provided bibliographic

Paul E. Ouboter (ed.), Freshwater Ecosystems of Suriname, vii–viii.

assistance. Shamita Sahdew contributed some fine drawings for the Chapters 4, 9, 12, 13, 15 and 16. The Translation Department of the Ministry of Education transformed the English writing in a seemingly acceptable product. The project was supported by a grant from the Beyerinck-Popping Fonds. The editor was inspired by several people to fulfil his task, of whom I especially wish to mention Shamita Sahdew, who incited me to start the project of this book, and Jan Mol, who from the beginning discussed almost all aspects with me. We also would like to thank the publisher of this book, who made the book look like a work of art.

Finally, we like to express our hope, that the publication of this book will urge scientists, both local and foreign, to conduct more research in Suriname.

Paramaribo, Suriname, April 1993

The president of the Board
of the University of Suriname
Prof. Dr. Baltus F.J. Oostburg.

The editor
Paul E. Ouboter

List of contributors

AMATALI, M., Hydraulic Research Division, Ministry of Public Works, Duisburglaan, Paramaribo
DE DIJN, B.P.E., National Zoological Collection, University of Suriname, P.O.B. 9212, Paramaribo, Suriname
HARIPERSAD-MAKHANLAL, A., Hydraulic Research Division, Ministry of Public Works, Duisburglaan, Paramaribo
HOOGMOED, M.S., National Museum of Natural History, P.O.B. 9517, 2300 RA Leiden, The Netherlands
LEENTVAAR, P., Sweelincklaan 10, 3723 JE Bilthoven, the Netherlands
MOHADIN, K., Department of Nature Preservation, Suriname Forest Service, P.O.B. 436, Paramaribo, Suriname
MOL, J.H.A., CELOS, University of Suriname, P.O.B. 9212, Paramaribo, Suriname
NOORDAM, D., Stichting voor een Schoon Suriname, Naarstraat 4, Paramaribo
OOSTBURG, B.F.J., University of Suriname, P.O.B. 9212, Paramaribo, Suriname
OUBOTER, P.E., National Zoological Collection, University of Suriname, P.O.B. 9212, Paramaribo, Suriname
PEETERS, G.M.T., National Zoological Collection, University of Suriname, P.O.B. 9212, Paramaribo, Suriname
TEUNISSEN, P.A., Laan v. Louiseshof 47, Paramaribo, Suriname
WASSCHER, M., National Museum of Natural History, P.O.B. 9517, 2300 RA Leiden, the Netherlands
WERKHOVEN, M.C.M., National Herbarium, University of Suriname, P.O.B. 9212, Paramaribo, Suriname

Introduction

PAUL E. OUBOTER

Since the arrival of the Spanish and Portuguese explorers and conquerors in the 15th and 16th century, Suriname has been a colony of various European nations. The Dutch, who were at first mainly traders (especially in negro-slaves), started to found trading-stations and plantations at several locations along the Guianan coast at the end of the 16th century. During most of the period between 1667 and 1975, Suriname was a Dutch colony. Consequently, most of the explorations were carried out by the Dutch. However, the first naturalists to study Suriname were all German and French, among whom Maria Sybilla Merian who arrived in 1699. In those days most naturalists collected plants and animals as a hobby, not professionally, and the focus was on taxonomy. The exploration of Suriname was taken up seriously in the 20th century, and more than 42 expeditions to the interior were organized (see reviews in Holthuis, 1959, Hoogmoed, 1973 and Ouboter *et al.* in press). Applied biological and ecological research was also begun in the early 20th century. Before World War II the focus was mainly on agricultural and medical entomology. Most of the studies of vegetation science, animal ecology and ecosystem functioning were only initiated after 1960. This is certainly true for aquatic ecological research which was significantly stimulated by the Brokopondo Research Project (1963–1967).

The following survey of research carried out on the components and ecology of freshwater ecosystems is merely a summary (Table 1). More detailed information will be presented in the chapters concerned.

"The Freshwater Ecosystems of Suriname" is a compilation of the research findings listed in the table. Some chapters are of a more general nature, while others will give a more detailed insight into selected topics. **Part I: Basic features** deals with the more general and basic chapters, whereas the more detailed chapters on specific subjects are treated in **Part II: Case studies**. **Part III** concerns **the human impact** on freshwater ecosystems.

Although much of the research carried out on aquatic ecosystems is, to a greater or lesser extent, touched upon in this book, many topics could not be covered, because no authors could be found to deal with the subject, or because there are no data available. The most serious gap in knowledge concerns

Paul E. Ouboter (ed.), Freshwater Ecosystems of Suriname, 1–6.

Table 1. Research on freshwater ecosystems and their biotic components in Suriname

Research topics	Scientist/author	Period
Vegetation		
Wetlands	Lindeman 1953	1948–1949
	Lindeman and Moolenaar 1959	1953–1955
	v. Donselaar	1965–1966
	Teunissen 1976	1974–1977
	Teunissen 1978	1972–1974
	Playfair 1992	1992
Macrophytes	v. Donselaar, 1968, 1989	1964–1966
Limnology		
General	Panday-Verheuvel 1982	1975–1983
	v.d. Lugt	1989–present
Phytoplankton	Haripersad-Makhanlal	1982–present
Brokopondo Reservoir	v.d. Heide, 1973, 1978	1964–1966
	Leentvaar 1975	1963–1968, 1972
Blackwater Rivers	Ouboter	1988–present
Animal systematics		
Macro-invertebrates	v.d. Land 1970	1967
Porifera	De Drago 1975	–
Gastropoda	Vernhout 1914	–
	Pain 1952	–
	Geijskes and Pain 1957	1938–1965
Crustacea	Holthuis 1959	1957
Odonata	Geijskes, e.g. 1959	1954–1965, 1971
	Belle, e.g. 1970	1955–1963
Hemiptera	Nieser, e.g. 1975	1969–1970
Trichoptera	Flint 1974	–
Diptera	Bonne and Bonne-Wepster 1925	1916, 1920–1923
	Geijskes 1946	1938–1965
	Bruyning 1959	1949–1955
Fish	Hoedeman, e.g. 1957	–
	Boeseman, e.g. 1968	1963–1964, 1971
	Nijssen, e.g. 1970	1966–1967
	Gery, e.g. 1972	–
	Mees, e.g. 1974	1965–1966, 1971–1972
	Isbrucker, e.g. 1981	–
Amphibians/reptiles	Goin 1971	1964
	Hoogmoed, e.g. 1973, 1979	1968, 1970, 1974–1975
Birds	Haverschmidt, e.g. 1968	1946–1974
Mammals, general	Husson, e.g. 1978	1962–1963
manatees	Dekker 1974	1967, 1971
chetacea	v. Bree	1973

Table 1. (continued)

Research topics	Scientist/author	Period
Animal ecology (incl. community composition)		
Odonata	Wasscher	1988, 1990
Diptera	de Kruijf 1972	1967–1969
	Panday, e.g. 1974	1972–1982
Fishes, Electrophoridae	Heiligenberg	?
Corantijn	Vari 1982	1979–1980
Callichthyidae	Mol	1987–present
Community composition	Ouboter	1987–present
Reptiles, caimans	Glastra 1983	1981
	Ouboter and Nanhoe, e.g. 1989	1982–1984
	Ouboter, in press	1986–present
	Peeters	1991–present
Mammals, otters	Duplaix 1980	1976–1978
manatees	Duplaix, Reichart, Bertram	1976–1978
Parasitology		
Malaria	Flu 1912	1910–1912
	v.d. Kuyp, e.g. 1950	1939–1961
	v. Thiel 1962	1960–1962
	Panday, e.g. 1977	1972–1982
	Oostburg	1957–present
	Rozendaal, e.g. 1990	1982–1986
Bilharzia	Heinemann 1971	1963–1970
	Tewarie-Mungra 1972	1960–1967, 1970–1971
Filaria	Bruyning 1953	1949–1955
	Oostburg 1974	1969–present
Environment		
Man-made lakes	Panday 1979	1972–1982
Pollution	Vermeer *et al.* 1974	1971
Mining	Teunissen	1990–present

freshwater fisheries. It seems that the Fisheries Division of the Ministry of Agriculture mainly collects data on marine fisheries, since it is the most important sector economically. The only data available on fresh waters are those from an inventory of the fishery potential of the Brokopondo Reservoir (Richter and Nijssen 1980). Other noticeable gaps in knowledge concern the communities of small waterholes (including tree holes and bromelias) and the functioning of most freshwater ecosystems on the level of nutrient recycling and energy supply and on the level of community structure.

It is hoped that this book will inspire scientists and initiate research to fill the gaps. Suriname is a country with enormous amounts of freshwater and with tropical ecosystems of a rare, untouched beauty and of an incredible complexity. The number of intriguing scientific questions remaining is tremendous. Let us hope that its potentials will be fully utilised and that Suriname will become one of the world's centres of ecological research in the tropics.

References

Belle, J., 1970. Studies on South American Gomphidae (Odonata). Stud. Fauna Suriname Guyanas 11(43): 1–158.

Boeseman, M., 1968. The genus *Hypostomus* Lacépède, 1803, and its Surinam representatives (Siluriformes, Loricariidae). Zool. Verhand., Leiden 99: 1–89.

Bonne, C., and J. Bonne-Wepster, 1925. Mosquitoes of Surinam – a study on neotropical mosquitoes. Meded. Kon. Vereen. Kolon. Instit. Amsterdam 21 (Afd. Trop. Hyg. no. 13): 1–558.

Bruyning, C.F.A., 1953. Some data on the Filaria vector *Culex quinquefasciatus* Say in Paramaribo. Doc. Med. Geogr. Trop. 5: 333–338.

Bruyning, C.F.A., 1959. Notes on *Wyeomyia* mosquitoes of Suriname, with a description of *Wyeomyia surinamensis* sp.n.. Stud. Fauna Suriname Guyanas 3(8): 99–146.

Dekker, D., 1974. On the natural history of manatees (*Trichechus manatus manatus*) from Suriname for the Amsterdam Zoo. Aquatic Mammals, Harderwijk 2(2): 1–3.

Donselaar, J. van, 1968. Water and marsh plants in the artificial Brokopondo Lake (Surinam, S. America) during the first three years of its existence. Acta Bot. Neerl. 17: 183–196.

Donselaar, J. van, 1989. Brokopondo Research Report, Surinam. Part III. The vegetation in the Brokopondo-lake basin (Surinam) before, during and after the inundation, 1964–1972. Nat. Wet. Studiekr. Sur. Ned. Ant., Amsterdam 124: 1–45.

Drago, I.E. de, 1975. Freshwater sponges of Suriname. Stud. Fauna Suriname Guyanas 15(57): 175–183.

Duplaix, N., 1980. Observations on the ecology and behavior of the giant river otter *Pteronura brasiliensis* in Suriname. Rev. Ecol. (Terre Vie) 34: 495–620.

Flint, O.S., Jr., 1974. The Trichoptera of Surinam. Stud. Fauna Suriname Guyanas 14(55): 1–151.

Flu, P.C., 1912. Rapport omtrent malaria-onderzoek in de binnenlanden van Suriname. 's Gravenhage, 124 pp.

Géry, J., 1972. Poissons characoïdes des Guyanes. I. Généralités, II. Famille des Serrasalmidae. Zool. Verhand., Leiden 122: 1–250.

Geyskes, D.C., 1946. Notes on the neotropical Anophelinae in Moengo, Surinam. Summary. Mosquito News 6(3): 113–118.

Geyskes, D.C., and T. Pain, 1957. Suriname freshwater snails of the genus *Pomacea*. Stud. Fauna Suriname Guyanas 1(3): 41–48.

Geyskes, D.C., 1959. The aeschnine genus *Staurophlebia*. Notes on Odonata of Surinam VII. Stud. Fauna Suriname Guyanas 3(9): 147–174.

Glastra, R., 1983. Notes on a population of *Caiman crocodilus crocodilus*, depleted by hide hunting in Surinam. Biol. Conserv. 26: 149–162.

Goin, C.J., 1971. A synopsis of the tree frogs of Suriname. Ann. Carnegie Mus. 43: 1–23.

Haverschmidt, F., 1968. Birds of Surinam. Oliver and Boyd, Edinburgh and London, 445 pp.

Heide, J. van der, 1973. Plankton development during the first years of inundation of the Van Blommenstein (Brokopondo) Reservoir in Suriname, S. America. Verh. Intern. Verein. Limnol. 18: 1784–1791.

Heide, J. van der, 1978. Stratification in the filling phase of the man-made lake Brokopondo in Suriname (S. America). Hydrobiol. Bull. 12(2): 85–98.

Heinemann, D.W., 1971. Epidemiologie en bestrijding van Schistosomiasis in Suriname. PhD thesis, Leiden, 107 pp.

Hoedeman, J.J., 1957. Notes on the ichthyology of Surinam (Dutch Guiana) 4. Additional records of Siluriform fishes (1). Beaufortia 6(71): 147–160.

Holthuis, L.B., 1959. Stomatopod Crustacea of Suriname. Stud. Fauna Suriname Guyanas 3(10): 173–191.

Hoogmoed, M.S., 1973. Notes on the Herpetofauna of Suriname IV. The lizards and amphisbaenians of Surinam. PhD thesis, Leiden. Junk, The Hague, 419 pp.

Hoogmoed, M.S., 1979. Resurrection of *Hyla ornatissima* Noble (Amphibia, Hylidae) and remarks on related species of green tree frogs from the Guiana area. Notes on the Herpetofauna of Surinam VI. Zool. Verhand., Leiden 172: 1–46.

Husson, A.M., 1978. The Mammals of Suriname. Zool. Monograph. Rijksmus. Nat. Hist., Leiden 2: 1–569.

Isbrücker, I.J.H., 1981. Revision of *Loricaria* Linnaeus, 1758 (Pisces, Siluriformes, Loricariidae). Beaufortia 31(3): 51–96.

Kruijf, H.A.M. de, 1972. Aspects of the ecology of mosquitoes in Surinam. Stud. Fauna Suriname Guyanas 13(51): 1–56.

Kuyp, E. van der, 1950. Contribution to the study of the malarial epidemiology in Surinam. PhD thesis, Amsterdam, 146 pp.

Land, J. van der, 1970. Kleine dieren uit het zoete water van Suriname. Verslag van een onderzoek in 1967. Zool. Bijdr., Leiden 12: 1–46.

Leentvaar, P., 1975. Hydrobiological observations in Surinam, with special reference to the man-made Brokopondo Lake. Stud. Fauna Suriname Guyanas 15(56): 1–173.

Lindeman, J.C., 1953. The vegetation of the coastal region of Suriname. The Vegetation of Suriname 1(1): 1–135.

Lindeman, J.C., and S.P. Moolenaar, 1959. Preliminary survey of the vegetation types of northern Suriname. The Vegetation of Suriname 1(2): 1–45.

Mees, G.F., 1974. The Auchenipteridae and Pimelodidae of Suriname (Pisces, Nematognathi). Zool. Verhand., Leiden 132: 1–256.

Nieser, N., 1975. The water bugs (Heteroptera, Nepomorpha) of the Guyana region. Stud. Fauna Suriname Guyanas 16(59): 1–308.

Nijssen, H., 1970. Revision of the Surinam catfishes of the genus *Corydoras* Lacép]ede, 1803 (Pisces, Siluriformes, Callichthyidae). Beaufortia 18(230): 1–75.

Oostburg, B.F.J., 1974. Wuchereriasis in Suriname. Acta Leidensia 41: 1–249.

Ouboter, P.E., P.C. Dixit, M.C.M. Werkhoven, J.H.A. Mol and M. Khodabaks, in press. Some aspects of environmental research in Suriname with special emphasis on biodiversity. In: Environmentally Sound Socio-Economic Development in the Humid Tropics. UNESCO, Paris.

Ouboter, P.E., and L.M.R. Nanhoe, 1989. Notes on the dynamics of a population of *Caiman crocodilus crocodilus* in northern Suriname and its implications for management. Biol. Conserv. 48(4): 243–264.

Pain, T., 1952. Notes on the *Pomacea* of Surinam, with special reference to *Ampullaria sowerbyi* Vernhout. Basteria 16(1–2): 30–32.

Panday, R.S., 1974. Mosquito ecology in relation to the transmission of pathogens in Surinam. PhD thesis, Paramaribo, 203 pp.

Panday, R.S., 1977. *Anopheles nuneztovari* and malaria transmission in Surinam. Mosquito News 37(4): 728–737.

Panday, R.S., ed., 1979. Proceedings of the symposium man-made lakes and human health. October 23–25, 1977, University of Suriname, Paramaribo, 73 pp.

Panday-Verheuvel, M.P., 1982. Factors influencing water quality and life in the Suriname River. Rapport P 83–1, Waterloopkundige Afdeling, Paramaribo, 16 pp.

Playfair, M., (ed.), 1992. Vegetatiekundig onderzoek voor het Multipurpose Corantijn Project. Paramaribo, 41 pp.

Richter, C.J.J. and H. Nijssen, 1980. Notes on the fishery potential and fish fauna of the Brokopondo Reservoir (Surinam). Fish. Mgmt. 11(3): 119–138.

Rozendaal, J.A., 1990. Epidemiology and control of malaria in Suriname with special reference to *Anopheles darlingi*. PhD thesis, Leiden, 171 pp.

Teunissen, P.A., 1976. Notes on the vegetation of Suriname I – Vegetation changes in a dammed up fresh water swamp in NW Suriname. Acta Amazonica 6(2): 117–150.

Teunissen, P.A., 1978. Ecosystemenkaart van het Surinaamse laagland (Kustvlakte en Savannegordel), schaal 1:200.000 (Reconnaissance map: Surinam lowland ecosystems (Coastal Plain and Savanna Belt), scale 1:200,000). Stinasu, Paramaribo.

Tewarie-Mungra, C.S.B., 1972. Schistosomiasis in Surinam – a clinico-pathological and epidemiological survey. PhD thesis, Amsterdam, 296 pp.

Thiel, P.H. van, 1962. Malaria problems arising from the construction of a reservoir in the interior of Surinam. Trop. Geogr. Med. 14: 259–278.

Vari, R.P., 1982. Environmental impact of the Kabalebo projekt, final report. Inventory, biology and ecology of the fishes in the Corantijn river system, Suriname. Working Group Ecology Kabalebo Projekt, Ministry of Development, Republic of Suriname, 129 pp.

Vermeer, K., R.W. Risebrough and A.L. Spaans, 1974. Pesticide effects on fishes and birds in rice fields of Surinam, South America. Environ. Pollut. 7: 217–236.

Vernhout, J.H., 1914. The non-marine molluscs of Surinam. I. Notes Leyden Museum 36: 1–45.

PART I: Basic features

Part I on basic features includes those chapters that give a more general account of the basic components, biotic and abiotic, of freshwater ecosystems.

The first two chapters set the abiotic boundaries of freshwater ecosystems in Suriname. *"The geographical outline"* (Chapter 2) describes the influence of geomorphology on hydrology and the physical and chemical characteristics of the water. The climate of Suriname and a hydrological description of the main rivers is presented in Chapter 3, *"Climate and surface water hydrology"*. The *physico-chemical parameters* of the water itself, *and phytoplankton composition* (combined by the word *"limnology"*) are elaborated on in Chapter 3, which compares characteristics at numerous localities all over the country.

Swamps cover about 60% of the Coastal Plain, which is inhabited by approximately 85% of the Surinamese population. Consequently, there will be many interactions between the human community and these wetlands. *"Vegetation and vegetation succession of freshwater wetlands"* (Chapter 5) describes these swampy areas and the threats caused by human interference.

The following chapters are concerned with an inventory of the living components, providing an overview of the knowledge gained from many localities all over Suriname.

Owing to the limited number of *aquatic macrophytes*, research efforts have never concentrated on this group. Chapter 6 contains a summary of the data available and will constitute the first survey of this group for the region.

In Suriname, information on *macro-invertebrates* has always been collected as a sideline. During research on the habitats of armoured catfish many data on macro-invertebrates (incl. zooplankton) were gathered as well. These will be presented in Chapter 7.

The *diversity of fish* in the tropics is much greater than in temperate regions. About 200 fish species from Europe are known. Even though knowledge of Brazil's ichthyofauna is not exhaustive, already more than 1,400 species have been described. Suriname does not possess the incredible variety of fresh water fish species of Brazil, but the number is still impressive, with more than 100 species in the larger rivers and at least 70 species counted in one small creek. The

Paul E. Ouboter (ed.), Freshwater Ecosystems of Suriname, 7–12.

diversity and species composition of the ichthyofauna of various river-sections and creeks are compared in Chapter 8.

An overview of amphibians and reptiles is lacking. Since the species of aquatic reptiles found in Amazonia and Suriname are almost identical, readers are referred to Best (1984). Two case studies on reptiles will be included in Part II.

Birds and mammals are the only animal classes whose systematics are completely known, mainly through the work of Haverschmidt (1968) and Husson (1978). With regard to birds, ecological research has included a very limited number of species. Spaans *et al.* (e.g. Spaans 1975; Spaans & de Jong 1982) have produced an extensive study on coastal birds, but the brackish coastal areas are beyond the scope of this book. Consequently, no chapter on birds could be included.

For mammals, the situation is more favourable as a thorough study of giant otters is available (Duplaix 1980), as well as some additional data on the swamp otters and manatees. Unfortunately, the author that was asked to deal with the subject, was not able to deliver her chapter in time.

Only Chapter 5 concerns the description of certain freshwater ecosystems (mainly swamps and swamp forests), all other chapters focus on certain components (e.g. aquatic macrophytes, fish), but do not describe the ecosystems themselves. Therefore, it seems desirable to give an overview of the freshwater ecosystems of Suriname in this introduction. They may provisionally be classified as follows:

Figure 1. Kuwini Creek (Interior).

1. Running water
   a. Brooklets
   b. Creeks (2nd and 3rd order) (Fig. 1)
   c. Rivers (higher order)
   Going from west to east the major rivers of Suriname are: the Corantijn (including Sipaliwini, Curuni, Lucie and Kabalebo rivers), Nickerie, Coppename, Saramacca, Suriname, Commewijne and Marowijne (including Tapanahoni, Palumeu, Ulemari, Litani and Lawa rivers) (Figs. 2 and 3).

Figure 2. Cusewijne River (Old Coastal Plain).

Figure 3. Sir Walter Raleigh Falls, Curuni River (Interior).

d. Canals
   There are many canals in the coastal area. Two of the larger ones are the Corantijn Canal in NW Suriname and the Saramacca Canal, connecting the Suriname and Saramacca rivers.

2. Stagnant open water
   a. Lakes
   In Suriname, the only natural lake is the small Nani Lake (NW Suriname) (Fig. 4). Many man-made lakes have been constructed in recent times. The largest of them is the Brokopondo Reservoir, covering 1560 km^2 (Fig. 5). Many exhausted bauxite mines have turned into small lakes.
   b. Lagoons
   Lagoons are defined here as very shallow areas of stagnant, open water, which may dry out during the dry season. They may hold seawater as well as freshwater. Many salt- and brackish-water lagoons are located just behind the coastline. The largest of them is Bigi Pan. Freshwater lagoons are very rare; under normal circumstances shallow, stagnant freshwater areas will change into swamps (i.e. covered by vegetation). In Suriname some freshwater lagoons exist, due to pollution by the bauxite industry (see Ouboter & De Dijn, Chapter 14).
   c. Puddles, waterholes, etc.
   Where the soil is impervious, puddles may form almost everywhere during the rainy season. In the high dryland forests of the interior puddles are rare. As the water level recedes, however, many waterholes are formed between the rocks of the riverbeds. These waterholes have their own specialized fauna and flora and are a kind of micro-ecosystems.
   d. Trenches

Figure 4. Nani Lake (Old Coastal Plain).

Figure 5. Brokopondo Lake.

Figure 6. Swamp in the Cusewijne area (Old Coastal Plain).

3. Wetlands
   a. Herbaceous and scrub swamps
   Across the country, swampy areas may border the rivers or creeks in certain localities. Extensive swamps occur only in the Coastal Plain. The largest and most inaccessible swamp is the Coronie Swamp. The Nani Creek and the Cusewijne River areas are also known for their large swamps (Fig. 6).
   b. Swamp forests
   Swamp forests border many of the rivers and creeks in the Savanna Belt and Coastal Plain. In the Interior the rivers have worn a narrow channel in the Guiana Shield and consequently swamp forests are rare.

This is as artificial a division as any. Many combinations and transitions are known. For instance, the Nani Creek and the Para River are surely in the category of running water. Without cleaning operations, these streams would be covered by floating meadows, finally developing a thick peat layer that could even hold trees. One could walk over the stream without ever noticing it. Without proper care, canals and trenches would also close rapidly, because of the growth of aquatic and semi-aquatic vegetation. These ecosystems may thus be classified as wetlands as well.

References

Best, R.C., 1984. The aquatic mammals and reptiles of the Amazon. In: H. Sioli (ed.), The Amazon. Limnology and landscape ecology of a mighty tropical river and its basin. Dr. W. Junk Publishers, Dordrecht: 371–412.

Duplaix, N., 1980. Observations on the ecology and behavior of the giant river otter *Pteronura brasiliensis* in Suriname. Rev. Ecol. (Terre Vie) 34: 495–619.

Haverschmidt, F., 1968. Birds of Surinam. Oliver & Boyd, Edinburgh, 445 pp.

Husson, A.M., 1978. The mammals of Suriname. E.J. Brill, Leiden, 569 pp.

Spaans, A.L., 1975. On the present breeding status of the scarlet ibis, *Eudonimus ruber*, along the North-Eastern coast of South America. Biol. Conserv. 7: 245–253.

Spaans, A.L. & B.H.J. de Jong, 1982. Present Status of some Colonial Waterbird Species in Surinam, South America. J. Field Ornith. 53(3): 269–272.

2. The geographical outline

DIRK NOORDAM

1. Introduction

Four major geographical zones can be distinguished in Suriname: the Interior (Precambrian Shield or Basement), the Savanna Belt (Zanderij Belt or Cover Landscape) and the Old and Young Coastal Plains. Each zone has its own typical combination of landscapes, soils and hydrological conditions, which will be discussed below (Fig. 1).

2. Geology

2.1. The Interior (Precambrian Shield or Basement)

The oldest part of Suriname is formed by Precambrian rocks, often indicated as the "Basement". These rocks are a part of the Guiana Shield, from which Suriname covers about 10%. The Guiana Shield forms a smaller counterpart of the Brazilian Shield, from which it is separated by the Amazonian geosyncline. Locally, this part of Suriname is often referred to as the "Residual Uplands", or the "Interior" (this term will be used here). It occupies about 80% of Suriname.

The shield gently slopes towards the north where it extends under the younger sediments and the continental shelf (Fig. 2). The boundary between Suriname and Brazil forms the watershed between the catchment of the Amazon and those of the Guianan rivers.

The shield is composed chiefly of a complex of granitoid rocks, gneisses, metasediments and metavolcanics, mostly dating from the Lower Proterozoic and older (mostly about 1900 million years BP). Very locally the basement is covered with sandstones and conglomerates which were formed during the middle Proterozoic, of which the Tafelberg is the principal representative. Dolerite dikes were formed during two periods, namely the Middle Proterozoic (NW-SE trending) and the Permo-Triassic (N-S striking). For more detailed information on the geology of the Precambrian Shield see De Vletter (1984) and Bosma *et al.* (1984).

Paul E. Ouboter (ed.), Freshwater Ecosystems of Suriname, 13–28.

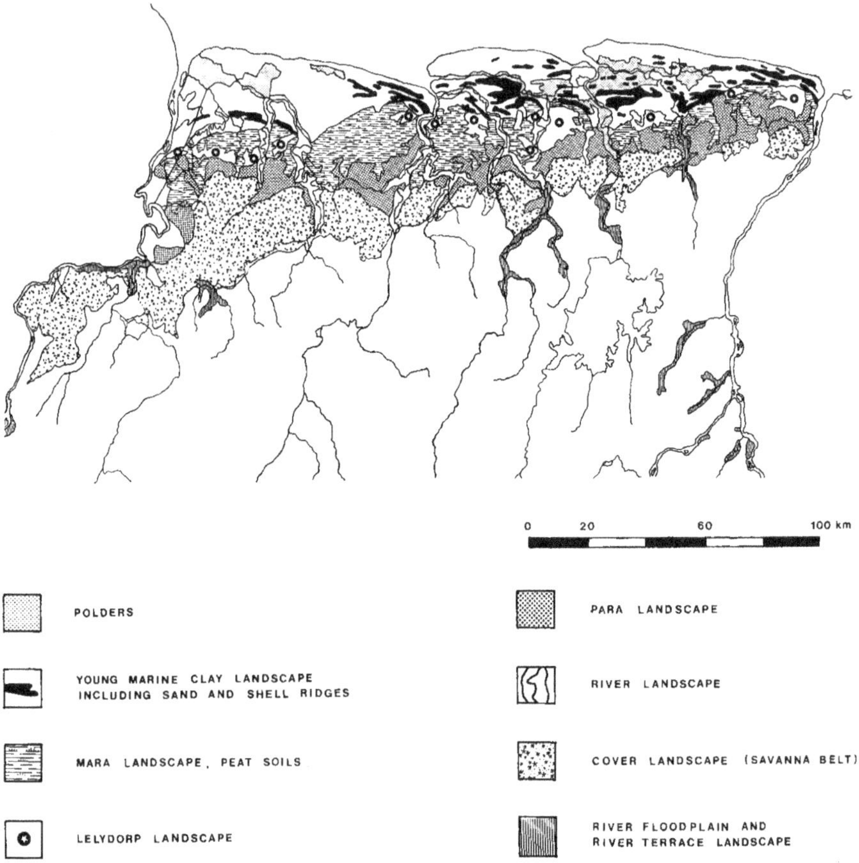

Figure 1. Geomorphological map of Northern Suriname (by S. Sahdew and P. Ouboter after SPS and OAS 1988, Suriname Planatlas).

No major geological changes have occurred during the Middle Proterozoic up to the Late Jurassic, when the Guiana sedimentary basin was formed and sedimentation started. The basin contains Upper Cretaceous, Tertiary and Quaternary sediments on top of the Precambrian basement, which dip gently and thicken towards the coast. They are thickest near the Corantijn rivermouth (1800 m), where the basin extends about 150 km inland. A detailed review of the stratigraphy and sedimentary history is presented by Wong (1984).

2.2. *The Savanna Belt*

The oldest sediments that appear at the surface are those of the Upper Coesewijne Formation. They were deposited under semi-arid conditions in the Pliocene, when braided rivers from the Interior deposited large amounts of

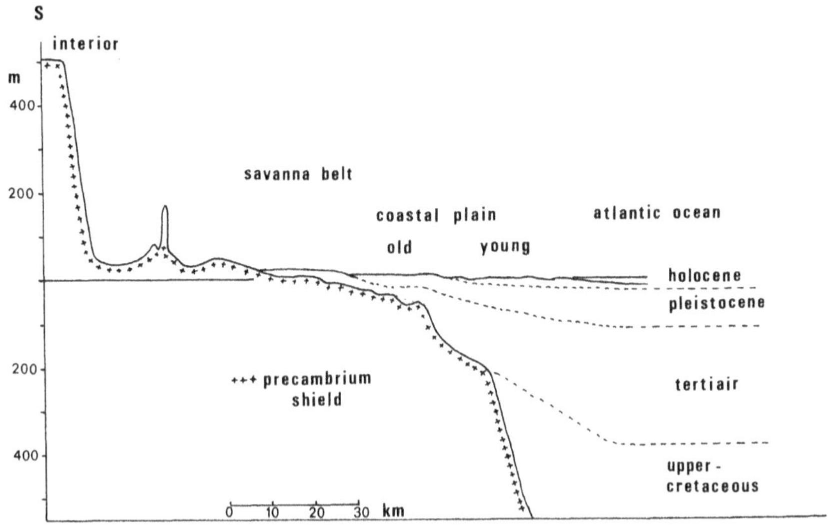

Figure 2. Simplified North-South section of the coastal area near Paramaribo (by J.H. Mol after Krook 1984).

predominantly sandy sediments to form a belt of alluvial fans along the northern edge of the basement. This belt, known as the Savanna or Zanderij Belt (Fig. 2), continues northward beneath the Pleistocene Coropina Formation. At the base of the upper Coesewijne Formation a clayey facies is found. The medium and coarse sandy layers of the Coesewijne Formation are important aquifers.

2.3. The Old Coastal Plain

During the Pleistocene, the Old Coastal Plain was deposited during two interglacial transgressions (Fig. 2). Clays where deposited as tidal flats and coastal swamps and sands as (semi-)parallel beachbars and offshore bars during transgressions. The deposits, known as the Coropina Formation, were strongly eroded during the regressive phases. A reconstruction of the geo- and pedogenesis has been presented by Veen (1970).

2.4. The Young Coastal Plain

The major part of the Holocene deposits forming the Young Coastal Plain consists of Amazon-derived clays (Fig. 2). Besides these, sands and shells have been deposited as (semi)parallel beach ridges. The deposits are known as the Demerara Formation. The oldest part was deposited during the post-glacial sea level rise (10,000–6,000 years BP) when clays were sedimented in a coastal mangrove swamp to tidal flat environment. The conditions of this environment were favourable for the formation of peat and pyrite. At the same time the erosion gullies in the Old Coastal Plain were filled with the same type of sediment. These clays were named Mara deposits by Brinkman and Pons (1968) in their pedo-geomorphological classification of the Coastal Plain of the Guianas. The deposits form a low-lying, discontinuous belt in the southern part of the Young Coastal Plain.

When the sea level rise came to a standstill, another type of sedimentation started, resulting in a rapid lateral accretion of the coast by sedimentation of clayey materials as mud banks alternated by phases with coastal erosion and the deposition of beach ridges (sands and shells). The coastal processes have been described by Augustinus (1978). These sediments deposited under more or less constant sea level conditions were named the Coronie deposits by Brinkman and Pons (1968). On morphological and pedological grounds these authors distinguish three phases of deposition, separated by periods of regression or erosion. The sedimentation phases are the Wanica (6,000–3,000 years BP), the Moleson (2,500–1,300 years BP) and the Comowine phase (1,000 years BP–present).

At some places deep water conditions and a stagnated drainage have resulted in the formation of thick peat layers.

2.5. Flood Plains and Riverbound Terraces

Besides the sediments of the Guiana basin also alluvial deposits have been formed along the rivers during the Quaternary Period. The Pleistocene deposits are found in terraces and vary in consistency from sand to clay, locally underlain by gravel layers. The terraces are mostly less than 2 km wide and they are found at several parts along the major rivers and creeks in the Interior. Two levels can be distinguished: a 15 m and a 30 m level.

The Holocene fluvial sediments have been, and still are deposited as levees and backswamps. They usually consist of silty clays at the levees and heavy clays in the backswamps. Along the Corantijn river, however, also sandy clays, clayey sands and sands can be found. Going inland the fluvial deposits become less extensive and discontinuous. They form isolated patches here. For an extensive description on the Quaternary geogenesis of the river valleys one is referred to de Boer (1972).

Except for the estuarine part where the bottom is usually formed by clay, the bottom of the rivers is covered with coarse sands while in the upper courses fine gravel may occur. Little sand is transported by the rivers in recent times.

3. Geomorphological Development and Geography

3.1. The Precambrian Shield

The Precambrian Shield is characterized by a series of step-like planation surfaces that can be followed over large distances. The older of these surfaces are thought to have developed under the influence of intermittent upwarping of the Interior, which gave rise to strong denudation, finally resulting in a more or less flat sloping surface toward the coast. In the next cycle of upwarping the same occurred but now at a lower level. These older surfaces, which date from the Tertiary Period, are now represented by their remnants: flat-topped, bauxite- or laterite-capped surfaces which escaped erosion because of their protective duricrust. They are separated from each other by a pronounced scarp. Isolated remnants of these older surfaces are found in the interior of Suriname, e.g. the Tafelberg, Brownsberg, Nassaugebergte and Bakhuysgebergte. Information on these so-called denudation surfaces in the Guianas is given by (e.g.) King *et al.* (1964). During the Quaternary the Late Tertiary surface was repeatedly attacked during alternating cycles of planation and linear erosion, occurring in respectively dry and humid periods. For the formation of the step-like build-up a very slow constant epeirogenic upward movement has been assumed. The resulting planation surfaces can be reconstructed by drawing planes through hill tops (hill top envelopes). Contrary to the Tertiary surfaces, which slope towards the sea, the Quaternary planation surfaces are clearly river bound. The process mechanism of the development of these lower planation surfaces has been described by Kips and Snel (1979). The Quaternary planation surfaces occupy large parts of the basement, which can be characterized as a multi-convex relief of low hills. The 30, 15 and 5 m planation surfaces can be considered as strath terraces, as fluvial deposits have been found on the hilltops.

The remaining landscape is predominantly low-lying with only a few mountain groups rising above the 250 meter contour, with summits exceeding 1000 meters. The border zone towards Brazil is somewhat higher, with altitudes generally between 250 and 500 meters above sea level, again with mountain groups with a height of approximately 1000 meters. This can be explained as the planation surfaces are bound towards the local erosion level, which increases going upstream.

Rock outcrops are only incidentally found, mostly in sula complexes, and scattered as inselbergs. Bornhardts (locally known as tebu), ruwares (rock plateaus) and boulder inselbergs are found. The latter two are usually hidden by the dense rain forest.

O'Herne (1969) made a study of the landscapes of Suriname using aerial photographs. He defines a landscape as "..an area, which as a result of its specific geological origin, morphologically forms a unit, characterized by a special rock formation and a variation typical of this area". He also states that "In a strongly weathered and densely forested country like Suriname, a landscape is characterized mainly by its relief and occasionally by its particular vegetation." For each of his landscapes O'Herne presents data on form and size of the hills, degree of dissection, altitude and the rock type on which it has developed.

Another approach to describe the landform of the Interior is that of Balsem *et al.* (no date) on their map of central Suriname. They emphasize the geomorphogenesis of the area presenting the various planation surfaces. These surfaces may comprise several landscapes as they have been described by O'Herne, while also one single landscape may be found on two or more planation surfaces.

The landform of the Precambrian Shield can shortly be described as follows. The flood plain and the 5 meter terrace are (almost) flat. The flood plain is periodically inundated at high river levels. The 15 meter terrace up to the 30 and 50 meter planation surfaces can be characterized as undulating to moderately steep, with increasing relief (15, 25 and 40 m respectively) and maximum slope (8, 18 and 30% respectively). The 100 meter planation and the older surfaces consist of steep hills, flat to rolling, steep sided plateaus and steeply dissected mountains, and locally elongated steep hills of the dolerite dikes.

3.2. The Savanna Belt Landscape

The Savanna Belt Landscape went through the same alternation of dry and humid conditions. The original depositional surface was flat and gently dipping to the north. Erosion has resulted in a strong dissection and locally proceeded so far that almost no Coesewijne material has been left. The present Cover ("Dek") Landscape is flat to gently undulating, with broad plateaus alternated by creeks and rivers. The river valleys are deep and here as well erosion terraces at different levels have been reported (Krook and Mulders 1971). In the western part of the country the Savanna Belt is found at a higher altitude and dissection is deeper. Steep to very steep slopes are reported with active erosion (Pleysier 1973). This is attributed to a recent lowering of the erosion basis.

Savannas on bleached, white sands form marked features within the landscape (hence, the name Savanna Belt) although they cover only 7%, the remaining part being mostly savanna forests.

3.3. The Old Coastal Plain

The Old Coastal Plain consists of a discontinuous belt of a large number of disconnected large and small surfaces. Two very large coherent parts are found within Suriname, one south of Paramaribo and one in the Wane Creek area. The original surface, comparable with the Young Coastal Plain, has been strongly attacked by erosion during the last glacial period, when large parts were completely eroded away, while in other parts deep erosion gullies have formed. Surface wash processes leveled the ridge landscape. The original tidal clay flats are now known as the Schollen (Islands), or Para landscape, while the part with the old sandbars is indicated as the Old Ridges, or Old Offshore Bar Lelydorp Landscape. The landscapes are low-lying and nearly flat with depressions.

3.4. The Young Coastal Plain

The Young Coastal Plain consists of extensive swamps and marshes, locally interrupted by roughly east-west striking ridges. The southern edge towards the Old Coastal Plain is often formed by very low lying deep swamps with a thick peat layer developed on Mara deposits. The same environment is found in the filled-up erosion gullies of the Old Coastal Plain.

The northern deposits have a slightly higher elevation, and less deep swamps and marshes are found. From the north to the south the originally saline deposits under a mangrove vegetation are gradually desalinized and a sequence of brackish to fresh environments can be distinguished. The peat layer gradually thickens, but usually very thick layers do not occur.

Ridges of variable width (25–400 m) rise 1–3 m above the neighbouring clay flats; they occur as elongated, usually narrow bodies, either individually or in bundles. Bundles are particularly abundant west of the major rivers. The most extensive ridge complexes were formed during the deposition of the Wanica phase. In the western part of the country only few ridges are found.

4. Soil Formation

4.1. The Precambrian Shield

A discussion on the pedogenetic processes and the polycyclic development of the soils of the Precambrian basement is presented by de Boer (1972); this largely also applies to the Tertiary and Pleistocene sediments.

The first cycle is represented by the often reworked sesquioxide nodules (laterite gravel, iron peas etc) from the various planation surfaces. They are the result of plinthitization, followed by hardening of the mottles. In the following period gleyzation of flat surfaces occurred, finally resulting in plinthite. These surfaces presently have a deep ground water table.

This period was followed by a period with illuviation of clay (lessivage), which occurred probably under drier conditions during the last glacial. The resulting orientated clay has been found preserved in nodules. Argillic horizons from that period were only found in soils with a low faunal activity, e.g. hydromorphic soils (de Boer 1972). Most of the remnants are, however, destroyed by the more recent soil formation and bioturbation.

Under the present humid conditions bioturbation and ferralitization are common processes on well drained soils. Another process considered active now on these soils is "appauvrisement" resulting in an increase in clay content from topsoil to subsoil. The hypothesis is that fine particles are brought to the surface by biological activity from where they are removed by surface wash processes. No-clayskins result from it, as the fine particles do not migrate downwards into the profile. Under hydromorphic conditions plinthitization occurs in older soils, while on the younger river deposits gleying can be observed.

Processes of local importance are podzolisation and the formation of Terra Preta. Podzolisation occurs on very poor, acid sandy soils under hydromorphic conditions. Terra Preta are found as patches, characterized by a deep black humic top layer. They have developed at the dwelling sites of pre-Columbian Indians. They are usually found near navigable waterways.

4.2. The Savanna Belt

The soils of the Savanna Belt as well have been subjected to clay illuviation in drier periods (Krook and Mulders 1971; Bennema 1982) and relics of clayskins have been found (De Fretes 1984). In the preceding, more humid climate podzolisation under hydromorphic conditions will have been active resulting in hardpans. These hardpans are found at a depth of 2 to 4 meters in the wet savannas and are considered fossilized features by Krook and Mulders (1971). Podzolisation is considered a presently ongoing process. Hardpans now develop only at the water courses at the foot of slopes with impeded drainage.

According to a theory developed by Lucas *et al.* (1982), podzolisation starts at the center of the plateaus and at the lower foot slopes, under the influence of impeded drainage triggering destruction and removal of clay followed by podzolisation. The areas with podzols will very gradually increase as a result of the worsening of the drainage conditions through the above process. In the final stage the whole plateau will be turned into podzols.

No extensive areas with signs of plinthitization of the plateaus in earlier humid periods have been found. Ferralitization and bioturbation are the common present-day processes on well drained sites with (sandy) loamy to clayey soils.

According to Poels (1987), a podzolisation process is active in the topsoil of all soils, made possible by the sandy textures, the poor nutrient status and their acidity. Organic matter is transported downslope by lateral flow or groundwater; it accumulates at the lower footslope, or is carried to the creeks. Hardly any iron or aluminium is, however, transported.

4.3. The Old Coastal Plain

Almost the same polycyclic development as described above for the Precambrian Shield has been presented for the Old Coastal Plain by Veen (1970), with the exception of the first cycle.

4.4. The Young Coastal Plain

The soil formation of the clay soils of the Young Coastal Plain has been described by Pons (1966) and Brinkman and Pons (1968). After a mudflat has been silted up high enough, the process of initial soil formation starts. This comprises desalinization, oxidation, mottling and water loss. These processes are summarized under the term "ripening". Through ripening, the initially soft and saline mud is gradually turned into a firm, non-saline acid clay soil with gley mottles, provided drainage is adequate. Generally the older claysoils will have the highest degree of ripening.

An exception has to be made for the oldest deposits of the Young Coastal Plain, the Mara deposits which, due to their low elevation at or below present sea level, never have been drained properly. They are largely still very soft, grey without mottles, and reduced.

Another exception is formed by the claysoils along the rivers, which have known different conditions during both geo- and pedogenesis e.g. the less saline environment and the stronger leaching as a result of a better drainage.

Finally, differences in parent material may have influenced soil formation. This is particularly the case when pyrite or potential acid sulphate soils (PoASS) are involved, which may turn into cat clays or acid sulphate soils. Outside the area with Mara deposits these PoASS are only found patchlike, usually correlated with estuarine deposits.

Peat accumulation is found on submerged terrains. Under the prevailing anaerobic conditions the mineralization of organic matter is slowed down considerably which results in an accumulation in the form of peat. Accumulation is more pronounced under conditions of low nutrient availability in soil and/or water, and under acid conditions. Little organic matter accumulates in an aquatic alkaline environment in contact with shell bodies. Besides water quality, also the vegetation type and the depth and frequency of inundation are important characteristics determining peat accumulation. Large areas of eustatic peat of varying thickness are found in the Coastal Plains of Suriname. Besides these, there are some vast areas in which peat grows above sea level. The best known is the Coronie swamp, where peat layers of over 4 meters have been reported. These domed low mounds of ombrogenous peat are the result of an impeded drainage, combined with a poor nutrient supply and a low pH (also see Teunissen, Chapter 5).

Also in the ridges, a progressive soil formation going from young to old can be observed. The originally pale sandy soils turn yellowish to reddish brown as a result of weathering and dispersion of iron oxides. Also, the formation of an accumulation horizon of organic matter starts as soon as vegetation has settled itself. On less well drained parts of sandy ridges gleying can be observed. Impeded drainage leads to ground water podzolisation, in particular on the older sandy ridges. However, podzolisation has been observed already at the slopes of a relatively recent ridge of the Comowine phase (Westerink 1989), probably under the influence of acid water from the neighbouring swamp, combined with the poor nutrient status of the medium sandy parent material.

A different soil formation occurs on the shell ridges where thick humic topsoils may form because of the presence of calcium carbonate.

5. Soil Conditions

5.1. The Precambrian Shield

The regolith, in which the soils of the Precambrian basement have developed, usually shows a deep and intensive weathering. Weathering depths for various rock types are presented by O'Herne (1965). For granites and granodiorites, the most common rock types, depths between 0–16 m are indicated, with maxima up to 28 meters. Basic rock types generally have developed a thicker regolith.

The major part of the basement is covered with (moderately) well drained brown to red ferralitic soils, usually with a loamy to clayey texture and a lighter textured topsoil. Often iron and/or quartz gravel is found, in particular on hill tops. The higher plateaus generally have a laterite or bauxite duricrust. Stripping in the recent dry period has locally produced shallow soils and bare rock land. Part of the (lower) rock inselbergs also date from these periods, but the majority is much older. Shallow, gravelly soils and bare parts are also found on steep slopes.

The lower footslopes and the creekvalleys are imperfectly to poorly drained. Grey to mottled sandy to loamy alluvial and colluvial materials, which are often stratified, are found here.

On the almost flat 5-meter terrace, less well-drained, mottled loamy to clayey soils are found.

Bleached soils have developed locally on poor sandy parent material, e.g. on coarse Pleistocene fluvial deposits and on some sandstones.

The flood plains along the rivers generally have poorly drained, very fine sandy to silty clay soils in the backswamps and better drained loamy to silty clayey soils on the levees. Locally, sandy to loamy soils can be found.

5.2. The Savanna Belt

The soils within this belt are often divided into the bleached white sands and the non-bleached, yellowish brown sands, loams and clays.

The white sands are internationally known as "Giant Podzols". They consist of several meters of white, medium to coarse quartz sand with a topsoil with varying amounts of highly mobile humus. The latter depends on the vegetation type (Heyligers 1963). A hardpan may be found, containing organic carbon and alumina.

The non-bleached soils have textures ranging from sand to clay. An increase in clay content is found mostly from topsoil to subsoil. The soils are very poor and very acid, and have moderate to good physical properties.

5.3. The Old Coastal Plain

In the southern part of the Old Coastal Plain, where the Old Para sea clay Landscape is found, the soils consists of stiff, compact, strongly mottled clays with an often slightly podzolized silty (clay) loam topsoil. The soils exhibit water stagnation on the stiff topsoil. Fertility is moderate and the soils are very acid.

In the Old Offshore Bar (Lelydorp) Landscape a rather complex soil pattern is found, ranging from well-drained, reddish brown, very fine sands or loams on the better drained ridges, to poorly drained (clay)loam over clay profiles or podzolized white fine sands at the plateaus. Fertility ranges from moderate to poor, and the physical properties vary from good in the sandy soils to poor in the clayloam and clay soils.

Throughout the Old Coastal Plain, erosion gullies can be found. They are filled up with the Holocene Mara deposits. The clays are soft and grey, and will turn extremely acid upon drainage (potential acid sulphate clays).

5.4. The Young Coastal Plain

As may be deduced from the description of the soil formation in the Young Coastal Plain, a wide variety of clay, peat, sand and shell soils occurs here. In general, it may be stated that the clay soils are very rich, with a variable pH and salinity. The physical properties of these soils, which have been described by Kamerling (1974), are usually moderate to bad.

On the ridges a toposequence of well to poorly drained soils is usually found, going from ridge top to footslope. The wider ridges may exhibit an impeded drainage at the central part, which can result in podzolization. At the footslopes, groundwater podzols may develop.

6. Hydrological Aspect of Land and Soil

The drainage of a piece of land is chiefly controlled by the climate, vegetation cover, lithology and relief. The relief, however, is the result of the actions which form the drainage pattern of a certain landscape. In general, these actions are in turn controlled by the characteristics of the regolith (lithology). Below, attention will be given to the lithology, in particular soil characteristics, as far as it influences the hydrology.

6.1. The Precambrian Shield

The hydrological factors in an area are, besides precipitation and evapotranspiration, run-off, percolation, ground water flow, lateral flow, deep flow and creek drainage. Run-off will only be an important factor in areas with bare rocks or shallow soils. Such areas cover minor parts of the Interior. For most soils under rain forest the infiltration capacity will be sufficiently high to prevent large-scale surface run-off, although some run-off may occur, in particular during rain storms and on steep slopes.

Percolation into the groundwater will in some soils be hampered, due to a clayey subsoil with a relatively low permeability. In this case lateral flow may occur. Other soil characteristics favouring lateral flow are the presence of sandy topsoils overlying loamy to clayey subsoils, and the presence of stonelines parallel to the surface. In many sloping soils, lateral flow will play an important part. In some soils percolation is so slow that some water stagnation occurs in the topsoil, resulting in epiaquic soil characteristics.

In sufficiently permeable soils, infiltrating water will effect a rise of the groundwater table, followed by a slow discharge into the creeks. It is assumed that groundwater flow indeed plays an more or less important part in most soils. Haug (1966) mentions that the forest creeks in his study area (Palumeu) carried water in the dry season as they are fed by springlets. Anonymous (1984) recorded a very low, but measurable flow in the Marechalls Creek during the long dry season of 1983. This creek discharges a small, predominantly schist area. Observations in the rainy season report on the other hand, a rapid increase of river and creek levels in the Interior after rainfall, indicating discharge through surface run-off and/or lateral flow. Anonymous (1984) mentions a time-to-peak period of about 13 to 28 hours and a concentration time of 51 to 72 hours, indicating a predominance of lateral flow processes over surface run-off.

As the discharge through run-off and lateral flow is much faster than by way of groundwater flow, and deep flow is usually not encountered, it may be expected that a higher drainage capacity (density) will be found in sloping areas with a slow percolation. So in an area with schists, a high drainage density may be found with small, round, low hills with colluvial footslopes without marshy bottomlands. On the other hand, granites produce a regolith which favors percolation and a slow release of water, which may result in a less dense drainage pattern, with rounded or elongated hills, and marshy bottomlands often without a clear colluvial footslope. At many places this generalized pattern is however not found because of the formation of planation surfaces, followed by linear erosion.

The drainage pattern is described as basically fine to coarse dendritic, with occasionally a rectangular pattern as controlled by diaclases and faults. Wensink (1968) mentions a rectangular drainage pattern with a definite SSW-NNE direction that can be followed over 18 km and comprises a number of creeks.

Haug (1966) describes a barbed wire pattern at a location where recent stream capture has occurred. The same author shows that the direction of faults and joints is reflected by the direction of the creeks and rivers in the area studied by him. The same holds for the whole Precambrian basement.

No figures on drainage density on various rock types are known to me.

Unlike rivers in the temperate zone, whose longitudal profiles can be characterized as a smooth curve with rapidly increasing steepness toward the headwaters, the thalweg of the rivers of Suriname, outside the estuarine area sector, shows a more or less straight profile with steps. In the headwater section a steep increase may be observed. The profile of the middle section, which comprises the bulk of the river stretches, is formed by numerous rapids (local name "sula") of some meters high and by somewhat higher falls that alternate with quiet reaches with only a very low gradient. Discussions on the origin and development of this river type are presented by, among others, de Boer (1972), Krook (1979), Kips and Snel (1979). A classification of rapids has been made by Zonneveld (1952).

According to Haug (1966), the watersheds may be very indistinct with swamps in the upstream parts.

River capture is reported for several parts of Suriname. An example is given by Haug (1966) who describes the capture of the upper part of the Jai Creek by the Tapanahoni River. This is related to planation during glacial periods.

6.2. The Savanna Belt

In his study of a small catchment in the Savanna Belt, Poels (1987) found that groundwater flow was the main contributor to discharge. Surface run-off and lateral flow were important in the valley bottom and on footslopes. There is, however, a decrease in permeability going from topsoil to subsoil, giving rise to some lateral flow. In other areas surface flow has been observed at places with a clayey subsoil, or at places where the kaolinitic substratum is close to the surface (Duynisveld and Van der Weert 1974). It was estimated by these authors that about 15% of the rainfall is drained into the rivers as direct flow.

Compared to the figures of the schist catchment (Anonymous 1984), the period of time-to-peak in the area studied by Duynisveld *et al.* is much quicker (6.5–15 hours). The systems react more quickly to rainfall.

On the other hand, the length of the dry period with no or very little discharge varies between 0 and 8 months, indicating a relatively rapid depletion of the water, or a deep flow. It has indeed been indicated that the Savanna Belt is an important source of recharge of an aquifer. According to Sevenhuysen (1977), the amount of water drained in this way can, however, be neglected.

The Savanna Belt has a relatively wide, dendritic drainage pattern. The creek valleys are usually wide and flat. The upper stretches may consist of shallow gullies, without a clear creek, carrying water during rainy periods.

Originally, drainage took place by consequent rivers but as a result of river capture the picture has been changed in the southern part of the belt, where subsequent rivers are now found, draining east to west. The northern part still has a more or less south-north drainage. Part of the water from the northern part is drained into the swamps north of the Savanna Belt.

6.3. The Old Coastal Plain

The stiff silty clays of the "schollen" in the Old Para (sea clay) Landscape exhibit a very low infiltration and percolation. Most of the rainfall is drained directly into the nearby depressions and swamps giving rise to an increase in waterlevel. Through these swamps, the water is very slowly drained into the rivers to which the swamps are connected by short creeks. Some swamps drain through more pronounced creeks (e.g. the Nani Creek).

More or less the same holds for the loamy to clayey plateaus of the Old Offshore Bar landscape, which also show predominantly a drainage by surface run-off. In areas with a sandy topsoil, lateral flow may be important, while the sandy and loamy soils of ridges may have a considerable groundwater flow.

Most of the water in these areas is drained toward depressions or swamps. Here as well, few distinctive creeks can be observed.

It has been mentioned that the sandy soils in some areas recharge the Zanderij aquifer, so some deep flow may occur as well.

6.4. The Young Coastal Plain

The vast swamps with clay and peat soils found here have hardly any percolation and surface drainage is also very slow. Some drainage into the few creeks or nearby rivers occurs. Most water is, however, removed by evapotranspiration.

For the Coronie Swamp a radial drainage pattern is indicated by Brinkman and Pons (1968), which is considered the result of the dome formed peat mound that is found here.

On the sand and shell ridges, groundwater flow is usually the most important component, although usually surface and lateral flow also occur in small quantities. The water is drained into the neighbouring lowlands (swamps). Drainage channels, if any, usually have an east-west direction.

7. A Few Remarks on Waterquality in Relation to Soil Conditions

7.1. The Precambrian Shield

The water draining from the shield will usually contain very few nutrients as the soils in these areas are very intensively weathered with a deep regolith. Areas with rock outcrops may influence the local water quality, but these areas are of limited extent.

The load of suspended matter is mostly very low as hardly any erosion takes place under natural conditions. At present it is sometimes slightly increased owing to human influences (villages, roads).

7.2. The Savanna Belt

A detailed study of the nutrient balance of the Savanna Belt has been executed by Poels (1987). He concludes that podzolisation strongly influences the water quality. The famous black water is the result of groundwater flow from podzol areas or lateral flow trough sandy topsoils, by which dissolved complexes of organic matter are drained into the creeks. Little or no iron or alumina is involved.

In areas with loamy to clayey soils the water is slightly brown.

The black water is very acid and very poor in nutrients. It has been observed that water from the Savanna Belt strongly influences the vegetation of the northern swamps into which it drains directly (Sevenhuysen 1977). A very poor vegetation type has developed here.

7.3. The Coastal Plains

In these areas with their moderate to high nutrient supply (Old and Young Coastal Plain resp.) a luxury vegetation has developed in marshes and swamps. Organic acids are produced by the destruction of organic matter in these environments, resulting in acid water. Most swamps produce brownish water.

In deep swamps where the vegetation is not rooted into the underlying soil, the vegetation may be poor as only rain water is involved.

Shell bodies will locally influence the water quality of nearby swamps by an increase of pH; the same is true for sandy ridges although the effect is much less pronounced.

A special case is formed by the potentially acid clays which are found in the Mara swamps and occur patch-like in the other part of the Young Coastal Plain. In the dry season the swamps may be drained sufficiently to oxidate the topsoil, thus producing strong acidity. These acid components will dilute into the water after recharge of the swamps in the rainy season, thus producing very to extremely acid conditions. Part of this water will be drained into creeks and rivers.

References

Anonymous, 1984. Micro-waterkrachtcentrale Marechallskreek. Hydrologische studie. Intern rapport Bureau voor Waterkrachtwerken, Ministerie v. Nat. Hulpbr.

Augustinus, P.G.E.F., 1978. The changing shoreline of Surinam. Thesis Utrecht.

Balsem, T., C.S. Bootsman, P.M. Dercksen, P.A. Kips, G.J. Rhebergen and A.R. Snel, n.d. Landforms, soils and use possibilities – Middle Suriname and Saramacca rivers. Map, scale 1:200,000. Soil Survey Dept. of Suriname.

Bennema, J., 1982. Acid soils of the humid tropics of South America, with special reference to the well drained soils on old alluvial sediments. In: Wienk, J.F. and H.A. de Wit (eds): Workshop on management of low fertility acid soils of the American humid tropics, Paramaribo, Suriname, November, 1981. IICA, San Jose, Costa Rica.

Bosma, W., S.B. Kroonenberg, R.V. van Lissa, K. Maas and E.W.F. de Roever, 1984. An explanation to the geology of Suriname. Meded. Geol. Mijnb. Dienst Suriname 27: 31–82.

Brinkman, R. and L.J. Pons, 1968. A pedo-geomorphological classification and map of the Holocene sediments in the coastal plain of the three Guianas. Soil Survey Papers No. 4, Soil Survey Inst., Wageningen, The Netherlands.

De Boer, M.W.H., 1972. Landforms and soils in eastern Surinam (South America). Doctoral thesis, Wageningen.

De Fretes, J.A. 1984. Micromorfologie van een Oxisol en een Ultisol uit Suriname. Internal report Dept. of Tropical Soil Science, Agricultural University, Wageningen.

De Vletter, D.R. (ed.), 1984. Synthesis of the Precambrian of Suriname and review of some outstanding problems. Meded. Geol. Mijnb. Dienst Suriname 27: 11–30.

Duynisveld, W. and R. van der Weert, 1974. Resultaten van een afvoeronderzoek in het Boven-Coesewijne gebied. Intern rapport no. 322, Afd. Bodemfysisch and Agrohydrologisch onderzoek, Landbouwproefstation, Paramaribo.

Haug, G.M.W., 1966. Verslag van het veldwerk in het stroomgebied van de Paloemeu. GMD informatie 17/66. Geol. Mijnb. Dienst Suriname.

Heyligers, P.C., 1963. Vegetation and soil of a white sand savanna in Surinam. The vegetation of Surinam, Vol.III. Thesis Utrecht. North Holland Publ. Cy, Amsterdam.

Kamerling, G.E., 1974. Bodemfysisch en Agrohydrologisch onderzoek in de jonge kustvlakte van Suriname. Verslagen van Landbouwkundige onderzoekingen no. 825.

King, L.C., D.K. Hobday and M. Mellody, 1964. Cyclic denudation in Surinam. Internal report on behalf of the Geological and Mining Service of Suriname.

Kips, P.A. and A.R. Snel, 1979. Landschap en bodem langs de midden-Suriname rivier. Report Dienst Bodemkartering no. 63.

Krook, L., 1984. Sediment petrographical studies in Northern Suriname. Meded. Geol. Mijnb. Dienst Suriname 28: 1–137.

Krook, L. and M.A. Mulders, 1971. Geological and related pedological aspects of the Upper Coesewijne Formation. Meded. Geol. Mijnb. Dienst Suriname 21: 183–208.

Lucas, Y., R. Boulet and A. Domeny, 1982. Acid soils of French Guiana. In: Wienk, J.F. and H.A. de Wit (eds): Workshop on management of low fertility acid soils of the American humid tropics, Paramaribo, Suriname, November, 1981. IICA, San Jose, Costa Rica.

O'Herne, L., 1965. Outline of the geomorphology of the basal complex of Surinam. Concept report Geol. Mijnb. Dienst Suriname.

O'Herne, L. 1969. A photo-geological study of the basal complex of Surinam. Meded. Geol. Mijnb. Dienst Suriname 20.

Pleysier, L.K., 1973. Land en bodem van het Avanavero gebied. Intern rapport no. 132, Dienst Bodemkartering, Suriname.

Poels, R.L.H., 1987. Soils, Water and Nutrients in a Forest Ecosystem in Suriname. Thesis, Wageningen.

Pons, L.J. 1966. Geogenese en pedogenese in de jong-Holocene kustvlakte van de drie Guyanas. Tijdschr. Kon. Ned. Aardr. Gen. 83: 153–172.

Sevenhuysen, R.J., 1977. Irrigatie uit een moeras. Een hydrologische studie van de Nannizwamp in Suriname. Doctoral thesis, Wageningen.

Veen, A.W.L., 1970. On geogenesis and pedogenesis in the Old Coastal plain of Surinam (South America).Publicaties van de Dienst Bodemk. nr. 32.

Wensink, J.J., 1968. The Emma Range in Surinam. Publ. v. h. Fys. Geogr. en Bodk. Lab. v. d. Univ. v. Amsterdam, no. 13.

Westerink, R.M., 1989. Excursiegids, excursie door Noord-Suriname 1989. Internal document, Instituut voor de Opleidingen van Leraren, Paramaribo.

Wong, Th. E., (1984). Stratigraphy and sedimentary history of the Guiana basin. Meded. Geol. Mijnb. Dienst Suriname 27: 83–90.

Zonneveld, J.I.S., 1952. Watervallen in Suriname. Tijdschr. Kon. Ned. Aardr. Gen. 69: 499–507.

3. Climate and surface water hydrology

MOEKIRAN A. AMATALI

1. Introduction

Suriname is drained by seven rivers towards the Atlantic Ocean. Three main groups can be distinguished, when the extent and shape of the drainage areas are compared, see Fig. 1.

The larger Marowijne and Corantijn Rivers, with drainage areas of approximately 68,700 km^2 and 67,600 km^2 respectively, represent the first group, and together they drain almost 58% of the country. These rivers are border rivers respectively in the East and the West.

The Coppename River (approx. 21,700 km^2) and the Suriname River (approx. 16,500 km^2) represent the second group, and together they drain approximately 24% of the country.

The Nickerie River (approx. 10,100 km^2), the Saramacca River (approx. 9,000 km^2) and the Commewijne River (approx. 6,600 km^2) represent the third group of rivers. These rivers together drain approximately 16% of the country.

It is obvious that the rivers of the first and second group mentioned above have a South-North flow, directly to the Atlantic Ocean. The rivers of the third group which are smaller ones deflect to the West on the Coastal Plain towards one of the larger rivers.

The courses of the rivers can be divided into two different parts, namely the courses through the hills and mountains of the Guiana Shield, and the estuaries traversing the Coastal Plain. Further intrusion of the tide is halted by the most seaward rapids, which mark the transition between the two parts. The Commewijne River and its tributary the Cottica River is the only river in Suriname which flows almost entirely through the Coastal Plain.

Hydrologically the country can be divided into three distinct areas. They are from the South to the North as follows, the Interior (Basement Area), the Savanna Belt, and the Coastal Plains. The largest one is the Interior covering 126,500 km^2 (Duba 1972) of the country, approximately South of 5° N Latitude. The infiltration and percolation in this area is comparatively small. As a consequence the surface run-off is relatively large, and the drainage density high.

Paul E. Ouboter (ed.), Freshwater Ecosystems of Suriname, 29–51.

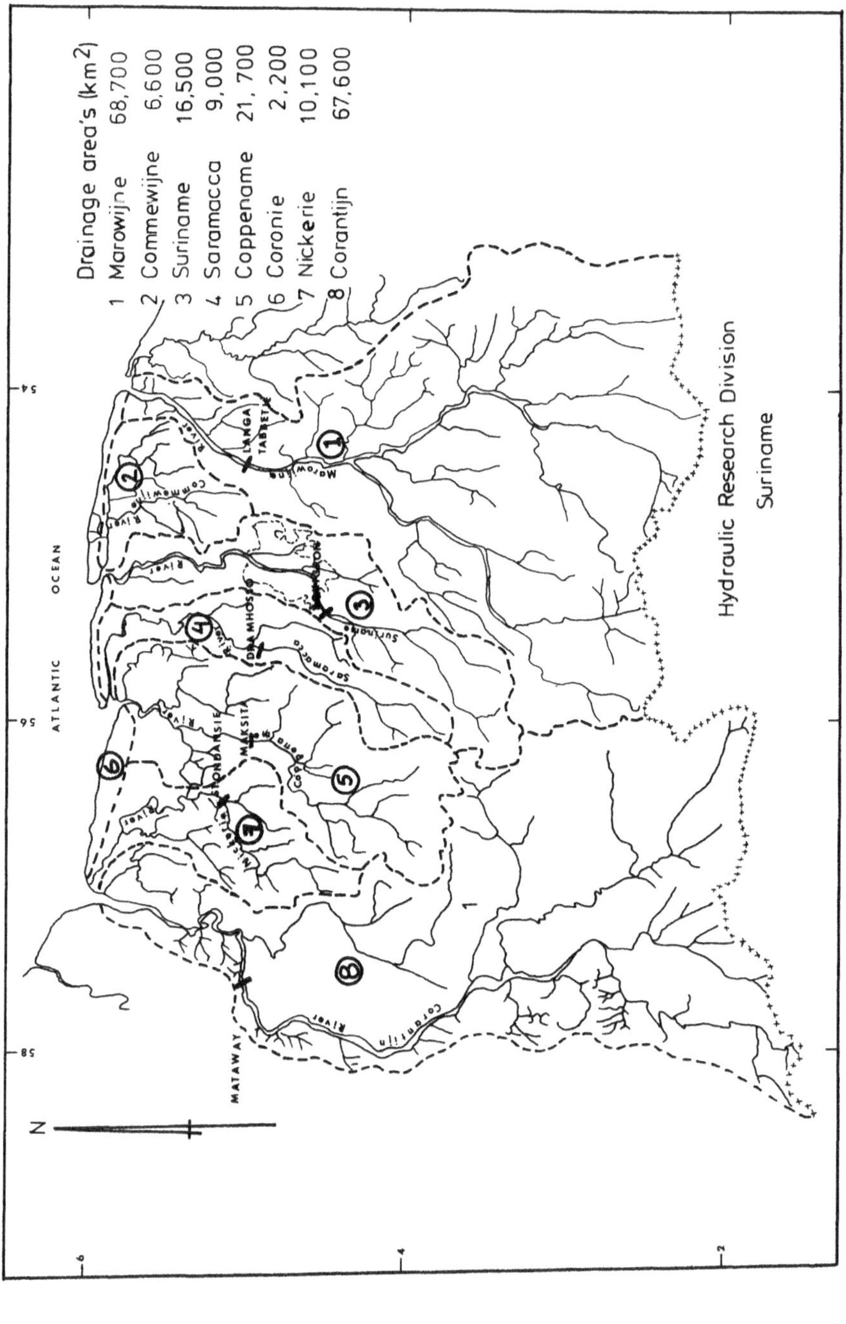

Figure 1. The main rivers of Suriname.

The Savanna Belt is the smallest of the three above-mentioned areas, with a total area of 8,500 km^2 (Duba 1972). This area is covered with coarse sands with interbedded sandy clay. One may expect that the infiltration and percolation is high, and the surface run-off small. The density of the drainage network of the Savanna Belt is less than the density of the drainage network of the Interior.

The Coastal Plains Area is spread over 21,000 km^2 (Duba 1972) of the country, and is low and flat. The surface run-off here is small, and the drainage network is not developed well. The water in the basin is stagnated and numerous swamps exist in this area. The major hydrological feature of this area is a vertical exchange of water via a rainfall - basin evapotranspiration mechanism. This results in a small run-off, and large input of water into and out of the Coastal Plains area.

At the upper reaches beyond the estuaries, where there is no tidal influence, the water level and currents in the rivers are influenced by the freshwater discharge. Consequently high waters correspond with peak flows, and low waters with low flows.

In the estuary the water level and currents are influenced by the tide, as well as by the freshwater discharge. Due to the tide, tidal inflow and outflow of seawater occur into and out of the estuary. The amount of water which moves into and out of the estuary is referred to as the tidal volume or tidal prism.

The tidal flow diminishes further upstream. Beyond a certain point upstream there is a continuous flow towards the lower reaches, and there is no reversal flow because of the freshwater discharge. This continuous flow varies in strength periodically, as far as the tide is noticeable. The location of the point at which the reverse tidal flow stops, depends upon the magnitudes of the freshwater discharge and the tidal range, and thus the tidal limit depends upon these two magnitudes.

The tide along the coast of Suriname is of the regular semi-diurnal type, having two high waters and two low waters during a tidal day (Nedeco 1968).

The tidal flow from the sea and the differences in density of the sea water and freshwater cause the salt-intrusion in the estuaries. On the other hand, the freshwater discharge of the rivers prevents further intrusion of the tide and saline waters. Tidal and salt-intrusion extend further landward, as the freshwater discharge decreases. So during periods of low flows (dry season) maximum intrusion occurs, and during peak flows (rainy seasons) minimum intrusion.

Silt originating from the Amazon basin is transported by the sea water along the coast of Suriname. These sediments are deposited at locations where the mixing of saline water and freshwater takes place (flocculation). As a result shallow banks and bars occur in the lower courses of the Surinamese rivers. On the other hand there is also a discharge of sediments, which originate from the Surinamese river-basins.

2. Climatological Aspects

2.1. General

Suriname lies in an area in which a North-East and South-East trade wind may occur. The Inter-Tropical Convergence Zone ("ITC-zone", also called the centre of the "Equatorial Trough") is the most decisive factor in creating the climate of Suriname. Two times a year the ITC-zone passes over Suriname, resulting in four seasons, namely

- the short dry season: early February to late April
- the long rainy season: late April to middle of August
- the long dry season: middle of August to early December
- the short rainy season: early December to early February.

The climate of Suriname is according to Köppen humid-tropical (Af), and has a mean monthly rainfall higher than 60 mm, during the driest months. However, the climate of a narrow strip along the coast, Coronie, and the South-West part of the country is drier. The long-term mean annual rainfall varies from about 1,450 mm (Coronie) to about 3,000 mm (Tafelberg) (National Planning Office of Suriname & Organization of American States 1988) (Fig. 2). The mean annual rainfall at Paramaribo, the capital of Suriname, which is located at about 10 km South of the coast, is about 2,200 mm. The maximum daily rainfall may rise up to 100–200 mm once every 2 or 3 years.

The mean temperature at Paramaribo is 27.1 °C. January is the coldest month with a mean temperature of 26.2 °C, while during the warmest months – September and October – the mean temperature is 28.2 °C. The mean of the daily maximum temperature is 30.5 °C and the mean daily minimum is 22.8 °C. Along the coast the daily fluctuations of the temperature is about 5 °C and is 10–12 °C in the Interior (National Planning Office of Suriname & Organization of American States 1988).

The mean windspeed is 1.3 on the scale of Beaufort. During the dry seasons the maximum windspeed occurs: a maximum of 1.6 Beaufort occurs in February, and a second peak in September and October. A minimum of 1.0 Beaufort occurs in January. During daytime windspeeds of 3–4 Beaufort generally occur, while in the evening and at night, especially in the Interior the windspeed drops (National Planning Office of Suriname & Organization of American States 1988).

2.2. Precipitation and Evapotranspiration

In the rainfall-run-off process the rainfall and evapotranspiration (= evaporation + plant transpiration) are important components. Four climatological stations are selected here where sufficient rainfall and pan-evaporation data are available, to give a rough idea of the precipitation and evapotranspiration over the whole country throughout the year. These stations

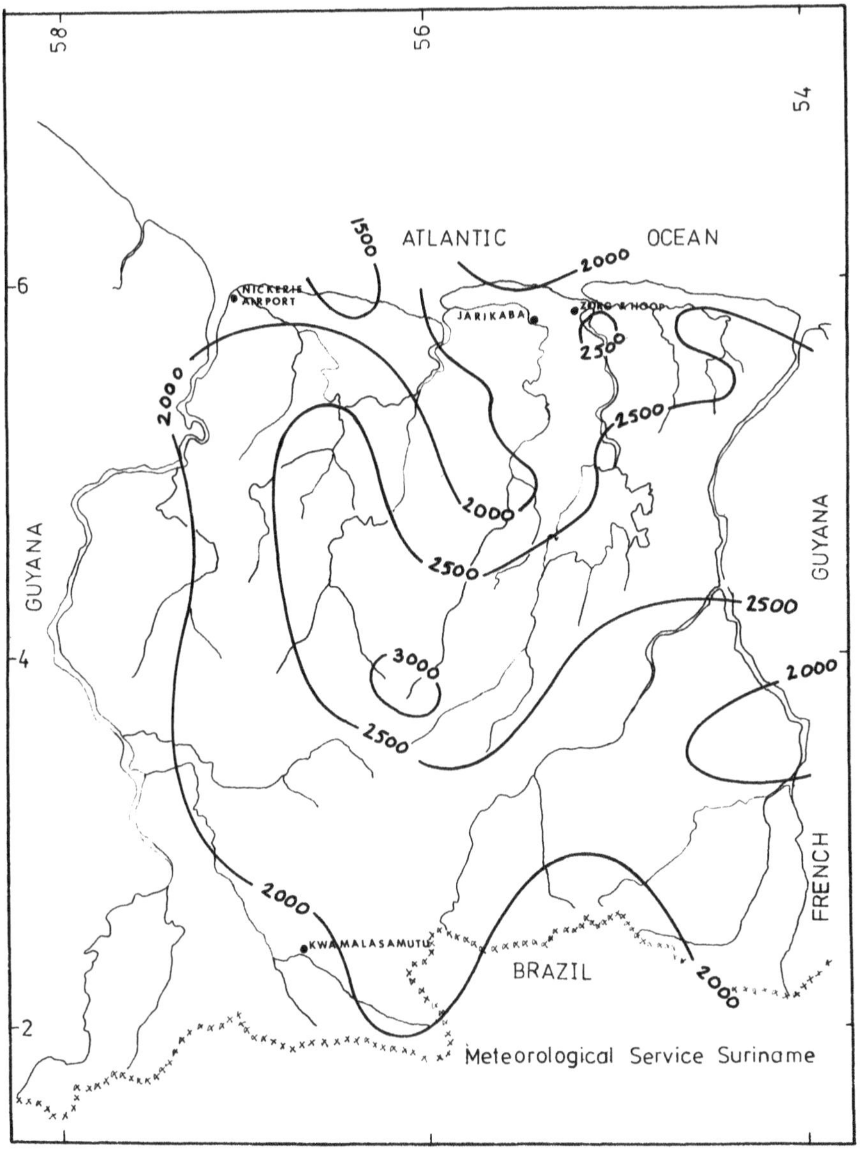

Figure 2. Average annual rainfall (in mm).

are Zorg & Hoop, Jarikaba, Nickerie Airport, and Kwamalasamutu. Zorg & Hoop is located near Paramaribo, Jarikaba about 20 km West of Paramaribo, and Nickerie Airport is near the outfall of the Nickerie River in North-West Suriname. Kwamalasamutu is located in the Southwestern part of the country, where the annual rainfall is less (see Fig. 2).

The long-term mean, maximum and minimum values, and the standard deviation of the monthly total and annual total rainfall of the above-mentioned stations are presented in Table 1.

According to recent analyses of rainfall data by C. Becker of the Meteorological Service of Suriname the precipitation in Suriname is decreasing at an average rate of about 4 mm a year since the past 100 years. Fig. 3 presents the trend of the annual precipitation of station Cultuurtuin in Paramaribo, as well as the 7-years moving average, and the years when the "El Niño" phenomenon occurred. It is obvious that in only approximately 20% of the years when the El Niño phenomenon occurred the annual rainfall was above the trend.

The long-term monthly total and annual total pan-evaporation are presented in Table 2. These values have been observed by means of a Class-A evaporation

Table 1. Longterm average, maximum and minimum precipitation totals (mm) at stations Kwamalasamutu, Nickerie Airport, Jarikaba and Zorg & Hoop.

Precipitation (mm) 1977-1990
Station: 144 Kwamalasamutu

	Jan	Feb	Mar	Apr	May	Jun	Jul	Aug	Sep	Oct	Nov	Dec	Annual
AVE	116.8	103.9	224.6	254.0	389.7	302.1	233.1	151.8	78.3	52.8	40.4	78.5	1,729.6
MAX	216.0	166.3	370.7	340.7	528.2	547.0	512.0	317.6	135.3	95.5	125.9	135.9	2,481.7
MIN	40.7	21.9	34.1	100.5	131.8	94.6	26.0	23.3	25.1	15.5	2.1	26.3	908.0
STDEV	56.5	50.9	120.0	100.5	130.1	127.2	128.6	82.8	35.4	22.6	42.3	35.6	464.0

Precipitation (mm) 1965–1990
Station: 209 Nickerie Airport

	Jan	Feb	Mar	Apr	May	Jun	Jul	Aug	Sep	Oct	Nov	Dec	Annual
AVE	167.2	93.7	98.8	173.3	246.1	264.1	229.2	166.1	83.2	58.7	83.4	156.0	1,754.1
MAX	427.1	297.1	258.4	454.0	480.0	396.0	381.9	281.9	342.8	181.8	224.0	329.5	2,215.8
MIN	37.3	15.7	2.6	20.0	50.4	192.5	130.6	35.6	10.4	0.5	6.1	28.9	1,366.5
STDEV	111.9	73.5	69.9	112.9	102.9	57.3	78.4	67.7	66.5	50.1	54.5	82.4	288.0

Precipitation (mm) 1969–1990
Station: 530 Jarikaba

	Jan	Feb	Mar	Apr	May	Jun	Jul	Aug	Sep	Oct	Nov	Dec	Annual
AVE	193.7	137.2	136.2	210.8	295.7	289.1	230.7	158.2	103.1	103.7	140.3	199.9	2,138.1
MAX	445.6	317.1	271.7	516.2	579.2	423.3	336.9	317.1	263.8	340.1	459.8	372.7	2,891.8
MIN	33.7	10.8	10.7	35.5	75.6	180.8	132.9	44.4	0.7	0.0	31.4	93.4	1,194.5
STDEV	130.9	84.4	71.3	130.0	107.2	72.8	49.5	80.7	69.2	74.4	95.3	67.0	410.1

Precipitation (mm) 1965–1990
Station: 604 Zorg & Hoop

	Jan	Feb	Mar	Apr	May	Jun	Jul	Aug	Sep	Oct	Nov	Dec	Annual
AVE	186.4	112.8	131.9	194.3	284.1	304.1	248.1	159.0	105.5	114.5	123.2	192.7	2,128.6
MAX	438.1	306.0	344.7	466.1	484.6	459.4	389.3	340.7	287.7	216.8	221.2	385.3	2,948.1
MIN	34.1	11.8	5.2	27.5	88.3	192.9	142.6	59.6	32.9	1.0	28.5	88.1	1,539.0
STDEV	127.0	72.7	76.1	112.7	96.4	74.5	67.6	72.5	58.3	64.2	54.8	72.6	348.6

Table 2. Long-term mean pan evaporation totals (mm) at stations Kwalasamutu, Nickerie Airport, Jarikaba and Zorg & Hoop.

Pan evaporation (mm) 1979–1986
Station: 144 Kwamalasamutu

	Jan	Feb	Mar	Apr	May	Jun	Jul	Aug	Sep	Oct	Nov	Dec	Annual
Mean	108.2	78.0	72.4	71.1	131.8	142.3	107.2	85.8	54.7	53.2	45.7	62.8	1,013.2

Pan evaporation (mm) 1973–1990
Station: 209 Nickerie Airport

	Jan	Feb	Mar	Apr	May	Jun	Jul	Aug	Sep	Oct	Nov	Dec	Annual
Mean	129.7	141.5	165.6	156.9	132.1	129.8	141.7	156.8	166.5	174.9	165.6	139.9	1,801.0

Pan evaporation (mm) 1975–1989
Station: 530 Jarikaba

	Jan	Feb	Mar	Apr	May	Jun	Jul	Aug	Sep	Oct	Nov	Dec	Annual
Mean	119.3	109.1	112.1	128.3	114.9	87.3	123.0	143.0	145.8	119.4	108.1	115.3	1,425.6

Pan evaporation (mm) 1973–1990
Station: 604 Zorg & Hoop

	Jan	Feb	Mar	Apr	May	Jun	Jul	Aug	Sep	Oct	Nov	Dec	Annual
Mean	93.3	97.6	126.8	118.6	128.2	118.4	132.4	140.1	119.9	104.2	101.7	123.8	1,405.0

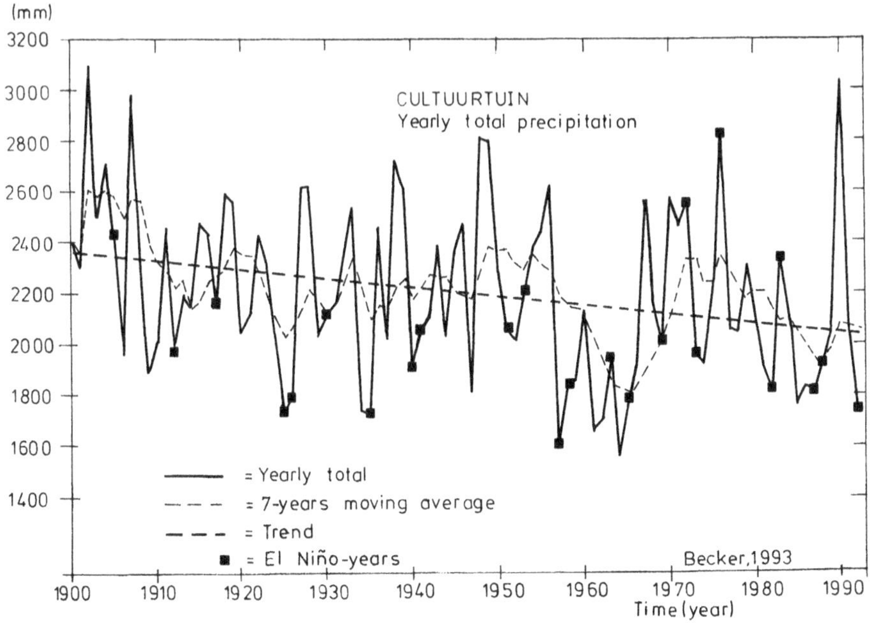

Figure 3. Yearly total precipitation at station Cultuurtuin.

pan, and give an idea of the evapotranspiration. One may estimate the potential evaporation by multiplying the pan-evaporation with the pan-coefficient, which is on average about 0.7 for the Class-A pan. Although the pan-evaporation does not give accurate estimates for the potential evapotranspiration (= maximum possible evapotranspiration in a given climatic condition, where the availability of water is not a limitation), one may conclude that the potential evapotranspiration in the Southwestern part of the country is less than the potential evapotranspiration in the Coastal Plain.

The precipitation surplus at the above-mentioned stations is presented by Figs. 4 and 5, and is equal to the total precipitation minus the potential evapotranspiration, assuming that the potential evapotranspiration here is equal to the potential evaporation. The precipitation surplus gives an indication of the availability of the precipitated water for the discharge. In Fig. 4 a pan-coefficient of 1.0 has been employed, and in Fig. 5 a pan-coefficient of 0.7. It should be mentioned that up to now only few investigations have been conducted in Suriname with regard to evapotranspiration.

Referring to Figs. 4 and 5 the precipitation surplus is the largest during the months of May, June and July, while during the months of September, October and November this parameter is the smallest. For the stations Zorg & Hoop and Jarikaba the values do not differ very much. For Nickerie Airport the values seem to be the smallest compared to the values of the other stations. For the first half year the values of the station Kwamalasamutu are larger than the values of the remaining stations.

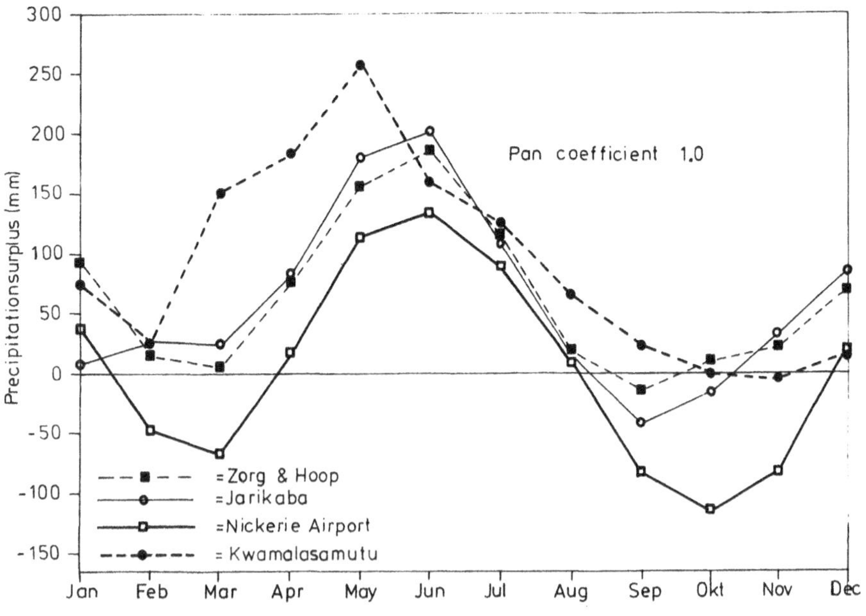

Figure 4. Precipitation surplus for four meteorological stations in Suriname, using a pan coefficient of 1.0.

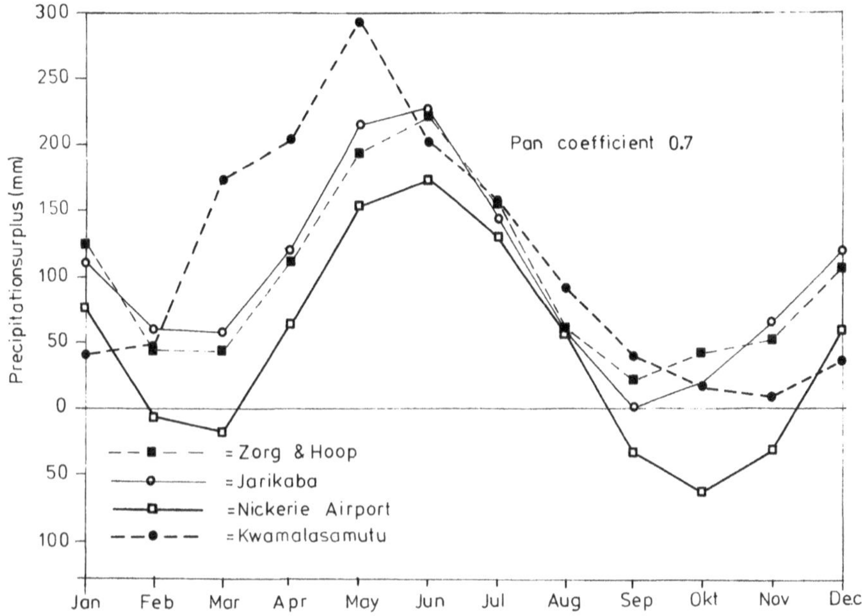

Figure 5. Precipitation surplus for four meteorological stations in Suriname, using a pan coefficient of 0.7.

3. The Rivers and Riverbasins

3.1. General

In this part a description is given of the specific features of the different Surinamese rivers and river-basins.

The locations along the river courses are indicated by kilometres along the thalweg. The zeropoint is in the rivermouth, where the depth reaches 15 meters below low water spring (Nedeco 1968).

The considered discharge measurement stations are beyond the tidal limit, since it is complicated and laborious to measure the discharge continuously in the tidal reaches.

The limit of 300 mg Cl/l is considered here for the salt-intrusion, which is equivalent to 500 mg salt/l. This limit is used here, since most crops are tolerant to water with a salinity of less than 300 mg Cl/l. According to the available data there is a relationship between the freshwater discharge and the salt-intrusion.

The silt content is also influenced by the tide and the freshwater discharge, since most of the silt originates from the Amazon basin, and is transported by the sea water along the Surinamese coast.

3.2. Marowijne River Basin

The sources of the Marowijne River, and most important tributaries are in the mountain ranges which form the border with Brazil, and in the central highlands. The catchment area of this river is about 68,700 km^2, including the catchment area on the right bank in French Guiana (Hydraulic Research Division 1969). 99.45% is in the Interior, 0.05% in the Savanna Belt, and 0.5% in the Coastal Plains (Duba 1972). Many falls and rapids are situated along the upper river courses and tributaries.

3.2.1. Estuary

The estuary of the Marowijne River is located downstream of the Armina Falls (km 114), about 90 km from the outfall. The mean tidal range is about 2 m in the rivermouth. The tidal volume of the Marowijne River is estimated at 200 million m^3 and is less than the tidal volume of the Corantijn River (Nedeco 1968; Hydraulic Research Division 1969).

The salt-intrusion in the Marowijne estuary is less than the salt-intrusion in the Corantijn estuary. During wet periods the 300 mg Cl/l limit is, according to available data at the Hydraulic Research Division (HRD), close to km 37 in the Marowijne River, approx. 20 km downstream of Albina (km 57). During dry seasons this limit is near km 59, several km upstream of Albina.

3.2.2. Freshwater Discharge

The mean discharge at the outfall of the river was estimated at 2,100 m^3/s. Peak discharges may exceed 9,000 m^3/s, and minimum discharges may be as low as 100 m^3/s (HRD 1969).

According to data collected over 29 years (1952–1980) the mean freshwater discharge of the Marowijne River at the station Langa Tabbetje (km 134) is 1,650 m^3/s. The recorded maximum is 5,670 m^3/s, which occurred in May 1976, and in November 1965 the minimum was recorded, namely 45 m^3/s. The drainage area of the river upstream of Langa Tabbetje is about 63,500 km^2, which is about 92% of the total catchment area of the Marowijne River (HRD 1982). Assuming a mean discharge of 1,650 m^3/s at Langa Tabbetje, and taking the catchment area into account, the mean discharge at the outfall is estimated here at 1,785 m^3/s.

The exceedence frequency of the monthly freshwater discharge at Langa Tabbetje is presented in Fig. 6 (HRD 1982). One can conclude that peak flows and high waters occur around May-June, and low flows and low waters around November. These figures coincide more or less with the pattern of the precipitation surplus (Figs. 4 and 5). There is clearly a relationship between the precipitationsurplus and the discharge here, with some lag-time.

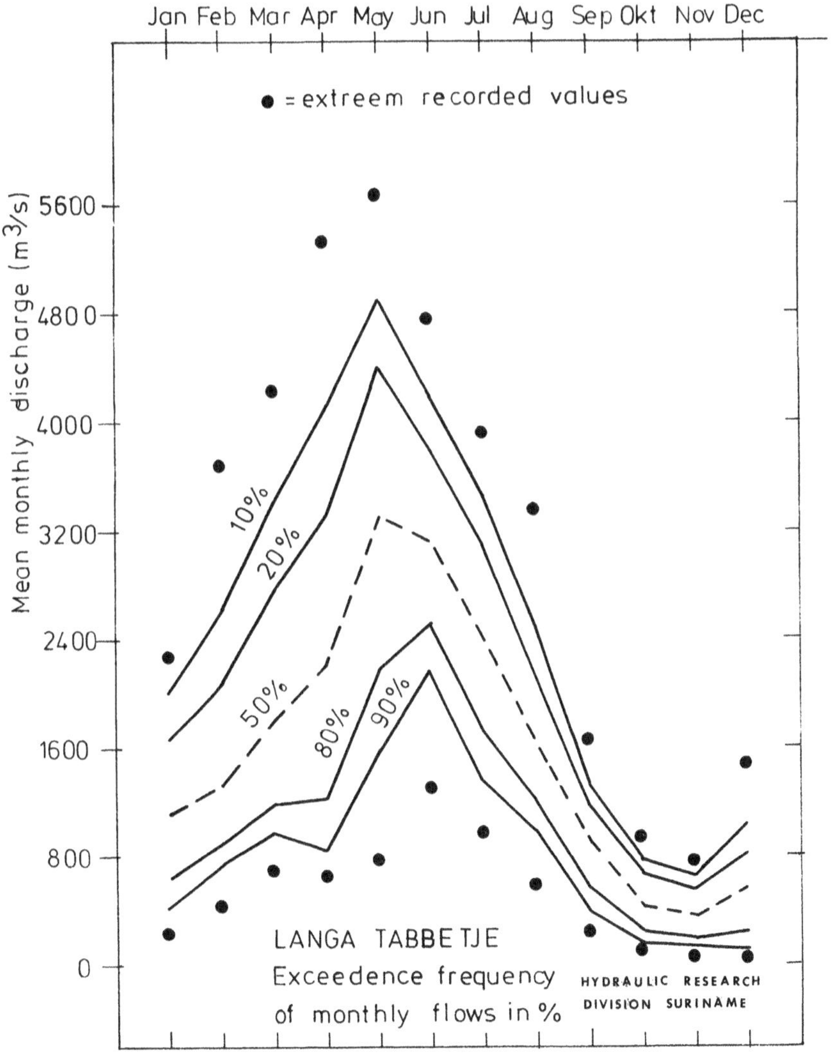

Figure 6. Exceedence frequency of the monthly freshwater discharge at Langa Tabbetje (Marowijne).

3.2.3. Sediment Transport

According to available data on silt content at the HRD, the silt content at Albina (km 57) varies from approx. 0.01 to 0.08 g/l, and it is expected that upstream the silt content decreases. Nedeco (1968) estimated the sediment discharge from the Marowijne River basin at 1.3 million tons per year.

3.2.4. Depths

The maximum depth along the thalweg at the lower reaches is approx. 21 m below the mean water level close to the French Guiana bank (km 47) and the minimum thalweg depth in this zone is approx. 4 m below mean sea level close to the outfall (km 15) (HRD 1969).

3.3. Commewijne River Basin

The sources of the Commewijne River are in the Hok-A-Hing Mountains range near the Brokopondo Storage Lake. The catchment area of this river is about 6,600 km² (HRD 1972). 36.4% is in the Interior, 11.5% is in the Savanna Belt and 52.1% in the Coastal Plain (Duba 1972). Of the rivers treated here, the Commewijne River is the only one without rapids in its upper reaches.

3.3.1. Estuary

The tidal range in the outfall is about 1.9 m and the tidal volume is estimated at 40 million m³ (Nedeco 1968). The location of the tidal limit is not exactly known yet.

According to the available data at the HRD, during dry seasons the 300 mg Cl/l limit is located near km 150 in the Cottica River, approx. 10 km downstream from Tamarin. This limit is approx. 5 km upstream from Alkmaar during the wet season.

3.3.2. Freshwater Discharge

The discharge at the outfall of the Commewijne River was estimated at 120 m³/s by Ringma (1961). Only a few discharge measurements have been done in the Commewijne River. According to discharge measurements at Destombesburg (km 115) during the period 8–21 December 1979, the mean discharge at that station is 51.3 m³/s, and varies between 28 to 97 m³/s (HRD 1983).

3.3.3. Sediment Transport

The silt content at Alliance (km 67.2) varies according to data of the HRD between approx. 0.07 g/l to 20 g/l. Nedeco (1968) estimated the sediment discharge from the Commewijne River at 0.06 million tons per year.

3.3.4. Depths

The minimum depth of the thalweg is near the outfall, and is approx. 6.6 m with respect to the mean sea level. The maximum depth is approx. 31.0 m with respect to the mean water level, and is located near Post Sommelsdijk (km 71) (HRD 1972).

3.4. Suriname River Basin

The sources of the Suriname River are in the Eilerts de Haan Mountain range. The length of this river is approx. 500 km, and the catchment area is approx. 16,500 km^2 (HRD 1988). 84.8% of the catchment is in the Interior, 4.5% in the Savanna Belt, and 10.7% in the Coastal Plain (Duba 1972). Along the upper river courses and tributaries, there are many falls and rapids. This is the only river in Suriname which is regulated. In 1964 a dam has been constructed for hydro-electric purposes at Afobaka (km 194). So an artificial lake, the Prof. van Blommenstein Storage Lake (Brokopondo Lake) exists between Afobaka and Pokigron (km 273), of which the area is about 1,600 km^2.

3.4.1. Estuary

The tidal intrusion depends upon the discharge from the Brokopondo Lake. Under normal discharge from the hydro-electric station, 250 m^3/s, the tidal influence can reach as far as Berg en Dal (km 164), while the mean tidal range in the river mouth is approx. 1.8 m. The tidal volume of the Suriname River is approx. 125 million m^3 (Nedeco 1968; HRD 1988).

The salt-intrusion also depends upon the discharge from the Brokopondo Lake. According to available data at the HRD the 300 mg Cl/l limit is near km 54, some kilometers downstream from Sluis Doorsteek during periods of peak flow; during periods of low flow this limit is near km 90, just upstream from Paranam.

3.4.2. Freshwater Discharge

The mean discharge at the outfall of the Suriname River was estimated at 440 m^3/s. Peak flow may be as large as 1,800 m^3/s and low flow as low as 220 m^3/s (HRD 1970).

According to collected data over 27 years (1952–1978) the mean freshwater discharge of the Suriname River at station Pokigron (km 273), which flows into the Brokopondo Lake is 224 m^3/s. The recorded maximum discharge at this station is 923 m^3/s (May 1976), and the recorded minimum is 5.8 m^3/s (December 1965) (HRD 1982).

The exceedence frequency of the monthly freshwater discharge at Pokigron (HRD 1982) shows that peak flows and high water occur around May–June and low flows and low water around November.

The average discharge at Afobaka (km 194) is 324 m^3/s according to the collected data over a period of 12 years (1965–1980). The recorded minimum is 214 m^3/s (July 1975), while the recorded maximum is 756 m^3/s (June 1976). The drainage area upstream of the Afobaka station is approx. 12,550 km^2, which is approx. 76% of the total catchment area of the Suriname River (HRD 1982). Assuming a mean discharge of 324 m^3/s, and taking the catchment area into account, the mean discharge at the outfall of the Suriname River is now estimated at 426 m^3/s.

3.4.3. Sediment Transport
According to available data at the HRD, the silt content at Nieuw Amsterdam (km 40) varies between approx. 0.01 to 20.0 g/l, and at Domburg (km 68) this parameter varies between approx. 0.001 to 1.5 g/l. The sediment discharge from the Suriname River is estimated at 0.25 million tons per year by Nedeco (1968).

3.4.4. Depths
Near the rivermouth there are shallow spots along the thalweg, less than approx. 5 meters below mean sea level. Upstream up to Paranam the depth increases to approx. 17 meters with respect to the mean water level. Near km 130 several km downstream from Phedra, there are also shallow spots along the thalweg (HRD 1988).

3.5. Saramacca River Basin

The sources of the Saramacca River are in the central highlands of the country. The river emerges from the Emma Mountains in the South-West, and from the Van Asch van Wijk Mountains in the South-East. The length of the river is approx. 460 km and the catchment area is approx. 9,000 km² (HRD 1987). 74.1% of the drainage area is in the Interior, 3.1% in the Savanna Belt and 22.8% in the Coastal Plain (Duba 1972). Similar to the other rivers of Suriname, there are numerous falls and rapids along the upper courses of this river.

3.5.1. Estuary
During periods of low flows the tidal influence reaches much further than Kwakugron (km 220) (HRD 1987). The tidal prism is estimated at 50 million m³ by Nedeco (1968).

According to collected data by the HRD the limit of 300 mg Cl/l is near km 37, approx. 6 km downstream from Carl François during periods of peak flow, and during periods of low flow this limit is near km 89, approx. 5 km upstream of Groningen.

3.5.2. Freshwater Discharge
The mean discharge of the Saramacca River at the outfall was estimated at 240 m³/s by Ringma (1961).

The mean discharge of the Saramacca River is according to collected data over the period 1961–1978 (18 years) 100 m³/s at station Dramhoso (km 285). The observed maximum discharge of this station is 493 m³/s (May 1976), and a minimum of 1.8 m³/s occurred in November 1964. The drainage area upstream from the Dramhoso station is approx. 3,520 km², which is 39.1% of the total catchment area of the Saramacca River (HRD 1982). Assuming a mean discharge of 100 m^3/s at the Dramhoso station, and taking into account the catchment area, the mean discharge at the outfall is estimated now at 225 m^3/s..

Taking the exceedence frequency of the monthly discharge at Dramhoso (HRD 1982) into account, one may conclude that peak flows and high waters occur around May–June, and low flows and low waters around November.

3.5.3. Sediment Transport

According to the available data at the HRD the silt content at Huwelijkszorg (km 53) varies between approx. 0.003 and 1.3 g/l and at Uitkijk (km 104) between about 0.003 and 0.8 g/l. The sediment discharge from the Saramacca River is estimated at 0.13 million tons per year (Nedeco 1968).

3.5.4. Depths

The minimum depth along the thalweg is approx. 3 m with respect to the mean sea level, and is located near the outfall. The depth increases upstream, and there are depths of approx. 24 meters below mean water level near km 145, some kilometers downstream from Locus Creek (HRD 1987).

3.6. Coppename River Basin

The sources of the Coppename River are in the Wilhelmina Mountains in the central highlands. The catchment area is approx. 21,700 km^2 (HRD 1971). 65.5% is in the Interior, 9.2% in the Savanna Belt and 26.3% in the Coastal Plain (Duba 1972). Similar to most other rivers of Suriname there are numerous falls and rapids along the upper reaches.

3.6.1. Estuary

The tidal range in the outfall is about 2 m, while the tidal prism is estimated at 75 million m^3 (Nedeco 1968). During periods of low flow the tidal influence can reach as far as Kwintikriki Falls (km 172) (HRD 1971).

According to data collected by the HRD the 300 mg Cl/l limit is near Boskamp (km 31) during periods of peak flow. During periods of low flow this limit is near km 83, about 12 km downstream from the outfall of the Wajambo River (km 95).

3.6.2. Freshwater Discharge

The mean discharge at the outfall was estimated at 490 m^3/s. Peak flows may exceed 2,200 m^3/s, and low flows may be as low as 6 m^3/s (HRD 1971).

At discharge station Maksita (km 175) the mean discharge is 319 m^3/s, based on available data over the period 1965–1980 (16 years). The maximum recorded discharge is 1,420 m^3/s (May 1976), and the minimum recorded discharge is 4.4 m^3/s (December 1969). The drainage area upstream from Maksita station is approx. 12,300 km^2, which is 57% of the total catchment area of the Coppename River (HRD 1982). Assuming a mean discharge of 319 m^3/s at Maksita, and taking the catchment area into account the mean discharge at the outfall of the Coppename River is estimated here at 500 m^3/s.

According to the exceedence frequency of the monthly flow at Maksita (HRD 1982) it is expected that generally peak flows and high water occur around May–June, and low flows and low water around November.

3.6.3. Sediment Transport

The silt content at Boskamp (km 32) varies according to data collected by the HRD between approx. 0.03 and 50 g/l, and at the Tibiti outfall (km 68) between approx. 0.02 and 1.3 g/l. Nedeco (1968) estimated the sediment discharge from the river at 0.25 million tons a year.

3.6.4. Depths

The minimum depth along the natural channel at the lower reaches is near the rivermouth, and is about 3.7 m with respect to the mean sea level. The greatest depth is approx. 22 m below mean water level, and is located near km 82, 13 km downstream of the Wajambo outfall (HRD 1971).

3.7. Nickerie River Basin

The source of the Nickerie River is situated in the Bakhuys Mountains. The catchment area of this river is approx. 10,100 km^2 (HRD 1970). Duba (1972) mentioned that 42.3% of the catchment is in the Interior, 23.6% in the Savanna Belt, and 34.1% in the Coastal Plains. Also in the upper reaches of this river there are numerous falls and rapids.

3.7.1. Estuary

During periods of low flow the tidal influence can be felt as far as Stondansie Falls (km 240) (HRD 1970). Nedeco (1968) mentioned that the tidal range is approx. 2.00 m at the outfall, and estimated the tidal prism at 10 million m^3.

According to data collected by the HRD the limit of 300 mg Cl/l may be further than the outfall of the Akwanza Creek during periods of low flow, while during periods of peak flow this limit is near Waterloo (km 28.2).

3.7.2. Freshwater Discharge

The mean discharge at the rivermouth was estimated at 160 m^3/s. The maximum discharge may exceed 500 m^3/s, while the minimum discharge may be as low as 2 m^3/s (HRD 1970).

At station Stondansie (km 240) the mean discharge of the Nickerie River is 91 m^3/s, based on collected data over the period 1963–1980 (18 years). In July 1971 a maximum of 450 m^3/s was recorded, and in May 1964 and February 1966 a minimum of 1.8 m^3/s occurred. The drainage area upstream from station Stondansie is approx. 5,160 km^2, which is 51% of the total catchment area of the Nickerie River (HRD 1982). Taking the catchment area into account, and assuming a mean discharge of 91 m^3/s at Stondansie the mean discharge at the rivermouth of the Nickerie River is estimated now at 178 m^3/s.

Peak flows and high water occur around June–July, while low flows and low water occur around November, according to the exceedence frequency of the monthly flow at Stondansie (HRD 1982).

3.7.3. Sediment Transport

According to data collected by the HRD the silt content at Klein Henar (km 52) varies between approx. 0.04 to 45 g/l. At Wageningen (km 73) it varies between approx. 0.03 to 6 g/l. The total annual sediment discharge from the Nickerie river is estimated at 0.1 million tons by Nedeco (1968).

3.7.4. Depth

A minimum depth of the natural channel of approx. 2.8 m with respect to the mean sea level is near the rivermouth. The greatest depth is approx. 17 m, and is recorded near Wageningen (km 73) (HRD 1970).

3.8. Corantijn River Basin

The sources of the Corantijn River are located in the Acarai Mountains near the boundary with Brazil, and in the central highlands. The catchment of this river is about 67,600 km^2, including the catchment area on the left bank of the Corantijn River which is in Guyana (HRD 1969). According to Duba (1972) 90% is in the Interior, 4.6% in the Savanna Belt, and 5.4% in the Coastal Plain. Along the upper courses of this river there are numerous falls and rapids.

3.8.1. Estuary

The tidal range in the outfall is approx. 2 m, and the tidal volume is estimated at 300 million m^3 (Nedeco 1968). During periods of low flow the tidal influence is up to Cow Falls, which is located at a distance of about 210 km from the outfall (HRD 1969).

According to collected data by the HRD, during peak flows the 300 mg Cl/l limit is near km 40, some km upstream from Springland, and during periods of low flows this limit is near Oroprecaro Creek (km 82), near Baboon Island.

3.8.2. Freshwater Discharge

The mean discharge at the outfall of the river was estimated at 1,800 m^3/s. Peak flows may exceed 8,000 m^3/s, and low flows can be as low as 30 m^3/s (HRD 1969).

Based on collected data over the period 1966–1980 (15 years), the mean discharge at the station Mataway (km 243) is 1,200 m^3/s. The maximum recorded discharge at this station is 5,370 m^3/s (July 1971) and the minimum recorded discharge is 31 m^3/s (January 1966). The catchment area of the Corantijn River upstream from this station is approx. 51,600 km^2, which is 76% of the total catchment area of the Corantijn River. Adopting a mean discharge of 1,200 m^3/s at the station Mataway, and taking the catchment area into account, the mean discharge of the Corantijn River is estimated now at 1,572 m^3/s.

According to the exceedence frequency of the monthly flow (HRD 1982), peak flows and high water occur during the months June–July, while low flows and low water occur during the months November–December.

3.8.3. Sediment Transport

The silt content at Springland (km 36) varies according to the data collected by the HRD between approx. 0.04 to 10.0 g/l. At Kwie-Kwie Creek (km 68.5), some kilometers downstream of Mc Clemen, the silt content varies from approx. 0.02 to 4 g/l. Nedeco (1968) estimated the sediment discharge from the Corantijn River at 1.2 million tons a year.

3.8.4. Depths

The minimum depth along the natural channel (4 m) is near the rivermouth. Further upstream it is deeper. The greatest depth is up to approx.30 m with respect to the mean water level, near Oreala (km 112) (HRD 1969).

4. The Water Balance and Climatic Impact

4.1. General

In hydrology, calculations are made of the water quantities which enter and leave the system (catchment). A water balance study may be carried out to estimate the various components of the hydrological cycle, employing the water balance equation, which is based on the law of continuity and conservation.

Water balance studies of Suriname have been done in the past by Ringma (1961) and Duba (1972). Although a water balance study involves thorough analyses, a simplified water balance of Suriname will be presented here to estimate the magnitude of the components of the hydrological cycle, and is limited to the long-term, mean annual water balance.

4.2. Areal precipitation

Based on the isohyets presented in Fig. 2, the long-term areal mean annual precipitation for the whole of Suriname is estimated here at 2,300 mm (rounded). Salati and Marques (1984) mentioned that the average annual rainfall for the Amazon region is 2,300 mm. These two values, which are estimated independently, do not differ.

4.3. Areal Pan-evaporation

The values of the long-term, mean annual pan-evaporation of the stations Zorg & Hoop, Jarikaba, Nickerie Airport and Kwamalasamutu are adopted here to estimate the long-term, mean annual pan-evaporation for the whole of Suriname. Here the weighted mean pan-evaporation of these stations is taken, where a weight of 0.17 (= 0.5/3) is given to the values of the stations Zorg & Hoop and Jarikaba, and a weight of 0.33 (=1/3) to the values of the stations Nickerie Airport and Kwamalasamutu. Thus, the long-term, mean annual pan-evaporation for Suriname is estimated at 1,279 mm.

4.4. Discharge

Referring to section 3 of this Chapter, the estimated discharges at the outfall of the rivers are as follows:

Marowijne River	1,785 m^3/s
Commewijne River	120 m^3/s
Suriname River	426 m^3/s
Saramacca River	225 m^3/s
Coppename River	500 m^3/s
Nickerie River	178 m^3/s
Corantijn River	1,572 m^3/s

Adding all these values, the Surinamese rivers together drain on average approx. 4,806 m^3/s freshwater into the Atlantic Ocean, which is equal to 757 mm/year.

4.5. Water Balance of Suriname

Neglecting the change in storage and all other components of the hydrological cycle which are not mentioned here, the long-term, mean annual actual evapotranspiration (= evapotranspiration which actually occurs) is equal to the residue long-term, areal mean annual precipitation (2,300 mm) minus the long-term annual discharge (757 mm). Thus the long-term, mean annual actual evapotranspiration is estimated here at 1.543 mm, which is about 1.2 times the long-term, mean annual pan-evaporation. Roche *et al.* (1991) mentioned that the actual evapotranspiration for Amazonian forests in French Guiana was evaluated at 1,470 mm a year. The difference between these two values for actual evapotranspiration is small (approx. 5%).

Referring to section 2.2 of this Chapter, one may conclude that for Surinamese conditions the potential evapotranspiration would be underestimated when a pan-coefficient of 0.7 is employed to convert the Class-A pan-evaporation to the potential evapotranspiration. A pan-coefficient for the Class-A pan of at least 1.2 should be employed for Surinamese conditions.

Approximately 67% of all the precipitated water on the Surinamese river basins returns to the atmosphere by evapotranspiration, and approx. 33% is drained off into the Atlantic Ocean. It should be noted that according to Salati and Marques (1984) for the Amazon region 46% of the precipitated water is discharged into the Atlantic Ocean, and 54% returns to the atmosphere by evapotranspiration. Moreover, it is agreed that the last mentioned ratio is between 60% and 65% for the continents (Roche *et al.* 1991).

4.6. Climatic Impact

It is concluded here that approx. 67% of the vapour forming precipitation on the Surinamese river-basins comes from evapotranspiration from the basin itself, and about 33% from the Atlantic Ocean by evaporation. Since the actual evapotranspiration is the greatest term of the water balance here, this term is very important for the hydrological cycle in Suriname.

Land-use and large scale deforestation in Suriname may affect the hydrological cycle and consequently the climate, since a great part of the precipitation on the Surinamese basins depends on the local evapotranspiration. Molion (1991) mentioned that large-scale land-use transformation from tropical humid forests to crop pasture fields reduce the evapotranspiration, and consequently the local rainfall. Run-off may also be reduced.

Summary

The seven main rivers of Suriname can be classified into three groups, namely the larger rivers (Marowijne and Corantijn), which are border rivers, the second largest rivers (Suriname and Coppename), and the smaller ones (Commewijne, Saramacca and Nickerie).

There are three hydrologically distinct areas, namely the Interior (approx. 126,550 km^2), the Savanna Belt (approx. 8,500 km^2), and the Coastal Plains (approx. 21,000 km^2).

The currents and water level in the upper reaches of the rivers are influenced by the freshwater discharge. In the estuary there is a combined influence of the tide and the freshwater discharge, upon which the currents and water level depend, as well as the salt-intrusion and transport of sediments.

The climate of Suriname is humid-tropic, with two rainy and two dry seasons per year. The long-term, mean annual rainfall varies from approx. 1,450 mm (Coronie) to approx. 3,000 mm (Tafelberg). At Paramaribo, the capital of the country, the mean temperature is 27.1 °C. The mean windspeed in the country is 1.3 on the scale of Beaufort.

The estimated precipitation surplus indicates that during the months of May-July much of the precipitated water is available for run-off, and during the months of September–October only little of the precipitated water is available.

The Marowijne River has the largest catchment area (approx. 68,700 km^2), the Corantijn River the second largest (approx. 67,600 km^2). The greatest part of the catchments of the rivers is in the Interior, except the catchments of the Commewijne and Nickerie Rivers. Of all the catchments only a small part (0.05% – 23.6%) is in the Savanna Belt.

In the West the tidal influence goes further landward, than in the East. The tidal range in the outfall of the main rivers is approx. 2.0 m. The Corantijn River has the greatest tidal prism (approx. 300 million m^3), the Marowijne River the second largest (approx. 200 million m^3), and the Nickerie River the smallest (approx. 10 million m^3). The tide is of the regular semi-diurnal type.

The minimum salt-intrusion is near km 30 – km 50 (about 10–30 kilometers upstream from the outfall), and occurs during periods of peak flows, while during periods of low flows, the maximum salt- intrusion occurs near km 60 – km 100 (about 40–80 kilometers upstream from the outfall).

The mean freshwater discharge at the outfall of the Marowijne River is the largest (approx. 1,785 $m^3/_s$), the second largest is at the outfall of the Corantijn River (approx. 1,572 $m^3/_s$), and the smallest at the outfall of the Commewijne River (approx. 120 $m^3/_s$). Peak flows and high water occur around June, and low flows and low water around November.

The silt content in the outfall of the main rivers is higher than further upstream, since sediments originating from the Amazon basin are transported by the sea water along the coast of Suriname. The sediment discharge from the Marowijne (approx. 1.3 million tons a year) and Corantijn basins (approx. 1.2 million tons a year) are the largest, and the sediment discharge from the Commewijne basin is the smallest (approx. 0.06 million tons a year).

The minimum depths along the thalweg are near the outfall of the main rivers, and vary from about 3 to 7 meters with respect to the mean sea level.

Further upstream the rivers are deeper. Maximum depths in the lower reaches vary between approx. 17 m to 30 m with respect to the mean water level.

About two thirds (± 67%) of the water which is precipitated on the Surinamese river basins, returns back to the atmosphere through evapotranspiration, and about one third (± 33%) is discharged into the Atlantic Ocean. Similar to the other river basins in the Amazon region, it is very likely that land-use and large-scale deforestation in Suriname could affect the climate significantly.

Acknowledgement

I wish to thank the staff and personnel of the Hydraulic Research Division Suriname for providing their kind assistance in gathering the hydrological data which is used here.

The kind cooperation of the Meteorological Service Suriname, especially of Mr. C. Becker, in providing climatological data and information for this chapter should also be mentioned here.

References

Adhin, H.S., 1982. Irrigation potential of Surinamese rivers. Proc. Congress Future of roads and rivers in Suriname and neighbouring region. University of Suriname and Delft University of Technology, Delft: 225–232.

Duba, D., 1972. Report on water balance studies, public water supplies and sewage, net-4, Suriname-2200. World Health Organization, 100 p.

Hydraulic Research Division, Ministry of Public Works, 1969. Hydrological data lower Marowijne river. Paramaribo, 85 p.

Hydraulic Research Division, Ministry of Public Works, 1969. Hydrological data lower Corantijn river. Paramaribo, 116 p.

Hydraulic Research Division, Ministry of Public Works, 1970. Hydrological data lower Suriname river. Paramaribo, 175 p.

Hydraulic Research Division, Ministry of Public Works, 1970. Hydrological data Nickerie river. Paramaribo, 140 p.

Hydraulic Research Division, Ministry of Public Works, 1971. Hydrological data lower Coppename river. Paramaribo, 122 p.

Hydraulic Research Division, Ministry of Public Works, 1972. Hydrological data Commewijne-Cottica river. Paramaribo, 88 p.

Hydraulic Research Division, Ministry of Public Works, 1982. Frequency analyses of revised monthly mean flows in the lower courses of the main rivers of Suriname. Paramaribo, 80 p.

Hydraulic Research Division, Ministry of Public Works, 1983. Onderzoek naar de irrigatiecapaciteit van de Commewijne en de Cottica. Nota 1: de metingen 1979/1980. Paramaribo, 70 p.

Hydraulic Research Division, Ministry of Public Works, 1987. Hydrological data lower Saramacca river (2nd ed.). Paramaribo, 147 p.

Hydraulic Research Division, Ministry of Public Works, 1988. Hydrological data lower Suriname river (2nd ed.). Paramaribo, 183 p.

Hydraulic Research Division, Ministry of Public Works. Hydrological annuals 1968–1980. Lower courses of the main rivers of Suriname. Paramaribo.

Molion, L.C.B., 1991. Climate variability and its effect on Amazonian hydrology. In: Water management of the Amazon basin. UNESCO/ROSTLAC, Montevideo: 261–274.
National Planning Office Suriname & Organization of American States, 1988. Suriname Plan Atlas. Washington, 51 p.
NEDECO, 1968. Suriname transportation study. Report on hydraulic investigations. The Hague, 293 p.
Ringma, S.A., 1961. Irrigatiecapaciteit van de Surinaamse rivieren. Bureau Landelijke Opbouw Suriname, Paramaribo, 82 p.
Roche, M.A. *et al.*, 1991. Water and salt balances of the Bolivian Amazon. In: Water management of the Amazon basin. UNESCO/ROSTLAC, Montevideo: 83–94.
Salati, E. and J. Marques, 1984. Climatology of the Amazon region. In: H. Sioli (ed.), The Amazon: limnology and landscape ecology of a mighty tropical river and its basin. Dr.W.Junk Publishers, Dordrecht: 85–226.
Shuttleworth, W.J. *et al.*, 1991. Post-deforestation Amazonian climate: Anglo-Brazilian research to improve prediction. In: Water management of the Amazon basin. UNESCO/ROSTLAC, Montevideo: 275–287.

4. Limnology: physico-chemical parameters and phytoplankton composition

ASHA HARIPERSAD-MAKHANLAL and PAUL E. OUBOTER

1. Introduction

A characterization of freshwater ecosystems can be made using many different factors (e.g. salinity, pH, eutrophication, colour, geomorphology, aquatic macrophytes, fauna). In 1950, Sioli presented a useful classification for the water types of Amazonia (primarily based on colour), a classification that is probably applicable to the whole of the Neotropics. He distinguishes clear water, white water and black water. Clear water has a very low content of nutrients, humic acids and sediment particles. White water carries an extensive load of sediments, colouring the water white to light-brown. White water is richer in nutrients than clear or black water. Black water is coloured tea brown due to dissolved humic acids. The presence of humic acids decreases the pH. The water is poor in nutrients, and the turbidity is low.

The limnological characteristics of Surinamese waters have been described for a limited number of localities by Leentvaar (1975) and Mohadin (1980). Until now, no one has tried to characterize the Surinamese freshwater ecosystems. The terrestrial ecosystems can be easily divided into four main areas: The Young Coastal Plain, the Old Coastal Plain, the Savanna Belt, and the Interior (see Noordam, Chapter 2). Here, our intent is to characterize the principal Surinamese freshwater ecosystems by their physico-chemical qualities and phytoplankton composition, and to see whether the aquatic ecosystems relate to the terrestrial ones.

Our data have been derived from various sources. In many localities of the Savanna Belt and Interior physico-chemical measurements were performed in the field by the junior author, using Hach test kits. These kits determine different parameters in a titrimetric or colorimetric way. These data were supplemented by data from Mohadin (1980), and analytical results of the WLA (Hydraulic Research Division) for several localities for which samples had been taken by different researchers (see Acknowledgements). The phytoplankton samples were also taken by various researchers to be analysed at the WLA (see Acknowledgements).

In this chapter we will only present a selected set of representative data, and

Paul E. Ouboter (ed.), Freshwater Ecosystems of Suriname, 53–75.

we will try to make a preliminary analysis. Unfortunately, we did not have the time to compare these results with data for Amazonia or other parts of the World.

2. Physico-chemical Parameters

Table 1 presents the results of the physico-chemical analysis of samples from localities all over the country. The position of these localities is shown in Fig. 1 and Fig. 2. The results are not always easy to compare, since different researchers performed different measurements. Still, a picture emerges that might be useful in the future to characterize aquatic ecosystems. A summary of the data is presented in Table 2.

2.1. pH and Humic Acids

The pH of the coastal and interior rivers is slightly acidic to almost neutral. Low pH values are characteristic of black-water rivers (Savanna Belt) and swamps, and are caused by humic acids. Humic acids are released in the water, directly from decaying plant material, or indirectly via the soil. In the first case, the water will only contain much humic acids if it is almost stagnant. This happens in swamps and small, almost stagnant forest creeks. Fast-flowing rivers could never have an increased load of humic acids only coming from plant material that has fallen into the water. The amount of humic acids released is negligible compared to the amount of water carrying the humic acids off. Black-water rivers, coloured by the high amount of humic acids, occur in areas with a podzolic soil (see Goulding *et al.* 1988), like the Savanna Belt. This soil serves as a reservoir for humic acids produced in the savanna forest which grows there. The humic acids are drained off to the creeks during the rainy season. The highest concentrations of humic acids are measured at the beginning of the long rainy season (unpubl. obs.). All rivers which drain the Savanna Belt in Suriname are black-water rivers carrying a high load of humic acids.

2.2. Conductivity

In general, conductivity is lower in the Interior than in the Coastal Plain. Enhanced values for the Coastal Plain, like for Bruynzeel (Suriname River), are caused by high tide, also increasing hardness, and salinity (Cl). A relatively low conductivity was measured for the Saramacca River (Uitkijk, Saronbrug).

Table 1. Physico-chemical parameters for water bodies in Suriname (see Fig. 1 and 2). Cond. – conductivity; Acid. – acidity; Alk. – alkalinity; Hardn. – hardness; H. acids. – humic acids; S. depth – secchi depth. Acidity, alkalinity and hardness are expressed in mg $CaCO_3$/L, all chemical substances in mg/L.

	pH	Cond.	Acid.	Alk.	CO_2	Hardn.	NH_4	NO_2	NO_3	PO_4	Cl	H. acids	O_2	% satur.	S.depth
Swamps															
2 Fen at Groenhart Road	4.0	77	–	–	12.2	7.52	0.12	0.01	0.05	0.10	16	–	2.14	24	max.
1 Blechnum-watrabebe swamp	5.5	46	–	–	9.0	8.95	0.17	0.01	0.05	0.18	14	–	1.96	23	1.0
3 Eleocharis swamp	5.1	40	–		4.0	6.80	0.01	0	0.06	0.02	13	–	4.32	53	max.
3 Swamp Prinsie	4.7	25	28	–	–	2.70	1.20	0	0	–	10	3.5	–	–	–
Coastal Plain															
7A Mac Clemen (Corantijn)	7.4	–	–	–	4.8	–	–	–	–	–	–	–	5.8	74	0.8
4A Uitkijk (Saramacca)	6.1	11	–	–	1.5	14.32	0.1	0	0.5	0.29	11	–	4.71	60	0.6
3A Bruynzeel (Suriname)	7.4	320	–	–	0.9	55.49	0.32	0	1.1	0.25	79	–	4.30	54	0.1
4B Saronbrug (Saramacca)	6.2	28	–	–	1.1	9.84	0.06	0.01	1.2	0.04	10	–	6.63	84	1.2
3B Paranam (Suriname)	7.5	47	–	–	–	8.95	0.09	0.01	0.94	0.06	10	–	5.36	68	1.0
2 Hooi Creek	6.2	41	–	–	6.3	5.37	0.05	0.01	0.23	0.08	11	–	4.10	49	max.
Savanna Belt															
13 Marataka	4.5	–	32	–	24.0	2.7	1.3	0	0	0	6.8	3.5	2.85	34	–
14 Cusewijne	5.2	–	41	5	26.3	9.74	1.10	0	0	0.22	7.78	3.29	2.65	33	–
4 Coropina Creek	4.5	34	35	3	19.0	3.27	0.83	0	0.01	0.12	5.73	2.97	1.26	16	0.9
15 Para	4.3	38	35	–	7.0	4.54	1.21	0	0	0.17	6.32	3.37	1.00	12	0.8
Transition to Interior															
11 Goliath Creek	5.7	–	31	14	38.4	–	0.91	0	0	–	7.2	2.5	2.02	24	–
16 Mindrineti	6.2	50	–	–	1.0	19.51	0.03	0.01	0.62	0.08	12		4.60	57	1.2
3C Berg en Dal (Suriname)	6.6	24	–	–	2.5	8.05	0.2	0.01	1.13	0.33	6	–	4.71	60	3.0

Table 1. (continued)

	pH	Cond.	Acid.	Alk.	$C0_2$	Hardn.	NH_4	$N0_2$	$N0_3$	$P0_4$	Cl	H. acids	02	% satur.	S.depth
Interior															
7B Corantijn	7.3	21	–	–	2.4	–	0.07	–	0.03	–	2.0	–	7.0	90	1.5
7 Maskita Creek	7.1	25	–	–	2.6	–	0.13	–	0.17	–	4.0	–	7.8	96	1.0
6 Sisa Creek	7.2	28	–	–	4.8	–	0.30	–	0.52	–	10.0	–	8.0	98	1.2
8A Kabalebo	6.7	–	–	–	4.5	–	0.09	–	0.05	–	9.0	–	6.6	85	1.2
6A Nickerie	6.5	–	28	51	28.6	43	0	0	0	0	4.7	1.6	7.85	99	–
12 Fallawatra	6.3	–	50	60	17.6	16	0	0	0	0	7.9	1.5	9.67	119	–
10 Browns Creek	6.5	23	–	–	2.0	2.68	0	0	0.25	0.08	9	–	6.76	78	max.
8 Mama Creek	6.6	28	–	–	2.0	11.46	0.08	0	0.38	0.15	8	–	6.53	78	0.7
9 Compagnie Creek	6.6	25	–	–	2.0	9.49	0.06	0	0.29	0.16	7	–	6.34	75	max.
12 Vier Gebroeders Creek	6.5	–	41	69	25.0	8.5	0.3	0	0	0	1.5	0.5	10.00	119	–
10 Sipaliwini	6.5	–	34	4	6.0	>1	0.5	0	0	–	1.0	1.8	7.75	89	–
9 Curuni	6.5	–	14	4	29.5	15.0	0.3	0	0	–	2.9	1.0	11.12	131	–
13 Pelulu Creek	6.8	–	13	1	–	0.8	–	0	0	0.02	1.1	1.2	–	–	–
11 Tapanahoni	6.5	–	5	>1	–	1.2	0.2	0	0	0.02	1.5	1.2	–	–	–

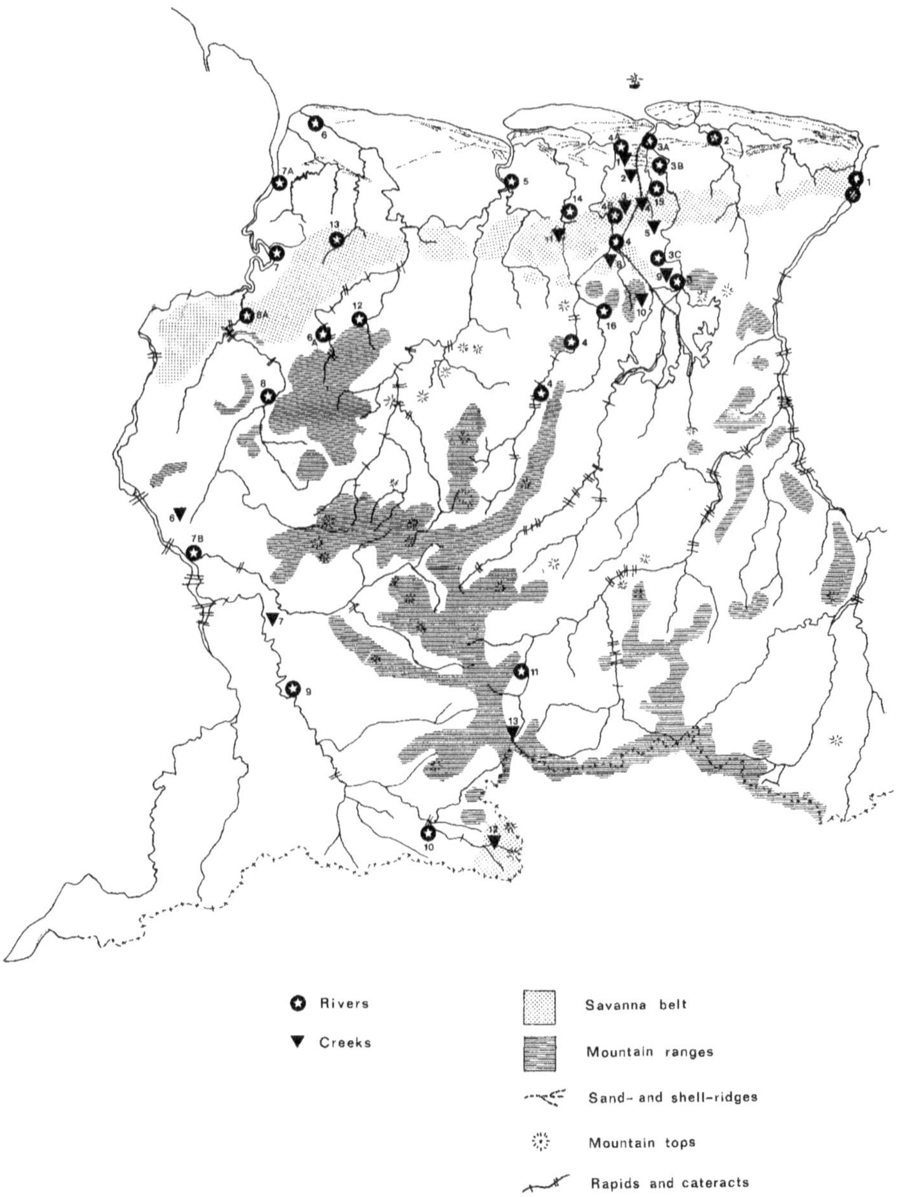

Figure 1. Sample locations of phytoplankton and physico-chemical parameters in rivers and creeks in Suriname. The numbers refer to the localities mentioned in the Tables 1, 3 and 4.

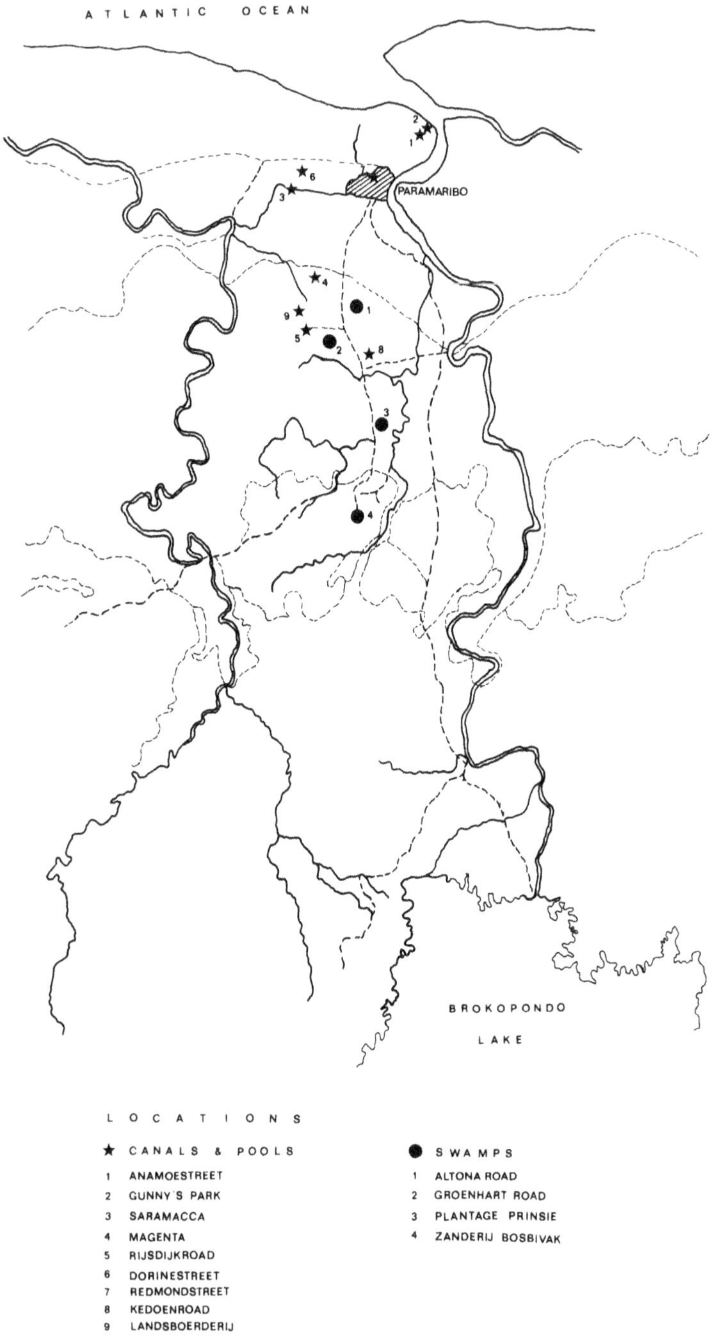

Figure 2. Map showing sample locations for swamps (circles) and canals and pools (stars) along the road Paramaribo – Zanderij. The numbers refer to the localities mentioned in the Tables 5 to 7.

Table 2. Summary of physico-chemical parameters for the Coastal Plain, Savanna Belt, Interior and Swamps in the Old Coastal Plain.

	Coastal Plain	Savanna Belt	Interior	Swamps (Old Coastal Plain)
pH and humic acids	pH medium (6.1-7.5)	pH low (<5) Usually more than 3 mg/l humic acids	pH medium (6.5-7.3) Less than 2 mg/l humic acids	pH low (<5.5) Often high contents in humic acids
Conductivity	Low to high	Low	Low	Low to high
Alkalinity	(no measure-ments)	Low	Low to medium	(no measure-ments)
Hardness	Low to medium	Low	Low to medium	Low (to high)
Nutrients	Low to medium	Low	Generally low	Low
Salinity	Low to brackish	Low	Low to very low	Low
Oxygen	Medium (49-84% sat.)	Low (<35% sat.)	High (66-131% sat.)	Usually low
Turbidity	Medium to high	Low	Low	Low

2.3. Hardness

Hardness is lowest in the black-water rivers draining the Savanna Belt, and in some of the rivers of the Interior. Some localities of the Interior show unexpected high figures for hardness. The Curuni River drains an area of migmatitic gneisses and amphibolites; the Nickerie and Fallawatra Rivers an area of charnockitic granulites. It is likely that dissolved solids from these rocks might cause the increased hardness.

It is remarkable that the Coastal Plain localities generally show a low hardness. It seems that the rivers of the Interior keep their oligotrophic character and low amount of dissolved solids until they are mixed with richer salt water (also see 2.4).

The swamps of the Old Coastal Plain carry soft water. Swamps in the areas of shell ridges may have hard water. Measurements of these swamps have not been included.

2.4. Nutrients

The nutrient load of all Surinamese waterbodies sampled is low. Ammonium is generally present in small quantities, the highest quantities being reported for the black-water rivers.

Nitrite is generally not present (Interior), or present in very small quantities.

Nitrate values are highest in the Coastal Plain, but here also low values were found (Hooi Creek), and relatively high values for some localities in the Interior (Sisa Creek, Mama Creek).

Phosphate values are generally low.

2.5. Salinity

The freshwater of Suriname contains on average between 10 and 20 mg/l Cl. Values below 2 mg/l were only found in the extreme south (Sipaliwini and Tapanahoni Rivers).

2.6. Oxygen

Oxygen is one of the most important water quality parameters for most aquatic organisms. Oxygen depletion may lead to massive fish mortality. Large differences in oxygen content occur between different types of rivers in Suriname.

The turbulence caused by rapids and cataracts in the Interior aerates the rivers sufficiently to raise the oxygen content to almost saturation, in some instances to supersaturation.

Completely different is the situation of the black-water rivers of the Savanna Belt. These are almost depleted of oxygen. This holds for all periods of the year (unpubl. obs.). Only shortly after rainfall the oxygen content of the top layer may increase temporarily. The low oxygen content of black-water rivers is caused by the following circumstances:

1. Little turbulence, and consequently little mixing of upper and lower water layers.
2. The water is poor in nutrients, and consequently poor in phytoplankton.
3. Because of the dark colour of the water, light is not able to penetrate deep, also causing a deficience in phytoplankton.
4. We suspect that photochemical oxidation of humic acids occurs, whereby oxygen is used from the water column.

Experiments to prove the action of photochemical oxidation gave results that were not unequivocal. Still, we believe that apart from little turbulence and phytoplankton another factor must be at work to explain the very low oxygen content of the black water.

Swamps have a low oxygen content as well, but here the values change over the day. Day and night oxygen is consumed by the decaying vegetation, but during the day, a considerable production of oxygen occurs by photosynthesis of algae. In a swamp near Prinsie we measured 0.5 mg/l O_2 at 8.00 hours. At 17.30 hours the oxygen content had increased to 2.9 mg/l. The values for the nearby black-water river were 0.75 and 0.70 mg/l, respectively. Many fish species living in swamps and black-water rivers are adapted to the low oxygen content (see Ouboter and Mol, Chapter 8, and Mol, Chapter 10). However, even fishes that have no apparent morphological adaptations, seem to survive well.

2.7. A Characterization of Surinamese Rivers and Creeks

On the basis of the results presented above, it is possible to make a preliminary classification of the Surinamese lotic environment. We did not include swamps, because the number of samples was to few to draw any conclusions. For more information on physico-chemical parameters of swamps, however, see Mol, Chapter 7.

Classification of Surinamese rivers and creeks

1. Rivers and creeks of the interior (above the falls).
   Characterized by:
   - "clear water"
   - the water is generally soft
   - oligotrophic
   - nearly saturated with oxygen
2. Coastal sections of rivers and creeks (below the falls).
   Characterized by:
   - "clear water"
   - the water is generally soft
   - oligotrophic
   - less oxygen

   Towards the sea there is an increase in turbidity, salinity and nutrient contents.
3. Black-water rivers which drain the Savanna Belt.
   Characterized by:
   - "black water"
   - the water is soft
   - the pH is low
   - oligotrophic
   - almost depleted of oxygen

How does this classification of rivers and creeks relate to the classification of terrestrial areas? There are many similarities. The area of the rivers and creeks of the interior completely covers the area of the Interior. The second category, however, is slightly different; it is composed of rivers coming from the Interior and entering the Savanna Belt and Coastal Plains, and creeks draining parts of the Coastal Plains (only a few). So, parts of the Savanna Belt are included here as well. On the other hand, the rivers draining the Savanna Belt itself, continue their influence almost to the sea, since they remain "black-water rivers" while flowing through the Coastal Plains.

3. Phytoplankton

3.1. Rivers (Table 3)

Three factors are of importance for phytoplankton species to adapt themselves to the running water of the large rivers:
- Turbidity. In the lower courses turbidity is high, resulting in very bad light conditions for phytoplankton.
- Oligotrophy. The rivers of the Interior are generally poor in nutrients. This is especially true during the rainy season when the volume of water increases.
- Current. Planktonic algae tend to dominate in stagnant or slow-moving waters; in the fast-flowing habitat of streams benthic species are more abundant.

Table 3. Phytoplankton sampled in rivers. 1 = Marowijne; 2 = Commewijne; 3 = Suriname; 4 = Saramacca; 5 = Coppename; 6 = Nickerie; 7 = Corantijn; 8 = Kabalebo.

Phytoplankton	1	2	3	4	5	6	7	8
Cyanophyta								
Chroococcus minutus			X					
Coelosphaerium sp.								X
Dactylococopsis acicularis			X					
Lyngbya sp.			X					
Microcystis sp.							X	X
Oscillatoria sp.				X				
Euglenophyta								
Euglena sp.			X					X
Phacus sp.			X					
Trachelomonas armata				X				
Trachelomonas cilindrica								X
Chlorophyta								
Ankistrodesmus acicularis			X					
Ankistrodesmus falcaltus var. *acicularis*			X					
Ankistrodesmus falcaltus								X
Ankistrodesmus spirales			X					
Ankistrodesmus sp.			X					
Coelastrum cambricum							X	
Crucigenia quadrata								X
Crucigenia sp.			X					
Dictyosphaerium pulchellum	X		X	X			X	X

Table 3. (continued)

Phytoplankton	1	2	3	4	5	6	7	8
Chlorophyta								
Eudorina elegans	X		X	X				X
Elakatorix gilatinosa			X					
Golenkinia sp.								X
Kirchneriella obesa	X						X	
Lagerhemia sp.								X
Mougeotia sp.			X				X	X
Oocystis sp.			X				X	X
Pediastrum duplex var. *gracilimum*							X	
Pediastrum duplex var. *rotundatum*								
Pediastrum tetras								X
Scenedesmus obliquus							X	X
Scenedesmus falcaltus							X	
Scenedesmus quadricauda							X	X
Scenedesmus sp.	X							
Selenastrum cambricum							X	
Selenastrum reticulatum							X	X
Selenastrum sp.							X	
Spirogyra rivularis								X
Spirogyra sp.							X	
Tetraedron gracile			X				X	
Ulotrix sp.			X					
Chlorophyta (Desmidiaceae)								
Arthrodesmus sp.			X					
Closterium contractum							X	X
Closterium keutzingii				X				X
Closterium venus	X			X				X
Closterium trifidium			X					
Closterium sp.	X							
Cosmarium floridanum							X	X
Cosmarium pseudoconnatum							X	X
Cosmarium sp.	X			X			X	
Cosmacladium sp.	X							
Euastrum monocylum	X							
Euastrum subtrilobulatum							X	X
Gonatozygon mucosum				X				
Micrasterias arcuata			X	X				
Micrasterias siolii								X
Micrasterias sp.				X				
Spondylosium sp.							X	
Staurastrum eriangularis			X					
Staurastrum leptocladum var. *cornutum*			X					
Staurastrum mutabilis			X				X	X
Staurastrum sebaldii							X	
Staurastrum talisheekens							X	
Staurastrum sp.	X						X	X

Table 3. (continued)

Phytoplankton	1	2	3	4	5	6	7	8
Chrysophyta								
Dinobryon sp.							X	
Bacillariophyta								
Achnanthes trinodus	X							
Actinocyclus normanni	X	X			X			
Asterionella formosa						X		
Asterionella japonica						X		
Biddulphia mobiliensis						X	X	
Biddulphia sinensis						X		
Bacillaria paradoxa	X							
Coscinodiscus asteromphalis			X			X	X	
Coscinodiscus curvalatus	X		X			X	X	
Coscinodiscus sp.		X			X	X	X	
Cyclotella stilorum						X		
Cyclotella menigheniana			X					
Cymbella turgida							X	X
Diatoma anceps						X	X	X
Diatoma sp.				X				
Eunotia asterionelloides				X			X	
Fragillaria virescens								X
Frustulia rhomboides	X				X		X	
Gomphonema longiceps							X	
Gomphonema olivaceum					X		X	
Gomphonema sp.			X					
Gyrosigma scalproides							X	
Melosira granulata	X		X	X		X	X	X
Melosira herzogii	X			X		X	X	
Melosira cf. *sulcata*			X					
Meridion sp.						X	X	
Navicula demerara							X	
Navicula jentzshii						X		
Navicula placentula							X	X
Navicula sp.							X	
Nitzschia sp.			X	X				
Pinnularia gentilis							X	
Pinnularia nobilis							X	
Pinnularia sp.	X			X				
Surirella beserata							X	X
Surirella bidentulata							X	X
Surirella didyma					X		X	
Surirella linearis			X		X		X	
Surirella robusta							X	X
Surirella tenera							X	
Surirella sp.	X	X		X	X			
Synedra ulna							X	
Stauroneis sp.								X
Tabellaria sp.							X	X
Pleurogsigma sp.					X			
Amphora sp.							X	

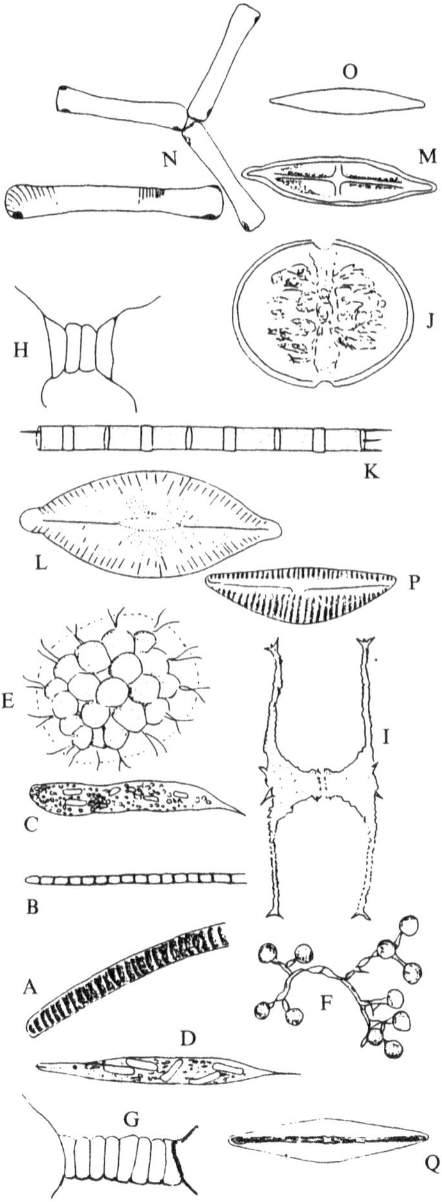

Figure 3. Some characteristic phytoplankton species from Suriname.
A. *Oscillatoria princeps* Vaucher & Gomont; B. *Oscillatoria* sp.; C. *Euglena oxyuris* var. *minor* Prescott; D. *Euglena acus* Ehrenberg; E. *Eudorina elegans* Ehrenberg; F. *Dictyosphaerium pulchellum* Wood; G. *Scenedesmus quadricauda* (Turp.) de Brebisson; H. *Scenedesmus quadricauda* var. *longispina* (Chod.) G.M.Smith; I. *Staurastrum leptocladum* var. *cornutum* Nordt.; J. *Cosmarium pseudoconnatum* Nordt. var. *borgei*; K. *Melosira granulata* (Ehrenberg) Ralfs; L. *Navicula demerara* Grun. ex Cleve; M. *Stauroneis nobilis* Schum; N. *Eunotia asterionelloides* Uherkovich; O. *Navicula* sp.; P. *Cymbella turgida* (Greg) Cleve Q. *Frustulia rhomboides* Ehrenberg.

In the running water of the large rivers, species of the phylum Bacillariophyta (Diatoms) appear to be the most common component of the phytoplankton community. Especially the benthic diatom species, which may be attached to all kinds of substratum (rock, silt, sand, mud) are widespread. In the fast-flowing border rivers, the Corantijn River in the west and the Marowijne River in the east, a large number of benthic and epiphytic diatoms were found, such as *Gomphonema*, *Surirella*, *Navicula* (Fig. 3L) and *Diatoma* species.

In the slower reaches of the large rivers, such as the tributaries, the species composition tends to be different from that of the main stream. For instance in the Kabalebo River, a tributary of the Corantijn River, the small green algae *Crucigenia quadrata*, *Selenastrum* sp., *Dictyosphaerium pulchellum* (Fig. 3F), *Kirchneriella obesa*, *Pediastrum tetras* may become more abundant than the diatoms. In general, the Chlorophyta (including desmids) are the second largest group in the rivers.

Many of the phytoplankton species occurring in the Suriname River probably originate from the man-made Brokopondo Lake. Two typical Brokopondo Lake species which are frequently found in the Suriname River are the desmids *Staurastrum leptocladum* var. *cornutum* (Fig. 3I), and the centric diatom *Cyclotella menigheniana*.

The Corantijn River has the highest number of phytoplankton species (more than 50). Rivers with a low plankton development are the Commewijne River with only 3 (diatom) species and the Coppename River with 9 (diatoms) species.

Insufficient data is available on seasonality. According to Leentvaar (1975) plankton development is strongest during the dry season, and at the beginning of the rainy season.

3.2. Creeks (Table 4)

In general, plankton development in creeks is low. Most likely, this is caused by the oligotrophic character of most creeks, as well as by the more shaded conditions.

Creeks are numerous in the Savanna Belt and the Interior. The few creeks that originate in the Coastal Plain are usually of the clear water type. Two examples are the Tawajari Creek and the Hooi Creek. The phytoplankton content of the Tawajari Creek is very poor. This may have been caused by the low pH (3.2) of this creek during the period of sampling (probably human induced; see Mohadin, 1980). In the Hooi Creek with a pH of 6.2, a few more species of the phylum Bacillariophyta (genera *Nitzschia*, *Navicula* and *Surirella*) were found, together with one filamentous green algae, *Mougeotia* sp.

The creeks draining the Savanna Belt are all of the black- water type (see above). These black-water creeks seem to be very unproductive habitats. The extreme poorness in anorganic ions and nutrients, the black colouring of the water (due to humic substances), and the acidic character (pH 3.5–5.4) explain the scarcity of phytoplankton in these creeks. The black-water creeks sampled are Cola Creek, Coropina Creek and Carolina Creek. The most frequently encountered phytoplankton species from these black waters are from the phylum Bacillariophyta (Diatoms) (genera *Navicula*, *Nitzschia*, *Surirella*, *Pinnularia*, *Cymbella*). The Desmids, which in general occur in very dilute, soft and unproductive water, are also found here (genera *Desmidium*, *Cosmarium*, *Bambusina*, *Staurastrum*).

Table 4. Phytoplankton sampled in creeks. 1 = Tawajari Creek; 2 = Hooi Creek; 3 = Cola Creek; 4 = Coropina Creek; 5 = Carolina Creek; 6 = Sisa Creek; 7 = Maskita Creek; 8 = Mama Creek; 9 = Compagnie Creek; 10 = Browns Creek.

Phytoplankton	1	2	3	4	5	6	7	8	9	10
Cyanophyta										
Anabaena sp.						X	X			
Anacystis sp.							X			
Coelosphaerium sp.						X				
Merismopedia glauca							X			
Microcystis sp.							X			
Oscillatoria sp.							X		X	
Euglenophyta										
Entosiphon ovatum				X						
Euglena sp.			X	X				X		
Trachelomonas volvocina				X						
Trachelomonas sp.			X	X		X		X		
Phacus sp.						X		X		
Chlorophyta										
Ankistrodesmus falcaltus						X	X			
Chlamydomonas sp.	X					X				
Coelastrum cambricum						X	X			
Crǔcigenia tetrapedia						X	X			
Dictyosphaerium pulchellum						X	X			
Errerela sp.		X								
Eudorina elegans							X			
Golenkinia sp.						X				
Kirchneriella lunaris							X			
Mougeotia sp.		X			X		X			
Oocystis sp.		X								
Pandorina morum							X			
Pediastrum duplex var. *gracilimum*							X			
Pediastrum duplex							X			
var. *rotundatum*							X			
Pediastrum tetras						X				
Scenedesmus obliquus						X	X			
Scenedesmus quadricauda							X			
Selenastrum westii						X				
Selenastrum sp.							X			

Table 4. (continued)

Phytoplankton	1	2	3	4	5	6	7	8	9	10
Chlorophyta										
Spirogyra rivularis							X			
Spirogyra sp.							X			
Tetraedron trigonum				X				X	X	
Ulotrix sp.						X	X			
Chlorophyta (Desmidiaceae)										
Bambusina sp.				X						
Closterium keutzingii						X	X			
Closterium sp.						X			X	
Cosmarium amoenum							X			
Cosmarium floridanum						X	X			
Cosmarium speciosa							X			
Cosmarium pseudoconnatum							X			
Cosmarium sp.			X	X						
Desmidium elegans				X						
Desmidium sp.			X		X				X	
Euastrum subtrilobulatum							X			
Euastrum sp.				X						
Micrasterias pusillum						X				
Pleurotaenium minutum									X	
Pleurotaenium sp.									X	
Sphaerozosma granulatum									X	
Staurastrum mutabilis						X	X			
Staurastrum sebaldii							X			
Staurastrum sp.				X						
Chrysophyta										
Dinobryon sp.				X						
Bacillariophyta										
Coscinodiscus sp.				X						
Cymbella turgida						X	X			
Cymbella sp.				X						
Diatoma sp.				X	X	X				
Eunotia indica						X				
Eunotia sp.			X	X						
Eunotia zygodon				X						
Fragillaria capucina							X		X	
Frustulia rhomboides				X		X	X	X		
Gomphonema longiceps							X			
Gomphonema olivaceum						X				
Gomphonema olivaceum var. *calcarea*							X			
Gyrosigma scalproides							X			
Melosira granulata						X	X			
Meridion sp.						X				
Navicula demerara						X	X			
Navicula placentula						X	X			
Navicula sp.	X	X		X		X				

Table 4. (continued)

Phytoplankton	1	2	3	4	5	6	7	8	9	10
Bacillariophyta										
Nitzschia sp.		X		X						
Pinnularia gibba				X			X			
Pinnularia subcapitata							X			
Synedra sp.			X	X	X					
Surirella beserata						X				
Surirella didyma						X	X			
Surirella elegans							X			
Surirella linearis						X				
Surirella robusta						X	X			
Surirella rob. var. *spendida*		X								
Surirella tenera							X			
Surirella sp.				X				X		
Stauroneis sp.						X	X			
Tabellaria sp.				X		X				
Tabellaria fenestrata							X			
Tribonema sp.		X		X						

The creeks in the Interior have the following characteristics: the water is soft and clear, pH between 6.2 and 7.2, the water is usually saturated with oxygen. As in the lowlands, the plankton community consists of Diatoms, Chlorophytes and Desmids. Two creeks in the west, Sisa Creek and Maskita Creek, contained a rich phytoplankton community, containing resp. 36 and 46 species (Table 3). Much less phytoplankton species were recorded in the eastern Mama Creek and Compagnie Creek (resp. 5 and 9), whereas the Browns Creek yielded no phytoplankton species at all.

3.3. Swamps (Table 5)

Swamps cover about 60% of the Coastal Plain. They are characterized by stagnant water, and a large amount of desintegrating vegetation (see Teunissen, Chapter 5). The swamps in the Young Coastal Plain are usually covered by water which is brackish, hard and eutrophic. These swamps will not be dealt with here. The water of swamps of the Old Coastal Plain and the Savanna Belt is usually soft, oligotrophic and stained with humic acids. Phytoplankton is usually not abundant. Desmids, which are characteristic of oligotrophic and acidic habitats, usually dominate. In a swamp nearby the Altona Road (Old Coastal Plain) the following Desmids were found: *Cosmarium* sp., *Desmidium elegans*, *Gonatozygon* sp., *Micrasterias laticeps* and *Pleurotaenium* sp. Besides the Desmids, some Chlorophyta were present: *Eudorina elegans* (Fig. 3E), *Mougeotia* sp., *Mougeotiopsis* sp. and *Spirogyra*. Euglenoid genera (*Euglena*,

Trachelomonas and *Phacus*) were also found, but in low concentrations (Fig. 3C and 3D). In a fen along the Groenhart Road (also Old Coastal Plain) the phytoplankton was remarkable poor in the number of species. The swamp at Plantage Prinsie is a clear-water habitat, where a variety of phytoplankton species was found. Some notable species are the desmids *Cosmarium* sp. and *Pleurotaenium* sp., and the Euglenophyta *Euglena* sp., *Phacus* sp. and *Trachelomonas* sp.

Table 5. Phytoplankton sampled in swamps of the Old Coastal Plain. 1 = Swamp to Altona road; 2 = Swamp to Groenhart road; 3 = Swamp to Prinsie (Plantage); 4 = Swamp to Zanderij (Bosbivak).

Phytoplankton	1	2	3	4
Cyanophyta				
Microcystis flos-aqua		X		
Microcystis sp.			X	
Oscillatoria princeps			X	
Oscillatoria sp.		X	X	X
Euglenophyta				
Euglena sp.		X	X	
Phacus sp.		X	X	
Trachelomonas sp.	X	X	X	
Chlorophyta				
Eudorina elegans	X			
Mougeotia sp.	X			
Mougeotiopsis sp.	X			
Oedogonium sp.		X		
Spirogyra sp.	X			
Chlorophyta (Desmidiaceae)				
Cosmarium contractum var. *rotundatum*				X
Cosmarium sp.	X		X	
Desmidium sp.				X
Desmidium elegans	X			
Gonatozygon sp.	X			
Micrasterias laticeps	X			
Pleurotaenium sp.	X		X	
Staurastrum sp.		X		
Staurodesmus sp.		X	X	X
Chrysophyta				
Dinobryon divergens				X
Bacillariophyta				
Coscinodiscus curvulatus		X		
Frustulia rhomboides				X
Frustulia sp.				X
Melosira sp.				X
Nitzschia sp.		X	X	X
Navicula sp.				X
Surirella beserata				X
Tabellaria sp.			X	

One characteristic species was seen in an oligotrophic clear-water swamp near Zanderij (Bosbivak) in the Savanna Belt: *Dinobryon divergens*. Besides this species an association of many Diatoms (e.g. *Frustulia* sp., *Melosira* sp.) and some Desmid members (*Desmidium* sp. and *Cosmarium contractum* var. *rotundatum*) was recorded.

3.4. Man-made Habitats

Man-made waterbodies, principally canals and pools, are mainly situated in the Coastal Plain. They are usually drainage systems along the road. The water is usually hard and eutrophic. In contrast to the soft and oligotrophic natural waters of the Old Coastal Plain, Savanna Belt and Interior, a rich algal flora develops in these habitats. Rich meaning that the number of algae is very high, the diversity of the species is usually low. The Euglenophyta is the most important group here.

3.4.1. Canals (Table 6)

A variety of Euglenophyta, including *Lepocinclis salina*, *L. fusiformis*, *Euglena ehrenbergii*, *Phacus platalea*, and a bloom of *Euglena oxyuris* may completely dominate a hard-water canal at the Anamoe Road. The canal will then be densely coloured by their tremendous numbers. In a hard, but more brackish, water canal near Leonsberg (Gunny's Park) a small number of filamentous blue-greens were found, including *Calotrix marchica*, *C. parietana* and *Rivularia* sp., and in addition five species of Euglenophyta. The Saramacca Canal carries soft and very turbid water. It is situated in an area with a lot of industrial activities (paints, wood, rice, washing powder), and is consequently strongly polluted. The most important plankton species are *Euglena acus*, *Lepocinclis ensifera* var. *javanica* and *Phacus longicauda*. Two species from the order Volvocales, *Eudorina elegans* and *Pandorina morum* are frequently found, as well as the Diatom *Melosira granulata* (Fig. 3K), and to some extent *Peridinium* sp. from the family of the Dinophyceae.

Table 6. Phytoplankton sampled in canals. 1 = Anamoe road; 2 = Gunny's Park; 3 = Saramacca canal (Kameelbrug); 4 = Magenta canal; 5 = end of Rijsdijk road.

Phytoplankton	1	2	3	4	5
Cyanophyta					
Calotrix braunii				X	
Calotrix marchica		X			
Calotrix parietana		X			
Lyngbya sp.			X		
Rivularia sp.		X			

Table 6. (continued)

Phytoplankton	1	2	3	4	5
Euglenophyta					
Euglena acus			X	X	
Euglena ehrenbergii	X				
Euglena oxyuris var. *minor*	X	X			
Euglena rostrifera		X			
Euglena sp.			X		
Euglena tiszae			X		
Leponcinclis ensifera var. *javanica*			X		
Lepocinclis fusiformis	X	X			
Lepocinclis salina	X				
Phacus anacoelus			X		
Phacus longicauda			X		
Phacus platalea	X	X			X
Phacus unguis		X			
Strombomonas anulata					X
Strombomonas ensifera javanica			X		
Strombomonas gibberosa			X		
Strombomonas verrucosa			X		
Strombomonas sp.			X		
Trachelomonas hispada					X
Trachelomonas volvocina					X
Trachelomonas zorensis					X
Chlorophyta					
Ankistrodesmus arcuatus				X	
Cladophora sp.					X
Kirchneriella lunaris					X
Scenedesmus dentriculatus					X
Scenedesmus ecornis					X
Scenedesmus quadricauda	X				
Tribonema sp.			X		
Eudorina elegans			X		
Pandorina morum			X		
Chlorphyta (Desmidiaceae)					
Closterium dianae				X	
Closterium moniliferum				X	
Closterium navicula					X
Closterium praeloneum				X	
Micrasterias mahabuleshwarensis				X	
Onyhonema leave				X	
Onyhonema unicatum				X	
Pleurotaenium minutum					X
Staurastrum sebaldii					X
Pyrrhophyta					
Peridinium sp.					X
Bacillariophyta					
Climacosphaeria monilifera				X	
Gyrosigma sp.		X			
Melosira granulata		X			

In two slightly acidic, soft-water canals a lot of Desmids occur. In the Magenta canal (pH 5.5) a large population of *Closterium* spp. was noticed (*C. dianea*, *C. moniliferum*, *C. praeloneum*) with few Diatom species, a blue-green alga and almost no Euglenophyta. This is in contrast with the algal flora of a canal at the end of Rijsdijk Road (pH 4.9), in which many Euglenophyta species were present.

3.4.2. Pools (Table 7)

Pools are usually shallow, and overgrown with aquatic macrophytes. Euglenophyta (*Euglena*, *Phacus*, *Trachelomonas*) together with a small association of blue-greens (*Lyngbya nordgaardii*), and Diatoms (*Navicula* spp.) are the typical species of the algal flora.

Table 7. Phytoplankton sampled in the pools. 6 = Dorinestreet (Rensprojekt); 7 = Hoek Redmond street (Rode Kruis avenue); 8 = Kedoen road; 4 = 's Landsboerderij.

Phytoplankton	6	7	8	9
Cyanophyta				
Lyngbya nordgaardii		X		
Rivullana dura				X
Schizotrix calcicola		X		
Euglenophyta				
Euglena acus		X		
Euglena clavata		X		
Euglena ehrenbergii	X			
Euglena limnophyla	X			
Euglena oxyuris	X			
Euglena rostrifera		X		
Euglena subehrenbergii	X			
Lepocinclis salina		X		
Phacus inflexus		X		
Phacus platalea			X	
Phacus suecicus				X
Phacus sp.	X			
Phacus venustrus		X		
Trachelomonas molesta	X			
Trachelomonas volvocina			X	
Trachelomonas sp.				X
Chlorophyta (Desmidiaceae)				
Cosmarium angulatum				X
Bacillariophyta				
Gomphonema augur	X			
Navicula gastrum	X			
Navicula sp.				X
Surirella sp.	X			

A rich flora of *Euglena limnophyla*, *E. ehrenbergii*, *E. oxyuris* (Fig. 3C), *Phacus* sp., and *Trachelomonas molesta* was reported from a hard-water drainage pool along the Dorinestreet (Rensproject) (pH 6.9). Some Diatom species (*Navicula gastrum* , *Surirella* sp., *Gomphonema augur*) were found in low concentrations.

The filamentous blue- green algae *Lyngbya nordgaardii* and *Schizotrix calcicola*, and the Euglenophyta *Euglena acus* (Fig. 3D), *E. rostrifera*, *Phacus inflexus* and *P. venustrus* were the most conspicuous representatives in a hard-water pool (pH 8.5) covered with many aquatic plants which is situated at the corner of the Dr. Sophie Redmond Street and the Rode Kruis Avenue.

In a soft-water pool (pH 4.2) at Kedoen road the algal flora was characteristically poor with only two Euglenophyta species.

In a small soft-water pool nearby 's Landsboerderij (pH 7.0), the Desmid *Cosmarium angulatum* may occur as dominant, together with some Euglenophyta (*Trachelomonas* sp., *Phacus suecicus*), and the Diatom *Navicula* sp.

Summary

The physico-chemical parameters are described for selected localities (including rivers, creeks and swamps) . For the rivers and creeks of Suriname the following classification was made. 1. Rivers and creeks of the interior (above the falls). These are characterized by "clear water" which is generally soft, oligotrophic and nearly saturated with oxygen; 2. Coastal sections of rivers and creeks (below the falls). The only difference with the first category is a lower oxygen content; 3. Black-water rivers which drain the Savanna Belt, characterized by "black water" which is soft, oligotrophic, acid and almost depleted of oxygen.

In the large rivers the phytoplankton is dominated by Diatoms (Bacillariophyta), since these are best adapted to the current. In slower reaches Chlorophyta may be more abundant. Creeks have generally a low development of phytoplankton. A few Diatoms and Desmids may be found in black-water creeks. The plankton community of creeks in the Interior could in addition contain some Chlorophytes as well, but some creeks almost lack plankton. Compared to the natural water bodies, man-made aquatic habitats are usually eutrophic, leading to a rich algal development (the number of species, however, is low). The most important group here is the Euglenophyta.

Acknowledgements

We like to thank here Mr. A. Sabiran of the Hydraulic Research Department for the determination of physico-chemical parameters in the laboratory and Mr. Crul, Mr. K. Mohadin and Mr. Reynink for collecting the phytoplankton samples.

References

Goulding, M., M.L. Carvalho and E.G. Ferreira, 1988. Rio Negro. Rich life in poor water. SPB Acad. Publishing,The Hague, 200 pp.

Leentvaar, P., 1975. Hydrobiological observations in Surinam, with special reference to the manmade Brokopondo Lake in 1964. Stud. Fauna Suriname Guyanas 15: 1–173.

Mohadin, K., 1980. Hydrobiologische verkenning van Noord-Suriname. Unpubl. report Instituut Opleiding Leraren, Paramaribo, 100 pp.

Sioli, H., 1950. Das Wasser im Amazonasgebiet. Forsch. Fortschr. 26: 274–280.

5. Vegetation and vegetation succession of the freshwater wetlands

PIETER A. TEUNISSEN

1. Introduction

In Suriname about 12,000 square kilometers (7% of the landsurface) is occupied by freshwater wetlands, mainly found in the coastal region, between the brackish coastal mangrovebelt and the interior highlands.

The Surinamese coastal region (about 20,000 km²) may be subdivided into four ecological landscapes (Teunissen 1988):

a. brackish wetlands along the coast with mangrove forest, salt water lagoons and brackish herbaceous swamps (10% of the coastal region surface);
b. freshwater wetlands consisting of herbaceous swamps (15%), swamp wood or low swamp forest (30%) and high swamp forest (15%);
c. cultivated areas: built-up areas, agricultural mainly impoldered land, recently abandoned plantations and mined out land etc. (10%);
d. marsh and highland areas mainly covered by high marsh forest and high dryland forest (20%).

In the Interior, wetlands are confined to narrow fringes of land along rivers and creeks.

In the area of freshwater wetlands, open water bodies are mainly confined to rivers, creeks and man-made canals. All other wetlands consist of swamplands: wetlands covered with emerged vegetation. Swampland soils are poorly to very poorly drained and are inundated either permanently or at least during the greater part of the year. Generally, swampland soils are covered by a layer of peat: slightly decomposed organic material, locally known as "pegasse" (Van Vuuren *et al.* 1977; Noordam 1988). During average dry seasons, this peat layer remains soaked with water (Lindeman 1953).

In Suriname, areas with imperfectly to moderately well-drained soils, which become inundated during rainy seasons but drain completely during average dry seasons, are known as marshland ("drasland") or seasonal swampland. Marshland soils, covered by a thin layer of litter instead of peat are further characterized by hog-wallow structures of the mineral soil. In Suriname these structures are known as "kawfutu" (literally: "cow feet") a pattern of knobs or polygons separated by a network of narrow gullies (Lindeman and Moolenaar

Paul E. Ouboter (ed.), Freshwater Ecosystems of Suriname, 77–98.

1959). Marshland vegetations are not discussed in this paper.

Swampland vegetations may be predominated by swamp grasses; sedges and forbs (herbaceous swamps), by treelets and trees, less than 20 m high (swamp wood or low swamp forest) or by higher trees forming several tree layers (swamp forest).

Floating herbaceous vegetations are known as floating meadows.

2. Environmental Factors: Soil, Water and Fire

2.1. Mineral Soils and Peat Layers

2.1.1. Mineral Soils

The mineral soils of all Surinamese swamps consist of Holocene clays. During the early Holocene (with a rising sea level), marine clays, known as "Mara clays" (Brinkman and Pons 1968) were deposited between the higher elevated Pleistocene erosion remnants of ridges (Lelydorp Landscape) and marine clay plateaus (Para Landscape). Mara clays were also deposited in large areas in between the Lelydorp and the Para Landscape areas, where erosion remnants are almost lacking. These extensive swamps form the Mara Landscape (Noordam 1988). In the very soft Mara clays, soil profiles are not yet developed (no physical, chemical or biological development). Mara clays are rich in pyrite (FeS).

North of the Lelydorp, Para and Mara Landscapes, during the later Holocene to the present, younger marine clays were deposited during a stable sea level, and now form the Nickerie Landscape. These clays were deposited also in between the ridges of the Kwatta Landscape (Noordam 1988). Except for the very recent deposits along the coast, the younger clays are physically, chemically and biologically well developed and show clear soil profiles (Noordam 1988).

2.1.2. Peat

Swampy areas are inundated most of the year, so a layer of peat is formed on top of the mineral soil. Peat layers accumulate during rainy seasons and decay during dry seasons when oxygen is entering the upper part of the peat layer. Peat layers growing up to sea level are known as eustatic peat.

Due to very slow drainage, in the central parts of extensive swamps (see also 2.2.1) the mean water and peat level is higher than along the swamp edges, giving the water and peat layer a convex form. Thus, a vast low mound of peat is formed which grows up above sea level until the increase in runoff from the slope balances the excess precipitation. This peat, fed by rain water is known as "ombrogenous peat". Examples of ombrogenous peat swamps are: the Coronie Swamp, with peat layers up to 2 1/2 m thick (Pons and Parsan 1964), the Upper Nani Swamp with up to 5 m thick peat layers (Sevenhuijsen 1976) and also the

Babun areas where access canals needed weirs to prevent these swamps from "running dry" (see 4.4), like the Esterslust-Berendslust Swamp and the Perica Swamp with peat layers up to 2 m thick (Vink 1964; Van Donselaar 1965a).

Besides eustatic and ombrogenous peat swamps, floating peat swamps (swaying swamps) can be recognized as well. Swaying swamps have floating peat layers, making swaying movements if entered by man (Lindeman, 1953). The best documented representative is the so-called Buku Swamp northeast of Mungo, drained by the Buku Creek into the Cottica River. The origin of these swamps is unknown. Possibly the floating peat layers started to grow after deep peat fires as floating meadows. The water layer beneath floating peat layers may be several meters deep. At the end of the 18th century revolting runaway slaves built the fortress "Buku" on a natural island in the middle of the "inaccessible" Buku Swamp. Other swaying swamps are known in parts of the Coronie and Nani Swamp (Lindeman 1953; Pons and Parsan 1964; Sevenhuysen 1976).

2.2. Water

2.2.1. Drainage Patterns

In the central parts of extensive swamp areas (such as the Nani, Coronie, Cusewijne, Commewijne and Perica Swamps) distinct creeks are lacking completely. This indicates drainage in such areas is very slow (commonly measured in millimeters per second) due to the high resistance by the vegetation. Slow drainage and the presence of a protective peat layer prevent the mineral soil from gully erosion and therefore from creek development (Buys 1988). In case such swamps are covered by a herbaceous vegetation growing on thick peat layers (Nani Swamp, Coronie Swamp), on aerial photographs the drainage pattern is visible in the vicinity of surfacing island-like sand and clay plateaus, which have a comet-like tail, pointing to the direction of the water flow. When very acid swamp waters, poor in nutrients, pass such islands, the swamp water may be enriched by nutrients provided by the islands, resulting in a downstream vegetation which on aerial photographs differs in texture and gray-tone from the surrounding swamp vegetation (Brinkman pers. comm.; Sevenhuijsen 1977). In this way, the extensive Coronie Swamp shows a radial drainage pattern (Brinkman and Pons 1968; Teunissen 1978). Since the famous Surinamese forester J.W. Gonggryp studied the first aerial photographs, this radial pattern was also interpreted as the impact of a meteorite fallen in the centre of the Coronie swamp (Bubberman 1981). However, in the upper Nani Swamp the numerous island-like old ridges and clay plateaus prevent a radial drainage pattern. There the pattern is adapted to the obstacles, and therefore much more winding. In the Nani Swamp comet tails are found in all directions, following the winding drainage pattern.

Drainage over peat and through vegetation may be very slow but a considerable amount of water may flow over such large areas. Closer to the edges of extensive swamps, water is draining quicker and only there the water flow is strong enough to cause peat and mineral soil erosion, creating small creeks (Buijs 1988).

2.2.2. Water Depths

Swamp vegetation types are related also to water depth. Shallow and deep swamps are defined here as follows:

a. shallow swamps with surface waters fluctuating between the peat surface (dry season) and 60–80 cm above the peat surface (rainy season) are found mainly in the youngest parts of the coastal region (Nickerie, Kwatta and Lelydorp Landscapes) and in the vast swamps of the Mara Landscape;
b. deep swamps with surface waters fluctuating between the peat surface (dry season) and 80–350 cm above the peat surface (rainy season). The deep swamps exist as relatively narrow fringes along creeks and rivers and in between the clay plateaus of the Old Coastal Plain (Para Landscape), close to the higher elevated sandy Savanna Belt (Cusewijne Formation).
   During rainy seasons, the water inflow in such areas from the Savanna Belt as well as the surface water runoff from the clay plateaus is high, while the swamp discharge towards the rivers is very slow, resulting in very high swamp levels. This is especially the case where discharging creeks are blocked with fallen logs, peat and floating vegetation (see: 4.1).

2.3 Fire

As a rule, a peat layer does not dry up during the long dry season (August through December). However, if the rains fail to appear in the next short rainy season (December through February) and the next short dry season is dry indeed, an extremely long dry season of 8 continuous months is created. During such periods, peat layers desiccate and peat fires may start. These fires may be self-generated by bacterial heating or lightening, but more often these fires are man-made (see 4.1).

Peat fires have been documented from the extremely dry periods of 1745/46, 1767/68, 1797/98, 1845/46, 1898/99, 1911/12, 1925/26, 1939/40 and 1963/64 (Bubberman 1973).

Peat fires destroy the rooting zone of trees completely, making living trees fall criss-cross onto the burning forest floor. Peat fires may transform extensive areas of high or low swamp forests as well as herbaceous swamps into temporarely open water areas (Fig. 1) which soon become covered with a newly established rooting and or floating vegetation. Peat fires may widen the profiles of rivers and creeks formerly running through peaty profiles in swamp forest areas. Widened profiles in combination with fallen logs decrease the flow speed of creeks and rivers. Next to this, ash and rotting debris create waters rich in nutrients. Together with lots of sunlight the situation is ideal to stimulate the development of peat-producing floating meadows which finally block the waterways almost entirely. This development was observed by the author after the 1964 peat fires in forests along the Upper Cottica River. The river became inaccessible soon after the accompanying swamp forests were destroyed by fires. Also the existence of floating meadows in other creeks like the Ricanau, Surnau, Para and Coropina, Cusewijne and Nani Creeks (Fig. 2) may be related to old peat fires in those areas.

Figure 1. Open-water area that developed after a peat fire.

Figure 2. Nani Creek partly covered by floating meadows.

3. Vegetation Types and Vegetation Succession

Detailed data on vegetation types and succession for certain areas are presented by Lanjouw 1936; Geyskes 1945, 1948; Lindeman 1953; Lindeman and Moolenaar 1959; Teunissen 1976, 1978, 1980, 1984, 1987, 1988, 1991b, 1992c).

3.1. Herbaceous Swamps

In the younger parts of the Coastal Plain (north of the Lelydorp, Para and Mara Landscapes) herbaceous swamps show a great variety of grasses, sedges and forbs. Although the species may be found mixed, certain species may predominate over large areas. In the slightly brackish- to freshwater areas these species are: "Adroen" (*Cyperus articulatus*), "Payagrasi" (also known as "Kunsuwiwiri") (*Typha angustifolia*) (Fig. 3), "Riet" (*Phragmites communis*), "Alesigrasi" (*Leersia hexandra*) and "Mokomoko" (*Montrichardia arborescens*). In slightly brackish- to freshwater areas "Brantimaka" (*Machaerium lunatum*) scrub may be found as well. "Prasorograsi" (*Cyperus giganteus*) may also dominate, but this species prefers swamps with permanently freshwater (Fig. 4).

The southern Mara clay swamps are poor in species and most of the time one single species predominates in these swamps. Where peat layers are thin or lacking (due to frequent peat fires), "Prapragrasi" (*Eleocharis interstincta*) dominates (Fig. 5); on peat layers predominating sedges are *Lagenocarpus guianensis* and/or *Rhynchospora gigantea*, both known as "Twatwagrasi".

Figure 3. Typha angustifolia swamp.

Figure 4. Swamp dominated by *Cyperus giganteus* near Galibi.

Figure 5. Swamp dominated by *Eleocharis interstincta.*

In the absence of grass fires, treelets invade herbaceous swamps and this announces the succession towards swamp wood.

3.2. Swamp Wood or Low Swamp Forest

The most common type of swamp wood, which in the absence of fire may develop in slightly brackish- to freshwater herbaceous swamps is "Watrabebe"-

Figure 6. Pterocarpus officinalis – Tabebuia insignis swamp wood.

"Zwamppanta" (*Pterocarpus officinalis – Tabebuia insignis*) swamp wood, most of the time dominated by the first species (Fig. 6).

Locally abundant is the "Mira-udu" (*Triplaris surinamensis*) swamp wood dominated by *Triplaris*. The area of this type of swamp wood, which probably developed after the intense peat fires of 1926 (Lindeman 1953) and 1964 (Van Donselaar 1965a), seems to be decreasing due to ongoing forest succession.

In the slightly brackish to freshwater swamps in the central part of the Young Coastal Plain, a low type of swamp wood is found which is relatively rich in species: "Zwamppruim"-"Zwampzuurzak" (*Chrysobalanus icaco – Annona glabra*) swamp wood (Fig. 7). The species composition of these fire adapted swamp woods shows a strong resemblance to the species composition of the undergrowth of "Morisi" (*Mauritia flexuosa*) palm forests. However, in "Zwamppruim"-"Zwampzuurzak" swamp wood, "Morisi" palms are lacking completely, probably due to the non-tolerance of the species to even periodic brackish conditions.

In the northwestern and northeastern part of the Young Coastal Region, where swamps are slightly brackish to fresh also "Kofimama" (*Erythrina glauca*) swamp wood is found. Under the most extreme environmental conditions the "Kofimama" tree (named "mother of coffee" because of its use as a shade tree in coffee plantations) may dominate. This vigorously growing tree survives temporary salt intrusions, severe peat fires and even extremely high

Figure 7. Chrysobalanus icaco – Annona glabra swamp wood.

Figure 8. Mauritia flexuosa palm forest.

water levels. The other name for "Kofimama" is "Immortelle": in case of peat fires, the fallen trees produce many more new trees by sprouting all over the stem; in case of lengthy high water levels, "Kofimama" trees develop floating corky respiratory roots from the stem at the water level, which may change into stilt roots after the water level recedes (Teunissen 1975).

In Suriname the distribution of "Kofimama" swamp wood seems to coincide with the distribution of abandoned Amerindian dwellings and precolumbian "ridged fields" (Bubberman pers. comm.). Incidentally "Kofimama" also escaped from abandoned coffee plantations elswhere in Suriname.

"Morisi" (*Mauritia flexuosa*) palm forests (Fig. 8) only occur in permanent freshwater swamps. "Morisi" palms are fire resistant and survive swamp vegetation fires. In swamps which burn frequently, "Morisi" palms may dominate and ultimately these palms may form pure stands.

It may be expected that during prolonged absence of fire all mentioned swamp wood types eventually develop into species-rich, high swamp forest. In case of peat fires swamp wood may be destroyed and taken over by herbaceous swamp vegetations.

3.3. High Swamp Forest

In the shallow (50–80 cm deep) freshwater swamps of the Coastal Plain, the high "Babun"-"Mataki"-"Pina" (*Virola surinamensis-Symphonia globulifera-Euterpe oleracea*) forest (Fig. 9), which is relatively rich in species, forms the vegetation at the final (climax) stage of succession. Characteristic for this type of swamp forest is the layer of "Pina" palms and the many broad-leaved Monocotyledons in the subgrowth. "Posentri" or "Poison trees" (*Hura crepitans*) may dominate in this forest type ("Posentri" forest) (Fig. 10). Also "Mataki" (*Symphonia globulifera*) may dominate ("Mataki" forest). Sporadically as is the case in the Jodensavanna area, this swamp forest is dominated by the "Krubara" tree (*Pentaclethra macroloba*): "Krubara" forest.

Figure 9. Virola surinamensis – Symphonia globulifera – Euterpe oleracea forest.

Figure 9. (continued) *Virola surinamensis* – *Symphonia globulifera* – *Euterpe oleracea* forest.

In the Old Coastal Plain, in up to 3.5 m deep swamps, the relatively species-poor "Watrabiri"-"Waratapa" (*Crudia glaberrima-Macrolobium acaciifolium*) forest is found (Fig. 11), a second type of climax swamp forest (Teunissen 1978, 1980). As may be expected, only a limited set of swamp forest tree species is adapted to the very high water levels. Characteristic of these forests is the rather

Figure 10. Hura crepitans forest.

Figure 11. Crudia glaberrima – Macrolobium acaciifolium forest.

open canopy with xeromorphic leaflets, probably adapted to times of lengthy inundation, during which treeroots may suffer from the lack of oxygen, preventing the intake of water (physiological dryness). "Pina" palms are completely lacking in this type of swamp forest. In "Watrabiri"-"Waratapa" forests there is no substantial subgrowth layer. Only the prickly "Keskesmaka" palms (*Bactris maraja*) and the broad-leaved Bambu grasses (*Olyra* spp.) may be locally abundant.

Along creeks in the Interior, "Babun"-"Mataki"-"Pina" swamp forest is found, enriched with a great number of species also found in the surrounding high dryland forest. This type of forest is known as "Babun"-"Mataki"-"Pina" creek forest. Creek forest along water courses in savanna areas is often reduced to "Morisi"-palm gallery forest due to frequent peat fires.

Peat fires may destroy high swamp forest (as well as swamp wood) and create temporarily open water areas which are soon covered with a newly established floating or rooting herbaceous vegetation.

The development from herbaceous swamps to climax swamp forest may take many hundreds of years or even longer if thick peat layers have to accumulate first.

4. Human Impact

The human impact on swamp vegetations will be briefly discussed, in the order of their importance.

4.1. Grass and Peat Fires

During every dry season, herbaceous swamps and savannas are set on fire, a common practice to keep these areas accessible to man and attractive to wildlife (Fig. 12). Also road sides are burned for weed control purposes, while slashed forests are burned for shifting cultivation purposes or for permanent forest conversion. Grass is set on fire intentionally for fun or unintentionally during campfires etc.

During extremely dry seasons, vegetation fires may become peat fires, destroying swamp wood and swamp forest vegetations. During such periods forest litter in dryland and marsh forest becomes bone dry and flammable and fires can spread from one swamp (or savanna) area to another.

Bubberman (1973) makes it plausible that grass (and peat) fires are related to human activities. During the extremely dry season of 1963/1964 Bubberman made several reconnaissance flights over the coastal area. He observed vegetation fires only in the populated parts of the Coastal Region, mainly along the East-West Connecting Road. No fires were observed in remote areas.

Peat fires must have been taking place at least since the arrival of man (Amerindians) in Suriname (10,000 y BP), which means since early Holocene times when the present Coastal Region was being deposited. During precolumbian times, Amerindians already practised agriculture on raised "ridged fields" in our coastal swamp areas (Janssen 1974).

Parts of the course of the Wane Creek in eastern Suriname appear to consist

Figure 12. Swamp fire.

of precolumbian canals, excavated right through higher elevated clay plateaus, bypassing creek sections which are blocked by floating meadows or completely filled up with peat deposits. Possibly these canals were constructed after peat fires, which made parts of the creek inaccessible for boats (see also 2.3).

The 1963/64 fires are the best documented (LBB 1965; Bubberman 1973; Fraeser *et al.* 1974; Teunissen 1992b). Between September 1963 and May 1964 5,000 km^2 of vegetation, among which 500 km^2 of swamp forest, were consumed by peat fires.

In 1965 vegetation plots were laid out in the burned Perica Swamp (Van Donselaar 1965a) and monitored during the next 10 years (Van Donselaar 1965b, 1966; Teunissen and Wildschut 1967; Teunissen and Werkhoven 1975). Observations have shown that during the first year the up to 2 meters deep open pools among the tree stems that had fallen criss-cross were eutrophic due to the ash and debris, but gradually these waters became oligotrophic again. In the first 10 years, the developing floating and rooting vegetation produced a new peat layer of about 12 cm thick.

Gonggryp and Burger (1948) estimate that 50% of the swamp vegetations of the Coastal Region were influenced by fire. Recent ecosystem mapping (Teunissen 1978) showed that only 25% of the Surinamese swampland is still covered by climax swamp forest, which means that at least 75% of the potential swamp forest area (now covered with swamp wood and herbaceous swamp vegetation) was once or more often destroyed by fire.

4.2. Agriculture

4.2.1. Impoldering

Drained agriculture on "ridged fields" in swamp areas was practised by Amerindians (see also 4.1). This system was already abandoned when the first Europeans arrived. After a period of practising agriculture on highlands, the invaders started to dyke the more fertile swamp clay areas, creating polders (Fig. 13). Since then, about 2,000 km² of polderland has been created. For the near future more polders (and wadoeks) are planned on the right bank of the Lower Saramacca River, also south of the East-West Connecting Road in Coronie and along the Nickerie and Corantijn Rivers. In areas impoldered for rice growing, swamp forest is usually removed with dozers and burned. If areas are impoldered for other purposes (bananas, cattle), swamp forest may be removed by poisoning or by raising the water level to drown the trees. Both methods are used to prevent loss of the organic matter and soil compaction.

Figure 13. New polders are created.

4.2.2. Dykes

Dykes are built through swamps, not only for the purpose of impoldering, but also for swamp retaining and drainage diversion, while dykes are also created where canals for irrigation or drainage are excavated. All have an impact on the surrounding swamp vegetation, as they change the water regime of the swamps considerably. Swamp levels are dammed up intensionally if a water storage basin has to be created to facilitate irrigation for a second rice crop.

To prevent drainage of swamp water into the Lower Nickerie River, during 1971 and 1972 a 33 km "drainage diversion dam" was built south of and parallel to this river between the Marataka ridges near the village of Cupido and the Nani Creek, to divert swamp water flowing in the direction of the Nickerie River through the dam's excavation canal towards the Nani water distribution station (Roest 1972; Sevenhuijsen and Buys 1975; Sevenhuijsen 1976, 1977). During the next rainy seasons of 1972, 1973 and 1974, the swamp water of the northern Nani basin was dammed up to a level approximately 20 cm higher than usual maximum levels. This slight rise had an impact over an area of 190 km^2. The 70 km^2 of swamp forest and swamp wood present in this area was severely damaged or killed by drowning (Teunissen 1976) (Fig. 14).

Figure 14. Pterocarpus forest dying due to the swamp water being dammed up.

The 80 km long irrigation canal of the Multipurpose Corantijn Project (MCP) was built between the Wakay pump station at the Corantijn River and the Nani distribution station in Nickerie (ILACO *et al.* 1978). Although swamp spillways were projected in the accompanying canal dykes to prevent forest damage caused by rising Nani swamp levels, these spillways were never constructed and forest damage is severe (Van der Lugt pers. comm.).

To supply rice fields with irrigation water for a second rice crop, numerous south-north oriented water supply canals (used also for swamp fishing purposes) were constructed into the coastal Buru Swamp between the Suriname and Coppename Rivers. Some of these canals are more than 10 km long and reach into the brackish coastal swamps, facilitating salt penetration into the freshwater area (Mohadin and Tjon Lim Sang 1980).

The construction of many more dams and canals are considered, as for instance damming of the Lower Cusewijne River and raising the existing southern retaining dyke of the Coronie-East Rice Polder, both to increase the drainage capacity of the upstream swamp areas (Teunissen 1993b).

4.2.3. Discharge of Agrochemicals

Discharge of agrochemicals from polders seems to have no (visible) impact on swamp vegetations. The impact on water quality and aquatic life was studied by Vermeer *et al.* (1974).

4.2.4. Introduction of Exotic Plants

"Bottle brush" (*Melaleuca leucodendron*) – introduced by French nuns (working in Saint Laurent du Maroni in French Guiana) for cajaputi oil production – is running wild in swamps around Albina, in swamps along the eastern Wane Creek and swamps bordering the sand ridges of Galibi. These fire resistant treelets form very dense, impenetrable stands.

Locally, as in the "Weg naar Zee" area, introduced fodder grasses such as "Kapuwa grass" (*Echinochloa polystachya*), "Paardegras" (*Echinochloa stagnina*) and Paragrass (*Brachiaria mutica*) invade coastal herbaceous swamps (and become a pest in eutrophic drainage canals all over the country).

4.3. Roads and Transportation Canals

In crossing swamps with improperly constructed roads and canal dykes the natural drainage pattern of swamp areas may change considerably. Roads without culverts or with culverts insufficient in number, dimension and level, and canal dykes without or with insufficient spillways, block and dam up the slowly running swamp as a result of which the woody swamp vegetation at the upstream side of the road is drowned and marshland conditions are created on the other side.

The Burnside-Wageningen road, part of the East-West Connecting Road, prevented drainage of the soutern swamp towards the sea. The sea water invaded the northern area (Teunissen 1973, 1978). During the reconstruction of this road, extra culverts were constructed to prevent further damming up of the southern swamp and to feed the northern area again with freshwater (Lubbers *et al.* 1985; Buys 1985; Van der Steege 1986).

The construction of a 20 km ore transport canal is being considered, right through the Esterlust-Berendslust swamp forest area on the east bank of the Suriname River (Teunissen 1991c).

4.4. Swamp Forest Exploitation

To open up the extensive "Babun"-"Mataki"-"Pina" swamp forest (see 3.3) areas for harvesting "Babun" (*Virola surinamensis*) trees (used for plywood production), the Suriname Forest Service (Vink 1965) and the Bruynzeel Suriname Timber Company blasted and dug canal systems in these areas to transport (floating) "Babun" logs to the main rivers. While constructing these 3 m wide canals it appeared to be necessary to build weirs, to keep enough water in the forest areas to float the "Babun" logs out. Neglect of weirs after logging activities may have improved the swamp drainage to marshland conditions and may have changed the swamp forest species composition by invasion of marsh forest species. On the one hand, this may have resulted in an increase of the species diversity, on the other hand, this could have a negative impact on the "Babun" tree production: more competition for "Babun" seedlings and a harder job for lumber men to get the "Babun" trees out (not enough water for floating, too much water for transport over land).

4.5. Mining

Before bauxite mining can take place in the Coastal Region (Paranam-Onverdacht-Lelydorp area and Curmotibo deep-seated deposit), mining areas have to be impoldered. The removed overburden is disposed in dyked spoil areas or as undyked spoil piles (about 20 km^2 in 1993), preferably overlying Mara swamp areas instead of highland areas which are suitable for agricultural land use. When pyritic Mara clays are spoiled, oxidation of these clays results in the generation of sulphuric acid, which may lower the pH of surrounding wetlands (Teunissen 1991a,b, 1992b,c; Van der Lugt in prep.). Also the discharge of pumped out brackish mine water into the surrounding wetlands has no visible impact on swamp vegetations.

During the period 1992–2002, swamp retaining dykes will be built for oil exploration and oil exploitation purposes in the coastal Buru Swamp. These dykes will successively enlarge the perimeter of the drilling area and protect it from flooding throughout the year.

4.6. Industry

Caustic wastewater of Paranam's Alumina Refinery and the discharge of caustic soda contaminated rainwater from the refinery's bauxite residue disposal areas (red mud lakes) changed about 2 km^2 of surrounding swamps (herbaceous swamps, swamp wood and swamp forest) into open lagoons, in which some salt tolerant coastal plant species have found a new habitat (Teunissen 1990, 1992a, 1993a; Rensch and Teunissen 1993) (see also Ouboter and De Dijn, Chapter 14).

5. Conclusions

The greatest impact on natural swamp vegetations in the Coastal Region of Suriname is as old as the Coastal Region itself: man-made peat fires, destroying large areas of climax swamp forest during extremely dry seasons. If these fires cannot be prevented in the future, a unique, exploitable tropical forest ecosystem is doomed to disappear.

Theoretically, other negative impacts on natural swamp vegetations are managable and can be prevented or minimized by better planning and proper engineering.

References

Abrahams, J.V., 1987. Voorstudie Irrigatiewater-voorziening uit de Coroniezwamp. Tussentijds verslag. EOF Project. Ministerie van LVV.

Abrahams, J.V., 1988. Concept Eindrapport Voorstudie Irrigatiewatervoorziening vanuit de Coroniezwamp. Ministerie van LVV.

Bubberman, F.C., 1973. De bosbranden van 1964 in Suriname. Nieuwe West-Indische Gids 49(3):163–173 + aanhangsel.

Bubberman, F.C., 1981. Forest and forestry in Suriname. Suralco Magazine 13(3): 1–16.

Brinkman, R. and L.J. Pons, 1968. A pedo-geomorphological classification and map of the Holocene sediments in the Coastal Plain of the three Guianas. Netherlands Soil Survey Institute, Soil Survey Papers #4, Wageningen. 40 pp.

Buys, J.N.J., 1985. Opmerkingen en alternatief m.b.t. hydrologische aspecten reconstructie van de weg Burnside – Wageningen. Ministerie van Openbare Werken, Paramaribo. 20 pp.

Buys, J.N.J., 1988. Hydrography. In: Suriname Planatlas. Maps 1: 1,500,000 and 1: 500,000 with explanation. Ed. National Planning Office of Suriname, Paramaribo / Organization of American States, Washington DC.

Donselaar, J. van, 1965a. Mogelijkheden tot onderzoek in de Perica-zwamp i.v.m. de regeneratie van baboen (*Virola surinamensis*). Nota Dienst 's-Landsbosbeheer, Paramaribo. 6 pp.

Donselaar, J. van, 1965b. Successie-onderzoek in de Perica-zwamp I. Het uitzetten van permanente proefvlakten. Rapport Dienst 's-Landsbosbeheer, Paramaribo. 7 pp. + bijlagen.

Donselaar, J. van, 1966. Successie-onderzoek in de Perica-zwamp II. Het controleren van permanente proefvlakten. Rapport Dienst 's-Landsbosbeheer, Paramaribo. 6 pp + bijlagen.

Fraeser, A.I., R.J. de Milde and C.J. Inglis, 1974. A survey of the area of potentially productive marsh and swamp forest in Surinam. Forestry Development in Surinam, Project Working Document No 18. Food and Agricultural Organization of the United Nations.

Geyskes, D.C., 1945. Plants of the Suriname coastland. Some observations on the littoral vegetation of Dutch Guiana. Journal New York Botanical Garden 46(550): 229–236.

Geyskes, D.C., 1948. Luchtfotografie en zwampbegroeiing in Suriname. Tijdschrift Koninklijk Nederlands Aardrijkskundig Genootschap LXV (4,5): 665–668.

Gonggryp, J.W. and D. Burger Hzn, 1948. Bosbouwkundige studien over Suriname. Publicatie Fonds Landbouwk. Exp. 32. Wageningen. 264 pp.

ILACO *et al.*, 1978. Multipurpose Corantijn Project. Voorbereiding fase I. Annex D: Ecologische aspekten. 6 pp. ILACO Suriname N.V., PROSUR N.V., Rustwijk & Rustwijk N.V. en SUNECON.

Janssen, J.J., 1974. Suriname's Amer-indians. Who were they and where did they come from. Suralco Magazine 1974. Also in: J. Fontaine (ed.), 1980. From Suriname's History. Fragments of a stirring past. Suralco/De Walburg Pers: 10–20.

Lanjouw, J., 1936. Studies on the vegetation of the Suriname savannahs and swamps. Nederlandsch Kruidkundig Archief 46: 823–851. Also as: Mededelingen Botanisch Museum en Herbarium Utrecht 33.

LBB, 1965. Jaarverslag Dienst 's-Landsbosbeheer 1964. Paramaribo.

Lindeman, J.C., 1953. The vegetation of the coastal region of Suriname. The Vegetation of Suriname I(1): 135 pp. Van Eeden Foundation, Amsterdam. 135 pp.

Lindeman, J. C. and S.P. Moolenaar, 1959. Preliminary survey of the vegetation types of Northern Suriname. The Vegetation of Suriname I(2): 45 pp. Van Eeden Foundation, Amsterdam. 45 pp

Lubbers, B.A., L.J.M. Houben and R.J. Sevenhuijsen, 1985. Ontwerp reconstructie van de weg Burnside-Wageningen, Suriname. TH Delft, Afdeling Civiele Techniek. The Netherlands. 93 pp.

Lugt, F.L. van der, 1990. Rapport Waterkwaliteit BMS mijngebied: januari–februari 1990. 11 pp. + bijlagen.

Mohadin, K., 1980. Hydrobiologische verkenning van Noord Suriname. Rapport Instituut voor de Opleiding van Leraren. 100 pp. + bijlagen.

Mohadin, K. and R. Tjon Lim Sang, 1980. Enkele ecologische aspekten t.a.v. de rijstbouw aan de rechteroever van de Saramacca Rivier. Intern Rapport STINASU/Dienst 's-Landsbosbeheer, Afd. Natuurbeheer, Paramaribo.

Noordam, D., 1988. Geomorphology and soils. In: Suriname Planatlas. Map 1: 1,000,000 with explanation. Ed. National Planning Office of Suriname, Paramaribo / Organization of American States, Washington DC. 2 pp.

Pons, L.J. and Parsan, 1964. De bodemgesteldheid van Oostelijk Coronie. Unpublished Report Dienst Bodemkartering, Paramaribo.

Rensch, L.K. and P.A. Teunissen, 1993. Surface water monitoring report. Paranam area. 1992. Report Suriname Aluminum Company. 9 pp + 28 pp Appendices.

Roest, C., 1972. Onderzoek naar de capaciteit van de Nani zwamp als reservoir voor irrigatiewater. Ministerie van Landbouw, Veeteelt en Visserij, Paramaribo.

Sevenhuijsen, R.J., 1976. Study of the hydrology and vegetation in the Nanni Swamp, Surinam. In: Report for the Year 1975 of the Netherlands Foundation for the Advancement of Tropical Research (WOTRO): pp 35–37.

Sevenhuijsen, R.J., 1977. Irrigatie uit een moeras. Een hydrologische studie van de Nannizwamp in Suriname. Thesis Agricultural University, Wageningen. 143 pp.

Sevenhuijsen, R.J. and J.N.J. Buys, 1975. Het Nanni zwamp als irrigatiewater reservoir. Rapport t.b.v. de Werkgroep Follow-Up Stondansie. 2 Volumes. Ministerie van Openbare Werken.

Steege, J.G. van der, 1986. Ecologische aspecten van de reconditionering van de wegverbinding tussen Burnside en Wageningen in Suriname. Rapport Dienst 's-Landsbosbeheer, Paramaribo. 42 pp. + bijlagen. Paramaribo.

Teunissen, P.A., 1973. Vegetatiekaart Nickerie-Burnside, schaal 1: 120,000. In: A.L. Spaans: Pleidooi voor de aanwijzing van het Coronie-zwamp als natuurreservaat. Rapport Dienst 's-Landsbosbeheer, Paramaribo.

Teunissen, P.A., 1976. Vegetation changes in a dammed up freshwater swamp in NW Suriname. Acta Amazonica 6(2): 117–150.

Teunissen, P.A., 1978. Reconnaissance Map Surinam Lowland Ecosystems (Coastal Region and Savanna Belt), scale 1 : 200,000. Foundation for Nature Preservation in Suriname, Paramaribo / De Walburgpers, Zutphen

Teunissen, P.A., 1980. Overzicht van de Surinaamse Laagland Ecosystemen met vegetatietabellen. Rapport Dienst 's-Landsbosbeheer, Paramaribo. 110 pp.

Teunissen, P.A., 1984. De vegetatie van Suriname. Instituut voor de opleiding van Leraren. 13 pp.

Teunissen, P.A., 1987. Surinaamse ecosystemen III en IV:. Zoetwater-ecosystemen / Graszwampen en zwampbossen. Instituut voor de Opleiding van Leraren, Paramaribo. 11 / 13 pp.

Teunissen, P.A., 1988. Vegetation Map, scale 1: 1,000,000 with explanation. In: Suriname Planatlas. Ed. National Planning Office of Suriname, Paramaribo / Organization of American States, Washington DC. 2pp.

Teunissen, P.A., 1990. Interim report on caustic soda pollution by SURALCO-BILLITON Joint Venture Alumina Refinery at Paranam (Suriname). Report BILLITON Maatschappij Suriname. 10 pp.

Teunissen, P.A., 1991a. Bauxietmijnen Onverdacht. Vegetatie-ontwikkeling stortarealen. Rapport BILLITON Maatschappij Suriname. 32 pp + bijlagen.

Teunissen, P.A., 1991b. Accaribo bauxietmijn / l'Esperance stortareaal. Environmental Baseline Study en Environmental Impact Assessment. Rapport BILLITON Maatschappij Suriname. 74 pp + bijlagen.

Teunissen, P.A., 1991c. Klaverblad-Kaimangrasi Bauxite Deposits. Environmental impact assessment of potential access routes. Annex of: Klaverblad - Kaimangrasi pre-feasibility study. EVP Mining International, Curacao, N.A. 35 pp. + attachments.

Teunissen, P.A., 1992a. Review surface water monitoring program 1977–1991. Report SURALCO Environmental Department, Paranam.

Teunissen, P.A., 1992b. Curmotibo mine. Pre-feasibility baseline study and Environmental Impact Assessment. Annex of: Curmotibo pre-feasibility study. Report EVP Mining International. 34 pp. + bijlagen.

Teunissen, P.A., 1992c. Lelydorp Bauxite Mines III. Environmental baseline study Flora and Vegetation. Report BILLITON Environmental Department. 36 pp.

Teunissen, P.A., 1993a. Preliminary species trial with salt tolerant coastal plants on a bauxite residue disposal area in Suriname. Report SURALCO Environmental Department, Paranam.

Teunissen, P.A., 1993b. Feasibility study for the improvement of three polders in Coronie East, Saramacca and Pad van Wanica, Reeberg and Rijsdijk. Annex 1: Assessment of environmental aspects. Lahmeyer, Sogreah and Sunecon. 30 pp.

Teunissen, P.A. and M.C.M. Werkhoven, 1975. Successie-onderzoek in de Perica-zwamp IV. Rapport Dienst 's-Landsbosbeheer, Paramaribo. 6 pp + bijlagen.

Teunissen, P.A. and J.Th. Wildschut, 1967. Successie-onderzoek in de Perica-zwamp III. Rapport Dienst 's-Landsbosbeheer, Paramaribo. 17 pp + bijlagen.

Vink A.T., 1964. Ditching with dynamyte. Report Surinam Forest Service, Paramaribo. 20pp + 10 pp Attachments.

Vuure, W. van, *et al.*, 1977. Reconnaissance Soil Map of Northern Suriname. Scale 1: 100,000. Soil Department, Paramaribo. Map - C 805.

WLA, 1978. Onderzoek Coroniezwamp. Vooronderzoek. Waterloopkundige Afdeling van het Ministerie van Openbare Werken en Verkeer. 17 pp + bijlagen.

6. Aquatic macrophytes

MARGA C.M. WERKHOVEN and GEERT M.T. PEETERS

1. Introduction

Aquatic macrophytes are of primary importance as producers in the freshwater ecosystems, where they are often the characterizing constituents. Hence, aquatic macrophytes cannot be ignored during research on these ecosystems. This chapter gives an overview of the species in Suriname and some annotations on their distribution, with special emphasis on the family of Podostemaceae. The most important freshwater habitats and their characteristic aquatic macrophytes will be discussed.

1.1. Definition

Defining the term 'aquatic macrophyte' or 'water plant' is not easy. There is no sharp boundary between land plants and water plants. Actually, water plants are nothing more than land plants adapted secondarily to the aquatic habitat (Bloemendaal *et al.* 1988). Many of them are capable of living both in water and on land, and accomplish a bigger or smaller part of their life cycle on land. So it is hardly surprising that one can find an abundance of definitions in the literature on aquatic macrophytes.

In this chapter, water plants will be defined as "plants which can complete their generative cycle when all vegetative parts are submerged or carried by the water (floating leaves), or which live usually submerged but only turn to generative reproduction when the vegetative parts die off by desiccation, or which float on the water surface in addition to which the roots are submerged while the other parts emerge from the water".

Actually this is the definition given in Bloemendaal *et al.* (1988), extended with the pleustohelophytes (i.e. floating plants with submerged roots while other parts emerge above the water).

Some groups of plants, which are considered water plants by many researchers on the subject, are excluded by this definition, viz.:

Paul E. Ouboter (ed.), Freshwater Ecosystems of Suriname, 99–112.

1. helophytes : plants rooting in the soil with the lower parts submerged, but whose leaves and flowers reach above the water surface (e.g. *Typha angustifolia*).
2. reptohelophytes : plants rooting in the shore or in floating plant remnants, and propagating on or just below the water surface by stolons from which air-leaves and flowers grow (e.g. *Hydrocotyle umbellata*).
3. pseudohydrophytes : plants able to grow submerged for a longer period and to increase vegetatively, but unable to reproduce generatively under these circumstances.

In this chapter we will restrict ourselves to the flowering plants (Spermatophyta) and ferns (Pteridophyta), while mosses (Bryophyta), stoneworts (Charaphyta) and other algae will not be discussed.

2. Enumeration of Species

So far, 68 species of aquatic macrophytes from Surinamese waters are known (Appendix 1). For the sake of completeness, the 2 species restricted to brackish waters, *Ruppia maritima* and *Ceratophyllum echinatum*, are included as well. In addition it should be noted that the distribution of *Nymphaea ampla* is strongly concentrated in the brackish-water area; the species is only occasionally found in freshwater habitats.

3. Notes on Distribution

Based on its geomorphological history, Suriname is divided into three distinctly different zones. From north to south, these are respectively the Coastal Plain, the Savanna Belt and the Interior (Stichting Planbureau Suriname 1988).

Table 1 gives the occurrence of aquatic macrophytes in these three zones. Interpreting this table, one should take into account that the Coastal Plain has been studied more extensively than the other two zones. Undoubtedly more species will be found in the latter in future.

For the time being, 38 species are known from the Coastal Plain, 7 (only!) from the Savanna Belt and 41 from the Interior. Restricted to the Coastal Plain are 20 species, to the Savanna Belt 1 species, and to the Interior 27 species. Of the species restricted to the Interior, 24 belong to the family Podostemaceae. The distribution of the Podostemaceae will be discussed in greater detail in the following paragraph.

Table 1. Surinamese water plants, occurrence. This table is based on information from Den Hollander & Berg (1983), Kramer (1978), Pulle *et al.* (1928 t.p.), Van Royen (1951, 1953, 1954), data from BBS (National Herbarium of Suriname) and U (Herbarium of the Department of Plant Ecology and Evolutionary Biology, Utrecht), and personal observations.
C.P.-Coastal Plain; S.B.-Savanna Belt; Int-Interior.

Species	C.P.	S.B.	Int.	Species	C.P.	S.B.	Int.	Species	C.P.	S.B.	Int.
Nymphaea ampla	■			*Utricularia pulcherrima*	■	■		*A. marowijnensis*			■
N. odorata	■			*Pontederia rotundifolia*	■	■		*A. nana*			■
Ceratophyllum echinatum	■			*Salvinia auriculata*	■	■		*A. penicillata*			■
Nymphoides indica	■			*Eriocaulon melanocephalum*		■		*A. petiolata*			■
Utricularia foliosa	■			*Utricularia gibba*	■		■	*A. richardiana*			■
U. inflata	■			*Eichhornia crassipes*	■		■	*A. secundiflora*			■
U. myriocista	■			*E. heterosperma*	■		■	*A. staheliana*			■
U. obtusa	■			*Mayaca fluviatilis*	■		■	*A. treslingiana*			■
Limnobium stoloniferum	■			*M. longipes*	■		■	*A. versteegiana*			■
Philotria granatensis	■			*Lemna aequinoctialis*	■		■	*Marathrum capillaceum*			■
Hydrocleys nymphoides	■			*L. valdiviana*	■		■	*Mourera fluviatilis*			■
Ruppia maritima	■			*Spirodela intermedia*	■		■	*Oserya minima*			■
Eichhornia azurea	■			*Ceratopteris deltoidea*	■		■	*O. perpusilla*			■
Pistia stratiotes	■			*C. pteridoides*	■		■	*Rhyncholacis cristata*			■
Wolffia brasiliensis	■			*C. thalictroides*	■		■	*R. dentata*			■
Wolffiela lingulata	■			*Cabomba piauhyensis*			■	*Tristicha trifaria*			■
W. neotropica	■			*Apinagia digitata*			■	*Weddellina squamulosa*			■
W. welwitschii	■			*A. divertens*			■	*Utricularia hydrocarpa*			■
Marsilea polycarpa	■			*A. flexuosa*			■	*Eichhornia diversifolia*			■
Azolla caroliniana	■			*A. guyanensis*			■	*Nymphaea blanda*	■	■	■
Nymphaea amazonum	■	■		*A. hulkiana*			■	*Bacopa reflexa*	■	■	■
N. rudgeana	■	■		*A. imthurnii*			■	*Tonina fluviatilis*	■	■	■
Cabomba aquatica	■	■		*A. longifolia*			■				

3.1. Podostemaceae

Podostemaceae are herbs, often annual, growing in cataracts and rapids, usually submerged, and frequently resembling algae or mosses. All species are attached with special organs to a stony substrate. Often the thallus is lichenoid. At low water levels, the plants die off after having produced aerial flowers and fruits.

There are 50 genera with approximately 275 species, mainly restricted to the tropics, but occurring especially in thc American and Asian tropics (Mabberley 1989). In Suriname, 150 collections of Podostemaceae have been made, belonging to 24 species and 7 genera (Table 2).

According to Van Royen (1951), each set of cataracts and rapids would have its own species, while most species have a restricted area. To test this hypothesis, the data on the distribution of Podostemaceae from Pulle (1906), Van Royen (1951, 1953, 1954), Den Hollander and Berg (1983), and from the herbaria in Paramaribo (Suriname) and Utrecht (the Netherlands) are compiled in Table 2.

The six basins with cataracts and rapids in their rivers, rivulets, creeks and tributaries are from east to west: Marowijne, Suriname, Saramacca, Coppename, Nickerie and Corantijn (Fig. 1). Figure 1 also gives the localities where the collections have been gathered. From these data it can be concluded that the upstream areas of the Marowijne, Coppename and Corantijn basins are under-collected. The scarce collection sites in the Nickerie basin are due to the few appropiate habitats for Podostemaceae.

If we take into consideration the number of species per river basin in Suriname (Table 2), it may be concluded that the basin of the Suriname River is the richest, with 16 species (66%), followed by the Marowijne basin (13 spp.), and the Coppename and Corantijn basins (9 spp.). However, it should be mentioned that the Marowijne and Corantijn basins cover the largest areas, and encompass most cataracts and rapids.

The most common species in Suriname are *Apinagia richardiana* and *Mourera fluviatilis* (Fig. 2), which have been collected in all basins. These two species constitute 50% of the collections. Looking at the size of the plants and the inflorescences (see Table 2), there seems to be no reason to suppose that the smaller species were overlooked during the collecting.

Notwithstanding the incompleteness of our data, we can conclude from Table 2 that four species (*Apinagia secundiflora, A. petiolata, Oserya minima, Rhyncholacis cristata*) are probably restricted to one river basin in Suriname. The hypothesis that most species are restricted to one river, cataract or rapid, stands for 16% of the species (4 spp.). One may assume that more, probably new, species grow in the rather inaccessible upstream areas that have not yet been investigated. Our preliminary conclusion is that some species appear to be quite rare in Suriname e.g. *Weddellina squamulosa*, and the above-mentioned four species.

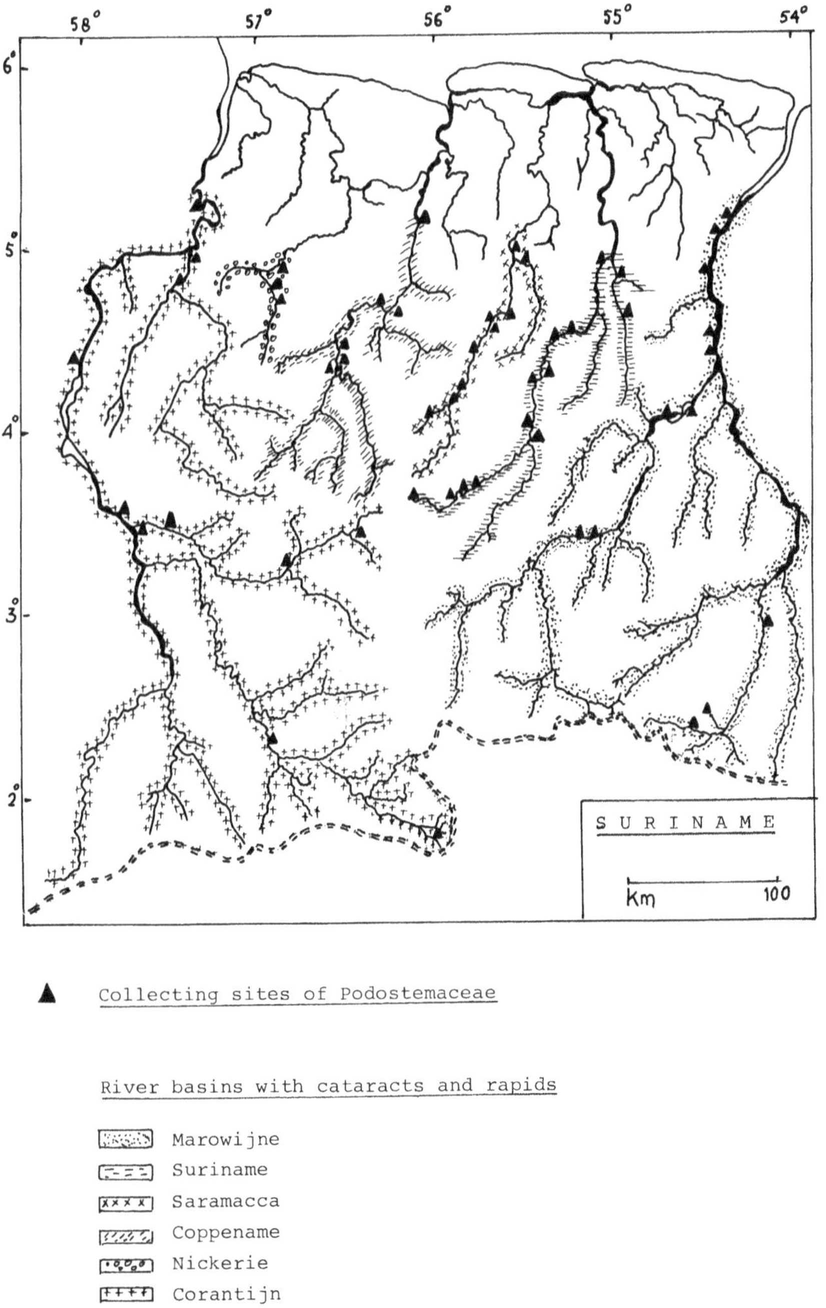

Figure 1. Map of Suriname, showing the river basins with cataracts and rapids, and the collection localities of Podostemaceae species.

Table 2. Surinamese Podostemaceae. Distribution, size, and number of collected Podostemaceae species in Suriname.

Species	Coll.	RIVER BASINS						infl.size	plant size	Distribution
		Mar.	Sur.	Sar.	Copp.	Nick.	Cor.			
Apinagia digitata	1	1						7	20	Fr.Gu./Sur.
Apinagia guyanensis	1	1						12	10	Fr.Gu./Sur.
Apinagia imthurnii	3	3						3	10	Fr.Gu./Sur./Guy.
Marathrum capillaceum	3	3						5	15	Sur./Ven.
Apinagia penicillata	3	1	2					1,5	7,5	Sur.
Apinagia versteegiana	4	2	2					1,5	15	Sur.
Apinagia flexuosa	6	1	3	1	1			4	25	Fr.Gu./Sur/Guy.
Apinagia divertens	3	1	1				1	0,5	2	Sur.
Oserya perpusilla	7	2		1	2		2	1,4	3	Fr.Gu./Sur.
Apinagia marowijnensis	3	1			2			2,5	15	Sur.
Apinagia longifolia	4	2			1		1	6	30	Sur./Guy.
Apinagia hulkiana	1		1					3,5	5	Sur./Guy.
Apinagia nana	2		2					0,2	0,4	Sur./Central Br.
Apinagia secundiflora	1		1					1	15	Sur.
Oserya minima	1		1					0,1	0,5	Sur.
Rhyncholacis cristata	1		1					3	30	Sur.
Weddellina squamulosa	5			1	2		2	12	80	N.Br./Sur./Guy./Col.

Table 2. (continued)

Species	Coll.	RIVER BASINS Mar.	Sur.	Sar.	Copp.	Nick.	Cor.	infl.size	plant size	Distribution
Rhyncholacis dentata	2		1		1			2,5	45	Sur.
Tristicha trifaria	4		2		1		1	0,2	10	Sur./From Mex. to Mad.
Apinagia staheliana	13		1		4	1	7	25	200	Sur./Guy.
Apinagia treslingiana	6		4				2	8	15	Sur.
Apinagia petiolata	1						1	4	30	Sur.
Apinagia richardiana	45	6	16	5	6	3	9	6	30	N.Br./Fr.Gu./Sur./Guy./Ven.
Mourera fluviatilis	30	7	4	9	3	2	5	60	200	N.Br./Sur./Guy./Ven.
Total	**150**									
no. of species per basin		13	16	5	9	4	9			
no of species restricted to one basin in Suriname		4	5	–	–	–	1			
no. of coll. per basins		31	43	18	21	8	29			

Coll. = Number of collections per species.
Mar. = Marowijne river.
Sur. = Suriname river.
Sar. = Saramacca river.
Copp. = Coppename river.
Nick. = Nickerie river.
Cor. = Corantijn river.
infl. size = Size of inflorescence, in cm.
plant-size = Size of plant in cm.

Fr.Gu. = French Guiana; Sur. = Suriname; Guy. = Guyana; Ven. = Venezuela; Br. = Brazil; Col. = Colombia; Mex. = Mexico; Mad. = Madagascar.

Figure 2. Mourera fluviatilis at Awaradam, Gran Rio, Suriname River basin.

4. Freshwater Habitats and Their Water Plants

Hitherto the ecology of water plants in Suriname has hardly been studied. Data on their ecology and habitat can only be found scattered in literature and herbaria. Moreover, hardly any attempt has been made to classify freshwater habitats in Suriname on the basis of the water plants present, or to distinguish water plant communities.

The following classification of freshwater habitats is a provisional one. It is based mainly on hydrological and morphological characteristics. Only the most frequent freshwater habitats of Suriname will be discussed.

4.1. Main River Channels

Because of their water depth, current velocity and turbidity, main river channels are, as a rule, not suitable for the growth of water plants. Only under the shelter of the shore, water plants are able to establish themselves. Particularly free-floating plants like *Eichhornia* spp, *Pistia stratiotes* and *Salvinia auriculata* may be mentioned, but also soil-rooting species with free-floating leaves like *Nymphaea* spp.

Figure 3. Extensive growth of *Salvinia auriculata* at Cusewijne River, April 1988 (photo J.A. Mol)

When the nutrient-content of the river water is sufficiently high, large masses of free-floating water plants may gather in the main river channel. For instance, almost every year at the end of the short dry season, large tracks of the Cusewijne River are covered by *Salvinia auriculata* (Fig. 3). During the subsequent long rainy season all of these floating plants are carried downstream, after which regrowth will start in the long dry season.

4.2. Rapids and Cataracts

In the Interior, where the rivers have worn their channels in the Guiana Shield, numerous rapids and cataracts occur. These form the characteristic habitat of the earlier mentioned Podostemaceae family.

4.3. Creeks and Rivulets

Creeks and rivulets may be colonized by water plants rooting in the soil. If favourable light conditions prevail, as in open savanna and swamp areas, richly variegated water plant vegetations may develop with species as *Cabomba aquatica, Nymphoides indica, Nymphaea* spp. and *Utricularia* spp. However, near cultivated areas, eutrofication will rapidly lead to dominance of free-floating species, while along the shore of the streams a strong development of grass matts may be seen. This can be illustrated by the Grote Borfelt Creek, a tributary of the Cusewijne River, which is densely covered by *Pistia stratiotes* and *Eichhornia crassipes* along with species like *Salvinia auriculata, Azolla caroliniana* and *Ceratopteris* spp.

If creeks are heavily shaded, as is usually the case in the densely forested areas, only a few water plants occur. *Tonina fluviatilis* is one of them.

4.4. Swamps

When swamps are flooded, water plants may develop among the characteristic helophyte vegetations. Most frequent in this habitat are *Azolla caroliniana, Salvinia auriculata, Spirodela intermedia, Utricularia* spp. and *Nymphaea* spp. (Teunissen 1980).

In swamp forests, water plants occur only sparsely after inundation (Teunissen 1980). The main reason for this is too much shade.

4.5. Man-made Aquatic Habitats

Most man-made aquatic habitats in cultivated areas, such as trenches, irrigation channels and puddles, are characterized by strong eutrofication and high light incidence. In these circumstances free-floating, light and nutrient loving water plants like *Eichhornia crassipes* and *Pistia stratiotes* grow luxuriously, together with such species as *Azolla caroliniana, Salvinia auriculata* and *Ceratopteris* spp.

A special case is the Brokopondo Reservoir. Van Donselaar (1989) gives an outline of the vegetation succession after the closing of the dam in 1964. Once the dam had been closed, the developing lake made a good habitat for *Eichhornia crassipes* and *Ceratopteris pteridoides*, viz. stagnant water with a high light incidence. The aforementioned species developed explosively and in the middle of 1966 they covered respectively 41,200 ha (=53% of the then lake area) and 17,000 ha (=22%). Other strongly expanding species were *Lemna valdiviana, Spirodela intermedia* and, along the shores, *Utricularia gibba*. After 1966 an intensive control with the herbicide 2,4-D caused a dramatic decrease of the floating vegetations. In 1972 *Eichhornia crassipes* was confined to the youngest (i.e. southern) parts of the reservoir, whereas *Ceratopteris pteridoides* had disappeared.

5. Conclusion

So far, research on aquatic macrophytes in Suriname has mainly concentrated on the inventory and the taxonomy of species. Our compilation shows that the species of the Coastal Plain are quite well-known. More data have to be collected on the species of the Savanna Belt and the Interior. Further investigations in these zones are strongly needed. The ecology of aquatic macrophytes in Suriname has hardly been studied. In the future, research should concentrate on this aspect, too.

Summary

A preliminary inventory of aquatic macrophytes in Suriname resulted in 68 species. Thirthy-eight species are known from the Coastal Plain, 7 from the Savanna Belt, and 41 from the Interior. Of these 41 species, 24 belong to the Podostemaceae. The hypothesis that each river or set of rapids has its own characteristic species of Podostemaceae scarcely holds true after our geographical analysis. In Suriname, only 4 species (=16%) are restricted to one river basin. Finally, some annotations are made on the freshwater habitats and their characteristic species in Suriname.

Appendix 1. Surinamese water plants, list of species.
This list is based on information from Den Hollander & Berg (1983), Kramer (1978), Pulle *et al.* (1928 - present), Taylor (1967), Van Royen (1951, 1953, 1954), and data from the National Herbarium of Suriname (BBS) and the Herbarium of the Department of Plant Ecology and Evolutionary Biology, Utrecht (U). The classification of the Spermatophyta is according to Stoffers (1982).

SPERMATOPHYTA
DICOTYLEDONAE
Ranunculales

Nymphaeaceae
Nymphaea
- *amazonum* Mart. et Zucc.
- ampla (Salisb.) D.C.
- *blanda* G.F.W. Mey.var.*fenzliana* (Lehm.) Caspary
- *odorata* Ait.var.*gigantea* Tricher
- *rudgeana* G.F.W. Mey.

Appendix 1. (continued)

Cabombaceae
Cabomba aquatica Aubl.
- *piauhyensis* Gardn.var.*piauhyensis*

Ceratophyllaceae
Ceratophyllum echinatum Gray

Rosales

Podostemaceae
Apinagia digitata v. Royen
- divertens Went
- *flexuosa* (Tul.)v. Royen
- *guyanensis* (Pulle)v. Royen
- *hulkiana* (Went)v.Royen
- *imthurnii* (Goebel)v.Royen
- *longifolia* (Tul.)v.Royen
- *marowijnensis* (Went)v.Royen
- *nana* Went
- *penicillata* (v.Royen)v.Royen
- *petiolata* den Hollander
- *richardiana* (Tul.)v.Royen
- *secundiflora* (Tul.)Pulle
- *staheliana* (Went)v.Royen
- *treslingiana* (Went)v. Royen
- *versteegiana* (Went)v.Royen
Marathrum capillaceum (Pulle)v.Royen
Mourera fluviatilis Aubl.
Oserya minima v.Royen
- *perpusilla* (Went)v. Royen
Rhyncholacis cristata v.Royen
- *dentata* v.Royen
Tristicha trifaria (Bory ex Willd.) Sprengel ssp. *trifaria*
Weddellina squamulosa Tul.f.*squamulosa*

Gentianales

Menyanthaceae
Nymphoides indica (L.)O.K.

Solanales

Scrophulariaceae
Bacopa reflexa (Benth.)Edwall

Lentibulariaceae
Utricularia foliosa L.
- *gibba* L.ssp.*gibba*
- *hydrocarpa* Vahl
- *inflata* Walter
- *myriocista* St.Hil.et Gir.
- *pulcherrima* Sylven
- *obtusa* Sw.

Appendix 1. (continued)

MONOCOTYLEDONAE

Alismatales

Hydrocharitaceae

Limnobium stoloniferum (G.F.W.Mey.)Chris.
Philotria granatensis (H.et B.)Marie-Victorin

Butomaceae

Hydrocleys nymphoides (Willd.)Buch.

Najadales

Ruppiaceae

Ruppia maritima L.

Liliales

Pontederiaceae

Eichhornia azurea (Sw.) Kunth
– *crassipes* (Mart.)Solms.
– *diversifolia* (Vahl)Urb.
– *heterosperma* Alexander
Pontederia rotundifolia L.f.

Commelinales

Mayacaceae

Mayaca fluviatilis Aubl.
– *longipes* Mart.ex Seubert

Eriocaulaceae

Eriocaulon melanocephalum Kunth
Tonina fluviatilis Aubl.

Arales

Araceae

Pistia stratiotes L.var.*obcordata* (Schleid.)Engl.

Lemnaceae

Lemna aequinoctialis Welwitsch
– *valdiviana* Phil.
Spirodela intermedia W.Koch
Wollffia brasiliensis Weddell
Wollffiella lingulata (Hegelm.)Hegelm.
– *neotropica* Landolt
– *welwitschii* (Hegelm.)Monod

Appendix 1. (continued)

PTERIDOPHYTA

Polypodiales

Parkeriaceae
Ceratopteris deltoidea Benedict
- *pteridoides* (Hook.)Hiern.
- *thalictroides* (L.)Brongn.

Marsileales

Marsileaceae
Marsilea polycarpa Hook.et Grev.

Salviniales

Salviniaceae
Salvinia auriculata Aubl.
Azollaceae
Azolla caroliniana Willd.

References

Bloemendaal, F.H.J.L., Th.C.M. Brock and C. den Hartog, 1988. Structuur van waterplanten en hun vegetaties. In: Bloemendaal, F.H.J.L. and J.G.M. Roelofs (eds.), Waterplanten en waterkwaliteit. Koninklijke Nederlandse Natuurhistorische Vereniging, Utrecht/Nijmegen, pp. 11–25.

Donselaar, J. van, 1989. Brokopondo Research Report, Suriname. part III. The vegetation in the Brokopondo-lake basin (Surinam) before, during and after the inundation, 1964–1972. Nat. Wet. Studiekr. Sur. Ned. Ant. No. 124, Amsterdam, 45 pp.

Hollander, G. den, and C.C. Berg, 1983. A new species of *Apinagia* (Podostemonaceae) and a key to the *Apinagia* species in Suriname. Acta Bot. Neerl. 32(3): 223–226.

Kramer, K.U., 1978. The Pteridophytes of Suriname. An enumeration with keys of the ferns and fern-allies. Nat. Wet. Studiekr. Sur. Ned. Ant. No. 93, Utrecht, 198 pp.

Mabberley, D.J., 1989. The plant book. A portable dictionary of the higher plants. Cambridge University Press, Cambridge, 706 pp.

Pulle, A., 1906. An Enumeration of the Vascular Plants known from Surinam, together with their Distribution and Synonomy. E.J. Brill, Leiden, 555 pp.

Pulle, A., J. Lanjouw, A.L. Stoffers and J.C. Lindeman (eds.), 1928–present. Flora of Suriname. E.J. Brill, Leiden.

Royen, P. van, 1951. The Podostemaceae of the New World. Part I. Meded. Bot. Mus. Herb. Rijksuniv. Utrecht 107: 1–151.

Royen, P. van, 1953. The Podostemaceae of the New World. Part II. Acta Bot. Neerl. 2(1): 1–20.

Royen, P. van, 1954. The Podostemaceae of the New World. Part III. Acta Bot. Neerl. 3(2): 215–263.

Stichting Planbureau Suriname, 1988. Suriname Planatlas. Department of Regional Development, O.A.S. Washington, D.C.

Stoffers, A.L. (ed.), 1982. Compendium van de Spermatophyta. Bohn, Scheltema & Holkema, Utrecht/Antwerpen, 310 pp.

Taylor, P., 1967. Lentibulariaceae. In: B. Maquire and coll., The Botany of the Guayana Highland. part VII. Memoirs of the New York Botanical Garden, 17(1): 210–227.

Teunissen, P.A., 1980. Overzicht van Surinaamse laagland ecosystemen met vegetatietabellen. Dienst 's Lands Bosbeheer (LBB) en Stichting Natuurbehoud Suriname (STINASU), Paramaribo, 110 pp.

7. Aquatic invertebrates of the Coastal Plain

JAN H.A. MOL

1. Introduction

The Coastal Plain of Suriname, located at 0–11 m above sea level, covers about 18,000 km² and is divided into the Old Coastal Plain with Pleistocene soils of clays and fine sands, and the Young Coastal Plain with Holocene soils of mainly clays with some beach ridges. The fine sands and clays of the Coastal Plain are probably mainly of Amazonian (Andean) origin, the material being transported by the Guiana stream to the Guiana coast (Reyne 1961; Krook 1979). Very poor drainage conditions in large areas lead to the formation of peat on top of clay flats in back swamps. The Coastal Plain is covered with herbaceous swamps, swamp forests and marsh forests (Lindeman 1953).

Large areas of shallow, standing waters create a dominant lentic environment in the Coastal Plain. Seasonal variations in rainfall lead to regular flooding of immense tracts of land. In the dry season the water retreats in small pools, canals, creeks and rivers. The chemistry of the Coastal Plain water bodies changes considerably during the seasonal cycle, mainly as a result of dilution by electrolyte-poor rainwater and the influence of penetrating sea water. Table 1 shows that the water at the collection sites in the Young Coastal Plain (Fig. 1) is fresh in the rainy season (163 ± 39.7 mg/l Cl), while in the dry season it has to be classified as oligohalinous or even mesohalinous (1,163 ± 592.4 mg/l Cl). Table 1 also shows differences in water chemistry between standing waters of the Old Coastal Plain and the Young Coastal Plain, to some extent a result of longer post-depositional weathering of Old Coastal Plain soils.

The clear-water streams of the Interior of Suriname and the black-water creeks of the Savanna Belt probably have little impact on the ecology of Coastal Plain waters. Surinamese rivers have no extended flood plains and thus can be considered "reservoir" rivers (Welcomme 1979). In cases where small flood plains occur (e.g. the Cusewijne, Saramacca and Nickerie Rivers) the contribution of the rivers to the water chemistry of the swamps is insignificant due to the low electrolyte content of both clear and black water streams (Sioli 1950, 1975; Leentvaar 1975).

It is well-known that fishes respond to the rise in water level in the rainy

Paul E. Ouboter (ed.), Freshwater Ecosystems of Suriname, 113–131.

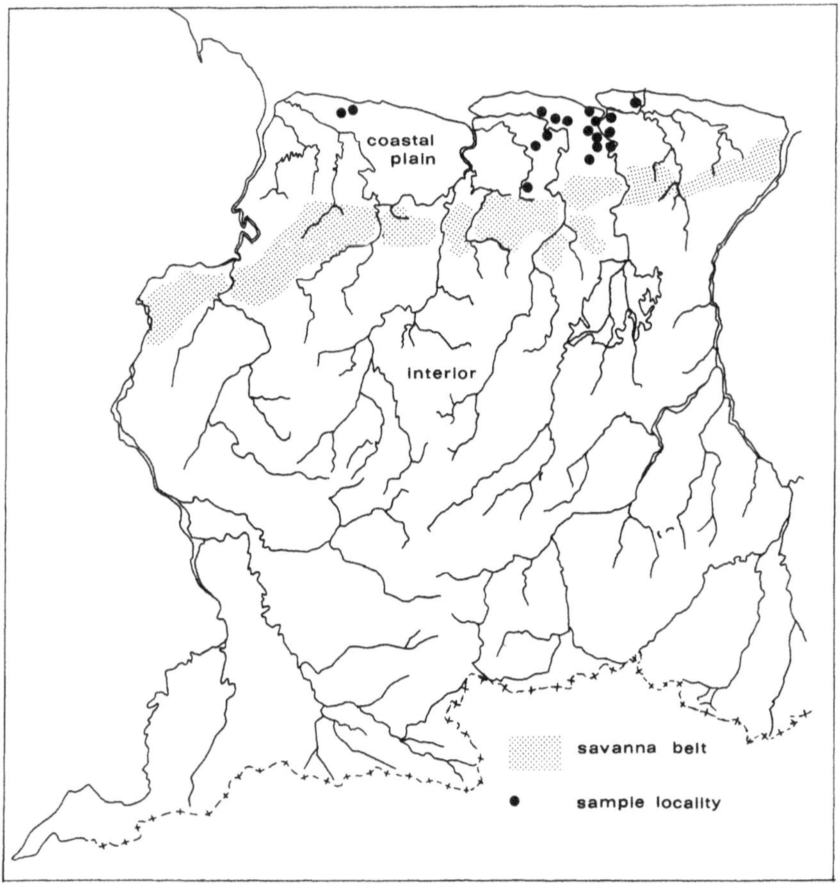

Figure 1. Map of Suriname with sample localities.

season, moving into the newly flooded aquatic habitats (Lowe-McConnell 1987). Most fish species of the Coastal Plain breed at the start of the rainy season (see Mol, Chapter 10). The high-water period is the main feeding and growing time and fish biomass increases rapidly (Goulding 1980; Lowe-McConnell 1987). It is therefore to be expected that the main food organisms of the fishes, the aquatic invertebrates, also react to the seasonal cycle of the Coastal Plain.

Aquatic invertebrates of the Coastal Plain of Suriname were collected by Van der Land (1970) in the short dry season of 1967. Some of the major taxa from his collection were extensively studied: Naididae (Harman 1974), Nematoda (Loof 1973) and Aeolosomatidae and Turbellaria (Van der Land 1970). Other taxa were only identified by specialists: Hydracarina, Odonata and Gastropoda (Van der Land 1970).

The present study aims at relating standing stock of aquatic invertebrates of

Table 1. Selected physico-chemical data (mean ± S.E.) of standing waters in the Coastal Plain of Suriname. Number of observations in parentheses.

Parameter	Young Coastal Plain		Old Coastal Plain
	rainy season	dry season	
pH	5.9 ± 0.17 (25)	7.2 ± 0.23 (17)	5.1 ± 0.21 (10)
conductance (μS/cm)	673 ± 125.1 (25)	1,755 ± 635.1 (17)	76.5 ± 13.38 (10)
Cl (mg/l)	163 ± 39.7 (25)	1,163 ± 592.4 (17)	15.4 ± 1.45 (10)
hardness (mg/l - $CaCO_3$)	125 ± 17.9 (24)	134 ± 26.0 (13)	11.0 ± 2.91 (6)
Ca (mg/l)	2.7 ± 0.37 (25)	21.5 ± 5.74 (17)	0.6 ± 0.20 (3)
Mg (mg/l)	16.4 ± 2.62 (25)	78.1 ± 36.85 (17)	0.8 ± 0.09 (3)
O_2 (mg/l)	2.0 ± 0.59 (23)	1.4 ± 0.41 (17)	1.9 ± 0.6 (7)
BOD (mg/l)	3.3 ± 0.24 (25)	4.3 ± 0.51 (13)	3.7 ± 0.75 (7)
PO_4 (mg/l)	0.2 ± 0.02 (25)	0.7 ± 0.24 (16)	0.1 ± 0.03 (7)
NO_3 (mg/l)	0.1 ± 0.04 (25)	0.3 ± 0.22 (14)	0.1 ± 0.04 (7)
NH_4 (mg/l)	0.2 ± 0.04 (25)	1.5 ± 0.69 (17)	0.1 ± 0.03 (7)
HCO_3 (mg/l)	74.1 ± 10.18 (25)	140 ± 22.4 (17)	7.3 ± 1.0 (6)
SO_4 (mg/l)	73.3 ± 10.35 (25)	137 ± 79.8 (17)	24.6 ± 0.65 (3)
K (mg/l)	8.5 ± 1.44 (25)	30.7 ± 14.44 (17)	2.1 ± 0.84 (3)

the water column (zooplankton) and the bottom substrate (zoobenthos) with the seasonal cycle of the Coastal Plain. In addition, densities of the invertebrate fauna in the root zone of the floating vegetation, an important habitat for aquatic invertebrates characteristic of the high-water period (Junk 1970, 1973), have also been studied and reported.

Samples of zooplankton, zoobenthos and the fauna of the root zone of the floating vegetation were collected in the main dry season (September–November) and the main rainy season (May–July) at 17 localities in the Coastal Plain (Fig. 1). Sample sites included both natural standing waters and man-made water bodies (e.g. paddy fields). Physico-chemical water parameters were measured following standard methods (APHA-AWWA 1976). They are summarized in Table 1. Zooplankton was collected by pouring 100 l of water through a 0.2 mm mesh-width plankton net. Rotatoria were sampled in one liter polyethylene bottles. The zooplankton samples yielded large numbers of aquatic invertebrates not belonging to the traditional zooplankton taxa (micro-crustacea), making a somewhat extended definition of zooplankton necessary. The overall shallowness of the Coastal Plain standing waters (62 ± 40.7 cm) allowed sampling of zoobenthos without special apparatus. At each site 500 ml bottom substrate up to a depth of 5 cm was collected (sample size 1 dm^2 of bottom substrate). The fauna of the root zone of the floating vegetation was collected by shoving a 3 dm^2 sieve with 0.106 mm pores beneath the floating vegetation and lifting it upwards out of the water. The mass of roots and rhizomes was rinsed and sieved three times in the laboratory.

2. Zooplankton

2.1. Composition

An overview of the zooplankton taxa collected in the standing waters of the Coastal Plain of Suriname is presented in Appendix 1. Due to the overall shallowness of the investigated water bodies the composition of the zooplankton samples shows some typical benthic taxa besides the pelagic fauna. The latter will be treated with the zoobenthos.

Compared with the approximately 250 species of Rotatoria recorded for Amazonia (Robertson and Hardy 1984), relatively few species (23) are present in the Coastal Plain of Suriname (Appendix 1). However, Robertson and Hardy (1984) point out that only 16 Amazonian species are numerically dominant in the pelagic, the remaining lot being sessile, illoricate or rare.

Cladocera and Copepoda are well represented in the plankton samples. The number of species of the Coastal Plain of Suriname (Appendix 1) is in the same range as reported for Amazonian lakes and rivers (Robertson and Hardy 1984). Unlike Rotatoria and Cladocera, which show a more pantropical or even cosmopolitan distribution, Amazonian Copepoda are more restricted in occurrence. The Copepoda of the Coastal Plain are not identified to the species level. Cyclopoid Copepoda are dominant but Calanoid and Harpacticoid species are also present. Ostracoda, Hydracarina, one Conchostracan species and the Malacostracan *Palaemonetes carteri* are present in smaller numbers.

Several orders of aquatic insects are abundant in the zooplankton samples (Appendix 1). Particularly Diptera larvae of the families Chironomidae, Ceratopogonidae and Culicinae compose a significant fraction of total zooplankton density. Chaoborinae are very rare. Adult and larval Coleoptera of the families Hydrophilidae, Dytiscidae and Noteridae regularly turn up in the samples. The presence of Odonata is readily apparent from the abundance of showy imagines: both Anisoptera and Zygoptera are well represented in the samples. Important families of Hemiptera from Coastal Plain standing waters are: Belostomatidae, Nepidae, Pleidae, Corixidae, Notonectidae, Gerridae, Mesoveliidae and Veliidae. Few Naucoridae, Hydrometridae, Gelastocoridae and Ochteridae are present. Scattered Ephemeroptera and Trichoptera are found, these orders being much more common in the Interior. No Plecoptera have been collected.

2.2. Standing Stock

Mean densities of zooplankton in the dry season and rainy season are summarized in Table 2. In both the dry season and rainy season, zooplankton densities differ widely between sample sites. This is evident for all taxa collected and is demonstrated by the large standard errors of Table 2. As most of the collection sites were sampled within one or two weeks in each season, the large differences in standing stock more probably indicate differences in local conditions than in timing of blooms. Robertson and Hardy (1984) also report large differences in population densities of zooplankton both between lakes and in a single lake in time. They cite Morgan *et al.* (1980) in that “in principle, densities can fluctuate between zero and the potential maximum in one and the same water body”.

The large differences in zooplankton densities between the samples make generalizations on the effect of flooding on the zooplankton standing stock difficult. Although all zooplankton taxa show higher densities in the high-water period, the difference in mean standing stock is only statistically significant for the Ostracoda and Diptera (*t*-test). However, when the probabilities yielded from the *t*-tests of the main taxa are combined (Sokal and Rohlf 1981), the overall difference between dry season and rainy season standing stock is significant.

The emerging pattern of highest densities of zooplankton organisms in the high-water period is not usually observed in Amazonian flood-plain lakes (Robertson and Hardy 1984). Low zooplankton densities in the high-water period are attributed by them to the dilution effect of growing water volumes, predation by planktivorous fish and, above all, the direct and indirect effect of the inflow of suspension-rich river water. In Surinamese Coastal Plain swamps inflow of sediment-laden river water is negligible due to the “reservoir” character of most rivers. Suspended matter in Coastal Plain standing waters originating from the bottom sediments of the swamps may even have a positive effect on zooplankton as it is much higher in caloric contents than the anorganic clay sediments of the Amazon River.

Overall zooplankton densities of Surinamese Coastal Plain standing waters varied between 16,260 ind/m^3 in the dry season and 25,480 ind/m^3 in the rainy season. These densities are somewhat low compared to those reported in the literature on tropical flood plains (Welcomme 1979; Robertson and Hardy

Table 2. Mean standing stock of zooplankton (ind/m³) in the Coastal Plain of Suriname.

Taxon	Dry season mean ± s.e. (n = 22)	Rainy season mean ± s.e. (n = 30)
Rotatoria	11,900 ± 3,420	14,400 ± 4,240
Cladocera	2,820 ± 2,269	7,409 ± 2,940
Copepoda	920 ± 484	1,764 ± 565
Ostracoda	44 ± 18	219 ± 65
Conchostraca	6 ± 4	92 ± 63
Malacostraca	5 ± 2	43 ± 28
Hydracarina	28 ± 14	43 ± 21
Diptera larvae	268 ± 99	701 ± 160
Coleoptera larvae	10 ± 4	53 ± 23
Odonata larvae	18 ± 6	28 ± 6
Ephemeroptera & Trichoptera larvae	4 ± 3	42 ± 33
aquatic Coleoptera	5 ± 3	19 ± 9
Hemiptera	33 ± 10	102 ± 34
Oligochaeta, Naididae	124 ± 39	387 ± 148
Hirudinea	–	3 ± 2
Nematoda	37 ± 24	95 ± 67
Gastropoda	6 ± 3	58 ± 28
Collembola	8 ± 2	8 ± 2
terrestrial Arthropoda	22 ± 4	11 ± 2

1984). Mean densities of Rotatoria, Cladocera and Copepoda are low compared to densities reported by Robertson and Hardy (1984) for three white-water and two black-water lakes in Amazonia. Especially Rotatoria were recorded in densities up to ten times as high as those of the Coastal Plain samples.

3. Zoobenthos

3.1. Bottom Sediments

Composition and density of benthic fauna are undoubtedly correlated to substrate type. Contrary to the extensive literature on zoobenthos from running waters with hard substrates (Hynes 1970), little information is available on the benthic fauna of tropical flood plains (Welcomme 1979). The mean sediment composition up to a depth of 5 cm at the collection sites is illustrated in Fig. 2. Soft mud and plant detritus dominate, but sand and shell fragments are present near old beach ridges. Samples of swamp bottom substrate consist almost entirely of detritus. Substrates in cultivated areas yield more clay and silt particles (e.g. paddy fields).

Reiss (1977) states that the littoral benthic fauna of Amazonian lakes and flood plains is also affected by water depth, oxygen supply (Table 1) and, possibly, H_2S concentration.

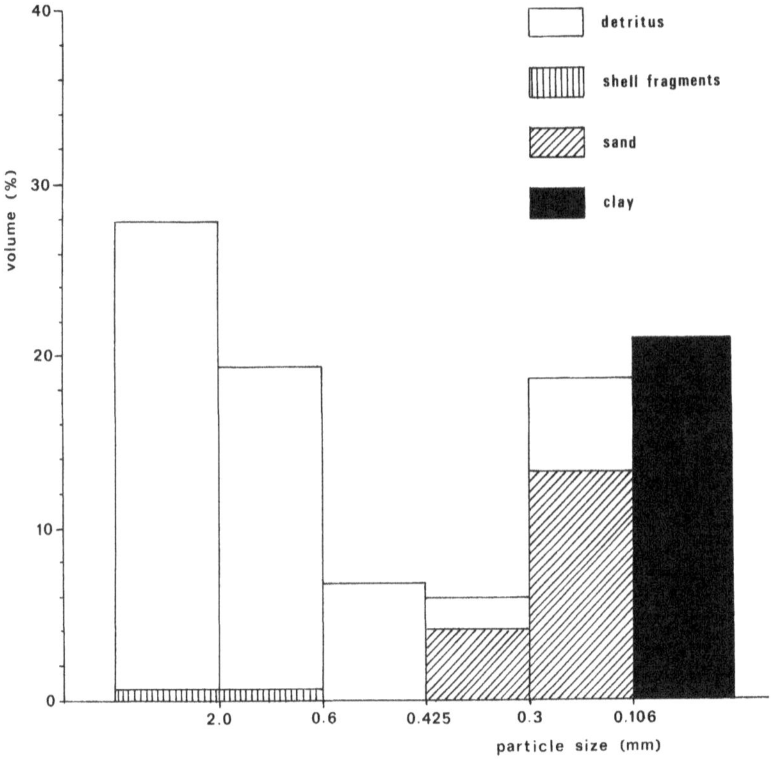

Figure 2. Mean sediment composition up to a depth of 5 cm at the zoobenthos collection sites in the Coastal Plain of Suriname.

3.2. Composition

The species composition of the major benthic taxa of the Coastal Plain (Appendix 2) is relatively well-known due to publications based on material collected by Van der Land in 1967 (Van der Land 1970).

Oligochaeta are well represented in the benthos samples. A listing of the Naididae and Aeolosomatidae of the Coastal Plain of Suriname is presented in Appendix 2. The genera *Pristina*, *Dero*, *Aulophorus* and *Aeolosoma* are always an important component of the zoobenthos. *Opistocysta* (Opistocystidae) is here reported for the first time for Suriname, although its presence is not surprising (Reiss 1977).

The freshwater Nematoda of the Coastal Plain collected by Van der Land are listed in Appendix 2. Nematoda often turn up in large numbers in the benthos samples.

Turbellaria are represented in the Coastal Plain benthic fauna mainly by the genera *Catenula*, *Stenostomum*, *Rhynchoscolex* and *Gyratrix* (Appendix 2).

Table 3. Mean standing stock of zoobenthos (ind/m²) in the Coastal Plain of Suriname.

Taxon	Dry season mean ± s.e. (n = 22)	Rainy season mean ± s.e. (n = 29)
Oligochaeta, Naididae	2,841 ± 1,052	4,359 ± 887
Oligochaeta, Aeolosomatidae	45 ± 24	1,196 ± 475
Nematoda	2,636 ± 1,396	1,900 ± 449
Turbellaria	1,027 ± 704	1,379 ± 301
Gastrotricha	273 ± 134	210 ± 105
Tardigrada	-	21 ± 14
Gastropoda	114 ± 93	341 ± 302
Diptera larvae	227 ± 106	286 ± 159
Ostracoda	409 ± 216	831 ± 273
Copepoda	300 ± 188	1,348 ± 553
Cladocera	59 ± 42	331 ± 122
Hydracarina	59 ± 33	14 ± 6

Other taxa of the microbenthic fauna of the Coastal Plain are Gastrotricha of the genus *Chaetonotus* and the Tardigrad *Hysibius augusti.*

Coastal Plain freshwater Mollusca are almost exclusively Gastropoda of the families Pomaceae, Planorbidae and Ancylidae. The genera *Pomacea* and *Biomphalaria* are especially abundant. The large Bivalve *Mycetopoda siliquosa* is not often collected.

Diptera larvae of the families Chironomidae and Ceratopogonidae and the Ostracod *Stenocypris* are often dominant in the samples.

3.3. Standing Stock

Mean standing stock of zoobenthos of the Coastal Plain in the dry season and rainy season are shown in Table 3. Again, as mentioned earlier for zooplankton, differences in recorded densities within one season are large for all taxa collected.

Total densities of zoobenthos are 7,990 ind/m² in the dry season and 12,216 ind/m² in the rainy season. Nematoda and Gastrotricha are collected in larger numbers in the dry season. Other important taxa are present in higher densities in the rainy season. Only Aeolosomid densities are significantly higher in the rainy season (t-test). Combining the probabilities of the major taxa (Sokal and Rohlf 1981) yielded no significant difference between overall zoobenthos densities in the high-water period and the low-water period. Bonetto (1975) considers the zoobenthos fauna the least affected of all communities of standing waters by seasonal fluctuations in water level.

Compared with the sparse information on zoobenthos of tropical standing waters summarized by Welcomme (1979), the densities recorded for Surinamese Coastal Plain water bodies are relatively high. Reiss (1977) reports maximum densities of 2,559 ind/m^2 in Amazonian flood-plain lakes during rising flood. However, Braun (1952) recorded a maximum of 12,000 ind/m^2 for Lago Jurucui in the mouth region of the Rio Tapajos.

4. Fauna of the Root Zone of the Floating Vegetation

4.1. The Floating Vegetation Habitat

Floating vegetation represents a characteristic feature of neotropical freshwaters. In Amazonia this floating vegetation often consists of grasses and thus the term "floating meadows" is coined for it (Junk 1970). Extensive floating meadows of aquatic grasses are not common in Suriname, but in Coastal Plain standing waters floating vegetation of other aquatic plants, among them some grasses, is abundant. The floating plants are especially well adapted to fluctuations in water level between dry and rainy season. They grow in large masses in small pools and canals and after flooding invade those areas that completely dry out during the low-water period (Carter and Beadle 1930). The most common species in the samples examined are *Salvinia auriculata*, *Pistia stratiotes*, *Eichhornia crassipes*, *Ipomoea reptans*, *Hymenachne amplexicaulis*, *Azolla* and different species of Lemnaceae. The submerged aquatics *Cabomba* and *Utricularia* are also present in the samples (also see Werkhoven and Peeters, Chapter 6).

The floating vegetation is a very special habitat which supports aquatic communities in the root zone, a terrestrial fauna among the leaves and flowers extending above the water's surface and semi-aquatic animals in the extensive transitional zone (Junk 1973).

The submerged root zone of the floating vegetation provides substratum for sessile biota. Among the roots are larger and smaller spaces of open water in which pelagic forms can live. Here, broad fluctuations in dissolved oxygen occur and in the center of dense stands of aquatic grasses a total lack of oxygen and H_2S formation is possible (Junk 1973). In the floating vegetation of Surinamese Coastal Plain standing waters H_2S was never recorded.

The parts of the floating vegetation above the water's surface provide a habitat for many insects whose larvae live in the root zone, e.g. Chironomidae and Odonata. Terrestrial fauna elements are also present: Collembola, Arachnida, Curculionidae and Formicidae (Junk 1973).

4.2. Composition

Most of the taxa present in the samples of the floating vegetation (Table 4) are already treated with zooplankton and zoobenthos. *Hydra* is frequently present and semi-aquatic Collembola are not rare either.

Many groups of terrestrial Arthropoda are represented in the samples, though often low in numbers: Arachnida, Isopoda, Coleoptera of the families Curculionidae, Staphylinidae, Chrysomelidae and Carabidae, Orthoptera, Homoptera, Thysanoptera, Blattidae and Formicidae.

4.3. Standing Stock

Table 4 shows the mean densities of the fauna of the root zone of Coastal Plain floating vegetation in the rainy season. Very conspicuous are the large numbers of benthic animals present in the floating vegetation, e.g. Naididae, Nematoda, Diptera and Ostracoda. It is possible that these animals are excluded to some extent from their benthic habitat due to oxygen depletion and H_2S formation near the bottom of the standing waters. Densities of Copepoda, Conchostraca, Odonata and aquatic Coleoptera are also very high compared with zooplankton standing stocks.

The total standing stock of the fauna of the Coastal Plain floating vegetation is 26,665 ind/m² in the rainy season. As this figure does not include the numerous small animals of e.g. Rotatoria and Turbellaria, the importance of the floating vegetation for secondary production in Coastal Plain standing waters must be very great. Yet, the densities recorded for the Surinamese Coastal Plain are low compared with the maximum densities reported by Junk

Table 4.. Mean standing stock of fauna from the root zone of the floating vegetation (ind/m²) in the Coastal Plain of Suriname.

Taxon	Rainy season mean ± s.e. (n = 25)
Oligochaeta, Naididae	6,000 ± 1,658
Hirudinea	40 ± 27
Nematoda	1,376 ± 657
Gastropoda	67 ± 24
Diptera larvae	5,536 ± 1,173
Coleoptera larvae	167 ± 36
Odonata larvae	151 ± 56
Ephemeroptera & Trichoptera larvae	31 ± 16
aquatic Coleoptera	33 ± 10
Hemiptera	196 ± 60
Ostracoda	2,394 ± 1,044
Copepoda	6,070 ± 1,672
Cladocera	3,906 ± 1,140
Conchostraca	468 ± 338
Malacostraca	1 ± 1
Hydracarina	72 ± 14
Hydra spec.	128 ± 72
Collembola	8 ± 3
terrestrial Arthropoda	21 ± 14

(1973) in the roots of floating meadows of aquatic grasses of white-water lakes and rivers in Amazonia. With Rotatoria and Turbellaria excluded, densities are between 50,000 and 700,000 ind/m² in Varzea lakes and around 50,000 ind/m² in floating meadows bordering white-water rivers. Petr (1968) found between 5,000 and 16,000 ind/m² in floating vegetation of *Pistia* and *Ceratophyllum* in Lake Volta.

It seems to me that the large differences in standing stock of the fauna of floating vegetation reported in the literature (Welcomme 1979) can be attributed at least partially to the species composition of the floating vegetation itself. Especially the depth and spatial structure of the root zone may be important in determining the faunal density that can be supported by the floating vegetation.

Summary

The composition and standing stock of aquatic invertebrates of the Coastal Plain of Suriname are discussed. An attempt is made to characterize the fauna of the three main habitats, bottom, water column and floating vegetation, and to evaluate the impact of the rain-induced seasonal cycle of high-water and low-water periods on the invertebrate standing stock.

Acknowledgements

I would like to take this opportunity to thank Mr. Sabiran and Mrs. Haripersad of the Hydraulic Research Division, Ministry of Public Works, for their help with, respectively, chemical analyses of water samples and identification of Rotatoria. Mrs. Tjon-Eng-Soe and Mr. Jewbali of the CELOS laboratory also assisted with chemical analyses of water samples. The Zoological Collection of the University of Suriname provided library facilities, especially with respect to the Surinamese fauna.

Appendix 1. Composition of freshwater zooplankton from the Coastal Plain of Suriname

Rotatoria		
1.	*Brachionus* spec.	++
2.	*Cephalodella exigua*	+
3.	*Cephalodella gibba*	+++
4.	*Collotheca ornata var cornuta*	+
5.	*Colurella adriatica*	++
6.	*Keratella* spec.	+
7.	*Lecane luna*	+
8.	*Lecane* spec.	++
9.	*Lepadella ovalis*	+
10.	*Lepadella patella*	++
11.	*Manfredium endactylotum*	+

Appendix 1. (continued)

Rotatoria (continued)	
12. *Manfredium grandis*	+
13. *Monommata grandis*	+
14. *Monostyla bulla*	++
15. *Monostyla bulla var constricta*	+
16. *Monostyla lunaris*	+
17. *Monostyla quadridentata*	+
18. *Monostyla* spec.	++
19. *Pedipartia gracilis*	+
20. *Philodina roseola*	+++
21. *Platyas quadricornis*	+
22. *Polyarthra* spec.	++
23. *Rotaria citrinus*	++
Cladocera	
1. *Alona costata* Sars 1862	+
2. *Alona* cf. *affinis* (Leydiy) 1860	+
3. *Alona* spec.	+
4. *Alonella dentifera* Sars 1901	+
5. *Chydorus* spec.	++
6. *Diaphanosoma brachyurum* (Lieven) 1848	+++
7. *Euryalona occidentalis* Sars 1901	++
8. *Grimaldina brazzai* Richard 1892	+
9. *Ilyocryptus spinifer* Herrick 1884	++
10. *Lathonura rectirostris* (O.F. Muller) 1785	+
11. *Macrothrix rosea* (Jurine) 1820	+++
12. *Moina* spec.	+++
13. *Scapheloberis* kingi Sars 1903	+
Copepoda	
1. Calanoida spec.	++
2. Cyclopoida spec.	+++
3. Harpacticoida spec.	++
Ostracoda	
1. *Cypridopsis vidua* (O.F. Muller) 1776	+
2. *Cyprinotus americanus* Cushman 1905	+
3. *Ilyocypris gibba* (Ramdohr) 1808	+
4. *Stenocypris* cf. *fontinalis* Vavra 1895	+++
5. *Strandesia bicuspis bicuspis* (Claus) 1892	+
Conchostraca	
1. Conchostraca spec.	++
Malacostraca	
1. *Palaemonetes carteri* Gordon	+++
Hydracarina[a]	
1. *Arrenurus* spec.	+
2. *Frontipoda* spec.	+
3. *Hydrachna* spec.	+
4. *Hydryphantes* spec.	+

Appendix 1. (continued)

Hydracarina[a] (continued)	
5. *Koenikea* spec.	+
6. *Mideopsis* spec.	+
7. *Neumania* spec.	+
8. *Piona* spec.	+
9. *Unionicola* spec.	+
Trichoptera[b]	
1. *Cyrnellus fraternus* (Banks) 1905	++
2. *Leptonema columbianum* Ulmer 1905	+++
3. *Leptonema sparsum* Ulmer 1905	+
4. *Macronema fraternum* Banks	+
5. *Macronema hageni* Banks 1924	+
6. *Macronema percitans* Walker 1860	+
7. *Macronema picteli* Banks 1915	+
8. *Nectopsyche gemma* (Muller) 1880	+
9. *Nectopsyche muhni* (Navas) 1916	+++
10. *Neoleptonema aspersum* Ulmer 1907	+++
11. *Oecetis punctipennis* (Ulmer)	?
12. *Plectromacronema comptum* Ulmer	?
13. *Smicridea bulbosa* Flint 1974	+
Odonata[c]	
1. *Lestes tenuatus* (Rambur)	+++
2. *Acanthagrion adustum* Williamson	+
3. *Acanthagrion ascendens* Calvert	+++
4. *Acanthagrion indefensum* Williamson	+
5. *Aeolagrion demararum* Williamson	+++
6. *Ceratura capreola* Hagen	+
7. *Ischnura ramburi credula* Hagen	+++
8. *Coryphaeschna adnexa* Hagen	+
9. *Coryphaeschna virens* Rambur	+++
10. *Coryphaeschna* cf. *viriditas* Calvert	+
11. *Leptobasis vacillans* Selys	+++
12. *Telebasis sanquinalis* Calvert	+
13. *Aeschnosoma auripennis* Geyskens	+
14. *Aeschnosoma forcipula* Selys	+
15. *Anax concolor* Brauer	++
16. *Aphylla albinensis* Belle	+
17. *Aphylla dentata* Selys	+
18. *Aphylla producta* Selys	++
19. *Aphylla simulata* Belle	+
20. *Cacoides latro* (Erichson)	+
21. *Cacoides mungo* (Needham)	+
22. *Cannaccria herbida* Gundl	+
23. *Cyanogomphus demerarae* Selys	+
24. *Cyanogomphus minutus* Belle	+
25. *Diastatops* spec.	+
26. *Erythemus* spec.	+++
27. *Erythrodiplax* spec.	+
28. *Lepthemus vesiculosa* (F.)	+
29. *Miathyria* spec.	+

Appendix 1. (continued)

Odonata[c] (continued)		
30.	*Micrathyria coropinae* Geyskens	+++
31.	*Misagria parana* (Kirkby)	++
32.	*Negomphoides cristatus* (Needham)	+
33.	*Negomphoides undulatus* (Needham)	+
34.	*Nephepeltia* spec.	+
35.	*Oligoclada* spec.	+
36.	*Orthemis* cf. *ferruginea* (F.)	+++
37.	*Planiplax phoenicura* (Ris)	++
38.	*Pontala flavescens* (F.)	+++
39.	*Staurophlebia reticulata* (Burmeister)	+++
40.	*Tramea* spec.	+++
Hemiptera[d]		
1.	*Ambrysus stali* La Rivers 1962	++
2.	*Ambrysus usingeri* La Rivers 1952	+
3.	*Belostoma denticolle* Montandon 1903	++
4.	*Belostoma harrisi* Lauck 1962	++
5.	*Belostoma micantulum* (Stal) 1858	+++
6.	*Brachymetra lata* Shaw 1933	+
7.	*Buenoa amnigenopsis* Nieser 1975	++
8.	*Buenoa amnigenus* (White) 1879	++
9.	*Buenoa incompta* Truxal 1953	+++
10.	*Buenoa salutis* Kirkaldy 1904	+++
11.	*Curicta doesburgi* De Carlo 1967	+
12.	*Cylindrostethus palmaris* Drake & Harris 1934	+
13.	*Gelastocoris flavus flavus* (Guerin-Meneville) 1835	++
14.	*Gelastocoris fuscus* Martin 1929	+
15.	*Hydrometra argentina* Berg 1879	+
16.	*Lethocerus annulipus* (Herrich Schafer) 1948	++
17.	*Lethocerus delpontei* De Carlo 1930	+
18.	*Limnogonus aduncus* Drake & Harris 1933	+++
19.	*Limnogonus hyalinus* (Fabricius) 1803	+
20.	*Limnogonus ignotus* Drake & Harris 1934	+
21.	*Limnogonus lotus* White 1879	+
22.	*Limnogonus lubricus* White 1879	+
23.	*Limnogonus visendus* Drake & Harris 1942	+
24.	*Martarega membranacea* White 1879	++
25.	*Mesovelia mulsanti* B. White	+
26.	*Neoplea absona* (Drake & Chapman) 1953	+++
27.	*Neoplea maculosa* (Berg) 1879	+
28.	*Nertha raptoria* (Fabricius) 1803	+
29.	*Nertha terrestris* (Kevan) 1948	++
30.	*Nertha unicornis* (Melin) 1929	+
31.	*Ochterus aeneifrons surinamensis* Nieser 1975	+
32.	*Ochterus perbosci* (Guerin-Meneville) 1843	+
33.	*Paraplea puella* (Barber) 1923	+++
34.	*Pelocoris impicticollis* Stal 1876	+
35.	*Pelocoris poeyi* (Guerin-Meneville) 1835	+++
36.	*Ranatra doesburgi* De Carlo 1963	+
37.	*Ranatra mediana* Montandon 1910	+++
38.	*Ranatra obscura* Montandon 1907	+

Appendix 1. (continued)

Hemiptera [d] (continued)	
39. *Rheumatobates crassifemur esakii* Schroeder 1931	+++
40. *Rheumatobates trinitalis* (China) 1943	++
41. *Tachygerris adamsoni* Drake 1942	+
42. *Telmatometra fusca* Kenaga 1941	++
43. *Tenagobia incerta* Lundblad 1928	+++
44. *Tenagobia schadei* Lundblad 1928	+
45. *Tenagobia socialis* (White) 1897	+++
46. *Trichocorixa orinocoensis* Sailer 1948	+++
47. *Trichocorixa reticulata* (Guerin-Meneville) 1857	+++
48. *Trichocorixa verticalis verticalis* (Fieber) 1851	++
49. *Trochopus epydros* Drake & Doesburg 1966	+
50. Veliidae spec.	+
Diptera	
1. Ceratopogonidae	+++
2. Chaoborinae	+
3. Chironomidae	+++
4. Culicinae	++
5. Psychodidae	++
6. Stratiomyidae	++
7. Tabanidae	+
Coleoptera	
1. Dytiscidae	+++
2. Elmidae	+
3. Gyrinidae	+
4. Haliplidae	+
5. Hydrophilidae	+++
6. Noteridae	+++
7. Ptilodactylidae	+

[a] references in Van der Land (1970)
[b] references in Van der Land (1970)
[c] after Flint (1974)
[d] after Nieser (1970a, 1970b, 1975)

+ rare; ++ not uncommon; +++ abundant.

Appendix 2. Composition of freshwater zoobenthos from the Coastal Plain of Suriname

Oligochaeta, Naididae[a]	
1. *Allonais paraguayensis* (Michaelsen) 1905	+++
2. *Aulophorus bimagnasetus* Harman 1974	+
3. *Aulophorus carteri* Stephenson 1931	+
4. *Aulophorus costatus* Marcus 1944	+
5. *Aulophorus furcatus* (Muller) 1773	+++
6. *Aulophorus pectinatus* Stephenson 1931	+
7. *Aulophorus tonkinensis* (Vejdovsky) 1894	+
8. *Chaetogaster langi* Bretscher 1896	+
9. *Dero digitata* (Muller) 1773	+++

Appendix 2. (continued)

Oligochaeta, Naididae[a] (continued)		
10.	*Dero magna* Harman 1974	+
11.	*Dero nivea* Aiyer 1930	++
12.	*Dero obtusa* d'Udekem 1855	++
13.	*Dero pectinata* Aiyer 1930	+++
14.	*Dero sawayai* Marcus 1943	+
15.	*Haemonais waldvogeli* Bretscher 1900	++
16.	*Nais raviensis* Stephenson 1914	+
17.	*Pristina aequiseta* Bourne 1891	+++
18.	*Pristina americana* Cernosvitov 1937	+
19.	*Pristina longidendata* Harman 1965	++
20.	*Pristina longiseta bidentata* Cernosvitov 1942	+++
21.	*Pristina macrochaeta* Stephenson 1931	++
22.	*Pristina menoni* (Aiyer) 1930	+
23.	*Pristina proboscidea* Beddard 1896	+
24.	*Pristina unidendata* Harman 1973	++
25.	*Slavina evelinae* (Marcus) 1942	++
26.	*Stephensoniana trivandra* (Aiyer) 1926	+
Oligochaeta, Opistocystidae		
1.	*Opistocysta* spec.	+
Oligochaeta, Aeolosomatidae[b]		
1.	*Aeolosoma beddardi* Michaelson 1900	+
2.	*Aeolosoma evelinae* Marcus 1944	++
3.	*Aeolosoma headleyi* Beddard 1888	++
4.	*Aeolosoma hemprichi* Ehrenberg 1831	++
5.	*Aeolosoma* cf. *hyalinum* Bunke 1967	+
6.	*Aeolosoma sawayai* Marcus 1944	+
7.	*Aeolosoma travancorense* Aiyer 1926	+++
Nematoda[c]		
1.	*Actinolaimus tripapillatus* (von Daday) 1905	+++
2.	*Alaimus simplex* Cobb 1914	+
3.	*Aporcelaimellus obtusicaudatus* (Bastian) 1865	+
4.	*Axonchium* spec.	+
5.	*Chronogaster andrassyi* Loof & Jairajpuri 1965	++
6.	*Chronogaster lissa* Loof 1973	+
7.	*Chronogaster serrulata* Loof 1973	++
8.	*Chronogaster* cf. *longicollis* (von Daday) 1899	+
9.	*Chrysonemoides limigenus* Siddiqi 1969	+
10.	*Cryptonchus abnormis* (Allgen) 1933	+
11.	*Idiodorylaimus annulatiformis* Loof 1973	+
12.	*Idiodorylaimus kreisi* Loof 1973	+
13.	*Ironus ignavus* Bastian 1865	++
14.	*Ironus longicaudatus* de Man 1884	+
15.	*Ironus paludicola* (W. Schneider) 1937	+
16.	*Mesodorylaimus flavomaculatus* (von Linstow) 1876	+++
17.	*Mesodorylaimus* spec. 1	+
18.	*Mesodorylaimus* spec. 2	+
19.	*Monhystera paludicola* de Man 1881	+

Appendix 2. (continued)

Nematoda[c] (continued)		
20.	*Mononchulus nodicaudatus* (von Daday) 1901	++
21.	*Mononchus tunbridgensis* Bastian 1865	+
22.	*Nothactinolaimus lacustris* Loof 1973	+
23.	*Nygolaimus* spec.	+
24.	*Oncholaimus vanderlandi* Loof 1973	+
25.	*Oxydirus tropicus* Thorne 1964	+
26.	*Panagrolaimus hygrophilus* Bassen 1940	+
27.	*Panagrolaimus thienemanni* Hirschmann 1952	+
28.	*Paraphanolaimus microstomus* (von Daday) 1905	+
29.	*Paraphanolaimus* cf. *behningi* Micoletzky 1923	+
30.	*Prismatolaimus eurylaimus* Loof 1973	+
31.	*Prodorylaimus depressus* Loof 1973	+++
32.	*Prodorylaimus hamatus* Loof 1973	+
33.	*Trobrilus longiformis* Loof 1973	++
34.	*Trobrilus vicinus* Loof 1973	++
Turbellaria, Lecithoepitheliata[b]		
1.	*Geocentrophora* spec.	+
2.	*Prorhynchus* spec.	+
Turbellaria, Kalyptorhynchia[b]		
1.	*Gyratrix hermaphroditus* Ehrenberg 1831	+++
Turbellaria, Catenulida[b]		
1.	*Catenula confusa* Nuttycombe 1956	+++
2.	*Chordarium evelinae* Marcus	+
3.	*Chordarium leucanthum* Marcus	++
4.	*Dasyhormus* cf. *lasius* Marcus	+
5.	*Myostenostomum* spec.	+
6.	*Rhynchoscolex nanus* Marcus	+++
7.	*Rhynchoscolex platypus* Marcus	++
8.	*Rhynchoscolex simplex* Leidy 1851	+
9.	*Stenostomum anatirostrum* Marcus	+
10.	*Stenostomum bicaudatum* Marcus	+
11.	*Stenostomum evelinae* Marcus	++
12.	*Stenostomum gladulosum* Kepner & Carter	+
13.	*Stenostomum grande* Child	+
14.	*Stenostomom hemisphericum* Nasonov	++
15.	*Stenostomum leucops* (Duges)	+++
16.	*Stenostomum paraguayensis* (Martin)	+
17.	*Stenostomum pseudoacetabulum* Nuttycombe & Waters	+
18.	*Stenostomum saliens* Kepner & Carter	+
19.	*Stenostomum tuberculosum* Nuttycombe & Waters	+
20.	*Stenostomum unicolor* Schmidt	++
21.	*Suomina evelinae* Marcus	+
Gastrotricha		
1.	*Chaetonotus* spec.	++

Appendix 2. (continued)

Mollusca, Gastropoda	
1. Ancylidae spec.	++
2. *Biomphalaria glabrata* (Say) 1818	+++
3. *Biomphalaria straminea* (Dunker) 1848	+++
4. *Drepanotrema anatinum* (d'Orbigny) 1835	+
5. *Drepanotrema cimex* (Moricand) 1837	?
6. *Drepanotrema surinamense* (Dunker & Clessin) 1884	++
7. *Neritina zebra* (Brugiere) 1799	+
8. *Pomacea doliodes* (Reeve) 1856	+++
9. *Pomacea glauca glauca* (L.) 1764	+++
10. *Pomacea glauca gevesensis* (Deshayes) 1838	+
Mollusca, Bivalvia	
1. *Mycetopoda siliquosa* d'Orbigny	+
Ostracoda	
1. *Cypridopsis vidua* (O.F. Muller) 1776	+
2. *Cyprinotus americanus* Cushman 1905	+
3. *Ilyocypris gibba* (Ramdohr) 1808	+
4. *Stenocypris* cf. *fontinalis* Vavra 1895	+++
5. *Strandesia bicuspis bicuspis* (Claus) 1892	+
Tardigrada	
1. *Hysibius augusti* (J. Murray) 1907	+

[a] after Harman (1974)
[b] after Van der Land (1970)
[c] after Loof (1973)

+ rare; ++ not uncommon; +++ abundant.

References

American Public Health Association, American Water Works Association, and Pollution Control Federation, 1976. Standard methods for the examination of water and waste water, 14th ed. Am. Pub. Health Association, Washington, 1193 pp.

Bonetto, A.A., 1975. Hydraulic regime of the Parana river and its influence on ecosystems. Ecol. Stud. 10: 175–197.

Braun, R., 1952. Limnologische Untersuchungen an einigen Seen in Amazonasgebiet. Schweiz. Z. Hydrol. 14: 1–128.

Carter, G.S. and L.C. Beadle, 1930. The fauna of the swamps of the Paraguayan Chaco in relation to its environment. I. Physico-chemical nature of the environment. J. Linn. Soc., Lond. (Zool.) 37: 205–258.

Flint, O.S., 1974. The Trichoptera of Surinam. Stud. Fauna Suriname Guyanas 14: 1–151.

Goulding, M., 1980. The fishes and the forest. University of California Press, Los Angeles, 280 pp.

Harman, W.J., 1974. The Naididae (Oligochaeta) of Surinam. Zool. Verhand., Leiden 133: 1–37.

Hynes, H.B.N., 1970. The ecology of running waters. Liverpool University Press, Liverpool, 555 pp.

Junk, W.J., 1970. Investigations on the ecology and production-biology of the "floating meadows" (*Paspalo-Echinochloetum*) on the Middle Amazon. Part I. The floating vegetation and its ecology. Amazoniana 2: 449–495.

Junk, W.J., 1973. Investigations on the ecology and production-biology of the "floating meadows" (*Paspalo-Echinochloetum*) on the Middle Amazon. Part II. The aquatic fauna in the root zone of the floating vegetation. Amazoniana 4: 9–102.
Krook, L., 1979. Sediment petrographical studies in northern Suriname. Ph.D. thesis University of Amsterdam, Amsterdam.
Land, J. van der, 1970. Kleine dieren uit het zoete water van Suriname. Zool. Bijdragen Leiden 12: 1–46.
Leentvaar, P., 1975. Hydrobiological observations in Surinam, with special references to the man-made Brokopondo Lake in 1964. Stud. Fauna Suriname Guyanas 15: 1–175.
Lindeman, J.C., 1953. The vegetation of the coastal region of Suriname. The vegetation of Surinam 1: 1–135.
Loof, P.A.A., 1973. Freshwater nematodes from Suriname collected by J. van der Land. Zool. Verhand., Leiden 129: 1–45.
Lowe-McConnell, R.H., 1987. Ecological studies in tropical fish communities. Cambridge University Press, Cambridge, 382 pp.
Morgan, N.C., T. Backiel, G. Bretschko, A. Duncan, A. Hillbricht-Illkowska, A. Kajak, J.F. Kitchell, P. Larsson, C. Leveque, A. Nauwerck, F. Sheimer and J.E. Thorpe, 1980. Secondary production. IBP 22, Cambridge University Press, Cambridge, 588 pp.
Nieser, N., 1970a. Gerridae of Suriname and the Amazon. Stud. Fauna Suriname Guyanas 12: 94–138.
Nieser, N., 1970b. Hydrometridae of Suriname and the Amazon. Stud. Fauna Suriname Guyanas 12: 139–151.
Nieser, N., 1975. The water bugs (Heteroptera: Nepomorpha) of the Guyana region. Stud. Fauna Suriname Guyanas 16: 1–308.
Petr, T., 1968. Population changes in aquatic invertebrates living on two water plants in a tropical man-made lake. Hydrobiologia 32: 449–485.
Reiss, F., 1977. Qualitative and quantitative investigations on the macrobenthic fauna of Central Amazonian lakes. Amazoniana 6: 203–235.
Reyne, A., 1961. On the contribution of the Amazon River to the accretion of the coast of the Guianas. Geologie en Mijnbouw 40: 219–226.
Robertson, B.A. and E.R. Hardy, 1984. Zooplankton of Amazonian lakes and rivers. In: H. Sioli (ed.), The Amazon. Limnology and landscape ecology of a mighty tropical river and its basin. Dr.W.Junk, Dordrecht: 337–352.
Sioli, H., 1950. Das Wasser im Amazonasgebiet. Forsch. Fortschr. 26: 274–280.
Sioli, H., 1975. Tropical rivers as expressions of their terrestrial environments. In: F.B. Golley and E. Medina (eds.), Tropical Ecosystems. Trends in Terrestrial and Aquatic Research. Springer Verlag, New York: 275–287.
Sokal, R.R. and F.J. Rohlf, 1981. Biometry (2nd ed.). Freeman, San Francisco, 859 pp.
Welcomme, R.L., 1979. Fisheries ecology of floodplain rivers. Longman, London, 317 pp.

8. The fish fauna of Suriname

PAUL E. OUBOTER and JAN H.A. MOL

1. Introduction

The ichthyofauna of the Neotropics is still badly known. Ecological studies of fish communities are possible, but many species that will be included in these studies are unnamed and can only be indicated by numbers (see for instance Goulding *et al.* 1988). The Surinamese ichthyofauna has been studied by numerous zoologists since Van der Stigchel (1947). During six years of fish collecting, between 1987 and 1993, we (probably) found no species new to science, and only a few that were not mentioned for Suriname before. This indicates a relatively well-known fish fauna (compared to other South American countries), enabling us to make ichthyofaunal comparisons between rivers and river-sections. This chapter will give an impression of the fish fauna of a selection of rivers and river-sections in different parts of the country and consequently with different environmental conditions. The ichthyofauna of the swamps and swamp forests of the Coastal Plain is not included in this study. However, the lotic environment is very dominant in Suriname, and probably few species are exclusively restricted to the standing waters of the Coastal Plain.

We are not the first to do an exercise like this: Eigenmann (1912) already made a comparison between several rivers for Guyana, at a time when the ichthyofauna of the Guianas was less well-known. The questions Eigenmann was trying to answer were quite different from ours, and consequently the results cannot be compared easily.

Our data are drawn from the specimens present at the National Zoological Collection of Suriname, which were mostly sampled during numerous trips and expeditions between 1987 and 1993. In addition, all available literature data were used (see References). Unfortunately, time did not permit us to include all Surinamese rivers and creeks for which data are available. Time also kept us from the examination of the extensive collections of fish from Suriname at the museums of Amsterdam and Leiden (the Netherlands). We limited our survey to those river-sections which were sampled by ourselves, or which were sampled thoroughly by numerous expeditions.

Paul E. Ouboter (ed.), Freshwater Ecosystems of Suriname, 133–154.

2. The River-sections

The terrestrial ecosystems of Suriname can roughly be categorized into three groups: 1. those of the Coastal Plain (Young and Old); 2. those of the Savanna Belt; 3. those of the Interior. The chemical composition of water of rivers and creeks originating in these areas differs greatly as was shown in Chapter 4. Fig. 1 shows the river-sections studied, five of which are in the Interior, three originate in the Savanna Belt and two are in the Coastal Plain. All river-sections of the Interior are located upstream of the first rapids and can be characterized as almost clear-water rivers (although usually some turbidity is visible). All three rivers originating in the Savanna Belt are black-water rivers. The two coastal river-sections are the downstream parts of interior rivers between the

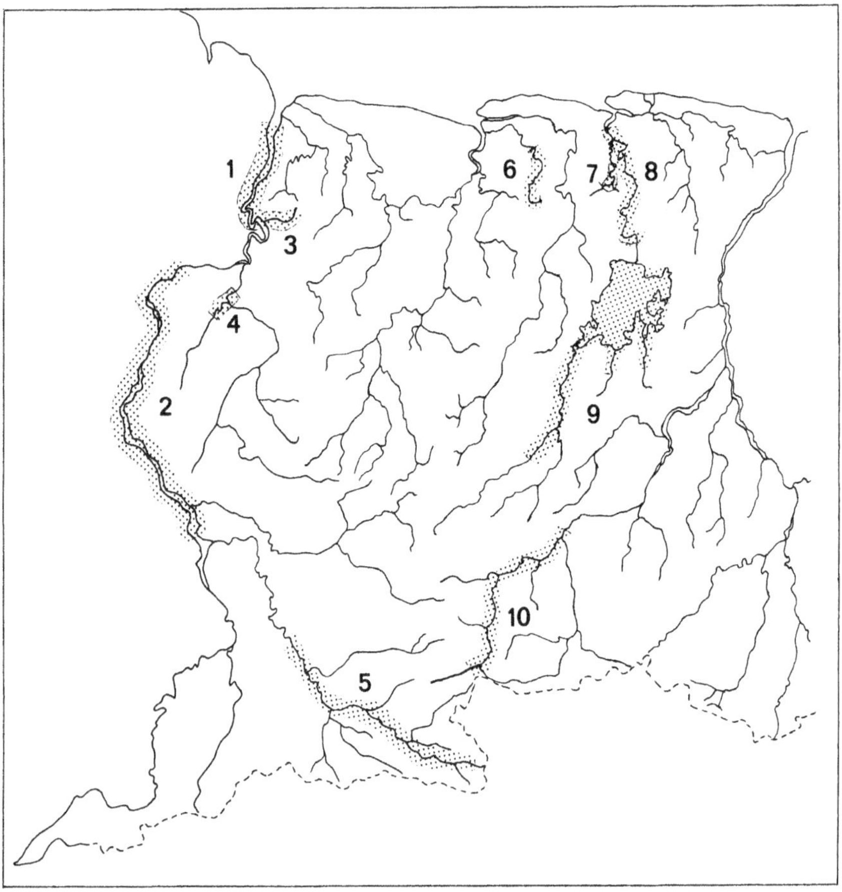

Figure 1. River-sections included in this study.
1. Corantijn downstream; 2. Corantijn midstream; 3. Kaburi Creek; 4. Kabalebo; 5. Curuni/Sipaliwini; 6. Cusewijne; 7. Para; 8. Suriname downstream; 9. Suriname midstream; 10. Tapanahoni

Savanna Belt and the mouth. These sections carry very turbid, fresh to slightly brackish water which is under a strong tidal influence. Physical and chemical differences are listed in Table 1 of Chapter 4.

Table 1. Composition of the ichthyofauna of Guyana, Suriname, French Guiana, the Rio Negro (Brazil), and Amazonia. Marine species included as well (*). In parenthesis the total number of species known to occur in Surinamese fresh-water habitats.

Family	Guyana	Suriname		Fr. Guiana	Negro	Amazonia
Potamotrygonidae	2	2		1	0	?
Osteoglossidae	2	0	(0)	2	3	3
Anostomidae	14	10		12	24	51
Characidae	78	62		63	103	304
Characidiidae	7	4		5	3	36
Crenuchidae	1	1	(1)	1	1	2
Ctenoluciidae	1	0	(0)	1	3	4
Curimatidae	6	9		11	15	55
Erythrinidae	4	4	(4)	4	6	5
Gasteropelecidae	2	2	(2)	2	2	7
Hemiodidae	3	8		8	12	23
Lebiasinidae	10	8		6	6	23
Serrasalmidae	17	10	(12)	13	20	35
subtotal	143	118		126	195	545
Apteronotidae	0	1		5	9	?
Electrophoridae	1	1	(1)	1	1	?
Gymnotidae	1	2	(2)	2	2	?
Rhamphichthyidae	14	11		10	24	?
subtotal	16	15		18	36	50
Ageneiosidae	4	3		2	4	11
Ariidae	13*	2		13*	0	0
Aspredinidae	8	5		6	1	0
Astroblepidae	0	0		0	0	15
Auchenipteridae	6	7	(9)	7	18	35
Callichthyidae	4	17	(17)	18	2	33
Cetopsidae	0	1		1	1	5
Doradidae	10	4		4	12	63
Helogenidae	1	1	(1)	1	0	2
Hypophthalmidae	1	1	(1)	1	2	3
Loricariidae	27	38		22	30	150
Pimelodidae	31	26	(28)	24	25	105
Scoloplacidae	0	0		0	1	1
Trichomycteridae	6	3		3	10	37
subtotal	111	108		102	106	460

Table 1. (continued)

Family	Guyana	Suriname	Fr. Guiana	Negro	Amazonia
Belonidae	1	1	?	3	3
Cyprinodontidae	8	4	6	1	?
Poeciliidae	5	4 (4)	4	1	?
Synbranchidae	1	1 (1)	1	1	1
Cichlidae	33	25 (26)	32	47	77
Nandidae	1	1 (1)	1	0	2
Lepidosirenidae	1	0	1	0	1
Megalopidae	1	1	?	0	?
Clupeidae	5	3	?	1	?
Engraulidae	0	9	?	5	?
subtotal	5	12	?	6	15
Anablepidae	2	1	?	0	?
Batrachoididae	0	1	?	0	2
Carangidae	0	1	?	0	0
Centropomidae	2	2	?	0	?
Echeneidae	0	1	?	0	?
Gobiidae	3	5	?	2	3
Mugilidae	4	1	?	0	?
Sciaenidae	14	10	?	10	?
Soleidae	3	3	?	1	?
Tetraodontidae	1	1	?	0	1
TOTAL	359	318	?	412	1163

3. The Ichthyofauna

Some representatives of the Surinamese fish fauna are illustrated in Fig. 2. The ichthyofauna of Suriname is mainly of Amazonian origin. A few marine families including species which enter the rivers of Suriname, are not listed for the Amazon (Table 1, Gery 1984) i.e. Ariidae, Anablepidae, Carangidae, Centropomidae, Echeneidae, Megalopidae, Pomadasyidae. All freshwater families known for Suriname are present in the Amazon. Of five Amazonian fish families no members are known to occur in Suriname; these are the Osteoglossidae, Ctenoluciidae, Astroblepidae (Andean), Scoloplacidae (only 1 species), and Lepidosirenidae. Of these, the Osteoglossidae, Ctenoluciidae and Lepidosirenidae occur in one or both neighbouring Guianan countries. Both Guyana (Rupununi, Essequibo), and French Guiana (Oyapock) have connections with the Amazon river-system. Suriname is completely separated from the Amazon system by mountains.

A further comparison is possible for the Guianas and the Rio Negro based on data from Eigenmann (1912) and Lowe-McConnell (1964) for Guyana, from Le

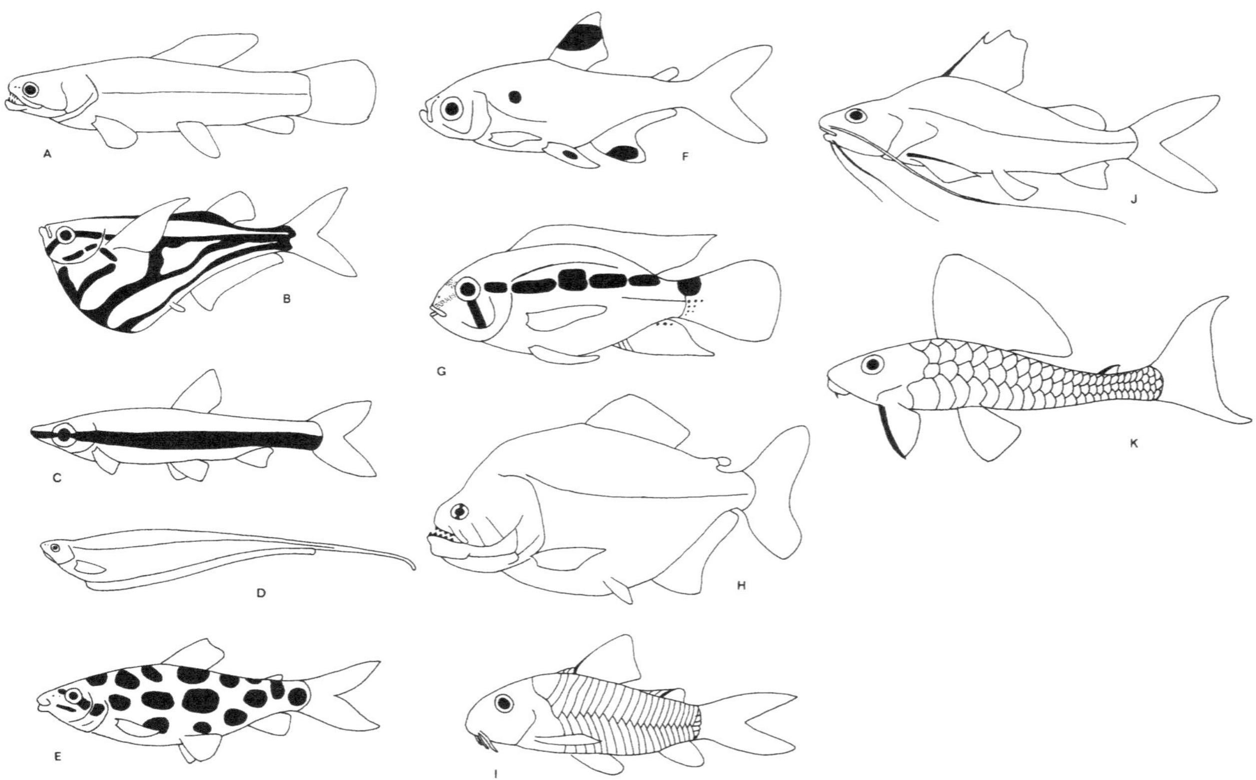

Figure 2. A. *Hoplias malabaricus* (Erythrinidae); B. *Carnegiella strigata* (Gasteropelecidae); C. *Nannostomus beckfordi* (Lebiasinidae); D. *Eigenmannia virescens* (Rhamphichthyidae); E. *Leporinus granti* (Anostomidae); F. *Pristella maxillaris* (Characidae); G. *Krobia guianensis* (Cichlidae); H. *Serrasalmus rhombeus* (Serrasalmidae); I. *Corydoras* sp. (Callichthyidae); J. *Pimelodus* sp. (Pimelodidae); K. *Hypostomus* sp. (Loricariidae)

Bail *et al.* (1984) and Rojas-Beltran (1984) for French Guiana, and from Goulding *et al.* (1988) for the Rio Negro. Comparisons with other areas of Amazonia or its periphery are hardly possible, because most authors omit part of the fish fauna (for instance the smaller species). Table 1 presents an enumeration of the number of species per family for the four areas and Amazonia-sec (peripheral drainage systems excluded). When interpreting the data, one has to bear in mind that it is a comparison between the ichthyofauna known for whole countries (Guyana and French Guiana), the data for several rivers and creeks in one country (Suriname), one large black-water river (Rio Negro), and the immense Amazon river-system. Still, we believe that it gives some insight in the ichthyofaunal composition of the region and the importance of the different families.

Compared to the ± 1200 species known from the Amazon and its tributaries (including the Rio Negro), the four case studies summarized in Table 1 reveal an impoverished fish fauna. Although "just one river", in general the number of species is highest in the Rio Negro. Apart from several marine families which do not enter the Rio Negro, the most obvious exceptions are the Aspredinidae, Callichthyidae, Characidiidae, Cyprinodontidae and Poeciliidae. Gery (1984) lists the Characidiidae as mainly peripheral. The Aspredinidae include many coastal species, which are not likely to enter the interior of Brazil as far as the Rio Negro. In Suriname all Poecilids are coastal as well, mainly dwelling in eutrophic environments. The Callichtyidae show a high diversity in both Suriname and French Guiana compared to the Rio Negro and Guyana. The largest genus of the family, *Corydoras*, was studied extensively in Suriname (Nijssen 1970). Future surveys for this group in Guyana will surely yield more species. Another family well-studied in Suriname is the Loricariidae, which not surprisingly, yield many species. Most other families compare well in numbers with the neighbouring Guianan countries; sometimes the number is a little lower. This is probably caused by the absence of some purely Amazonian species, which were able to enter the other Guianas, not Suriname.

3.1. Comparison Between the River-sections

The fish species recorded for the 10 river-sections are listed in Appendix 1 (A-F). Several species found in the Para River were certainly, others might have been introduced (indicated by an asterisk in Appendix 1). Their introduction probably resulted from the actions of traders in aquarium fish which were active in the upstream area of Coropina Creek in the sixties. It is uncertain whether *Mesonauta festivus* is really an exotic. Old inhabitants of villages in the area, remember catching the species when they were still young. On the other hand, the limitation of *M. festivus* to the Para river-system is suspicious.

Species that are reported from Suriname for the first time, outside the Para area, are *Leporinus despaxi* and *L. lebaili* (Anostomidae), three Loricariids from the Tapanahoni River, *Ancistrus cirrhosus*, *Exastilithoxus* sp.,

Table 2. Ichthyofaunal composition of 10 river-sections in the Coastal Region, Savanna Belt and Interior of Suriname.

	Lower courses		Black water			Interior midstream			Interior upstream	
	Cor.low	Sur.low	Kabu	Cus	Par	Kaba	Cor.mid	Sur.mid	Tap	Sip
Cypriniformes										
Characoidei	27	17	39	22	45	59	70	40	38	44
Characoidea	20	15	31	22	38	50	58	35	36	41
Gymnotoidea	7	2	8	0	7	9	12	5	2	3
Siluriformes	22	15	14	10	12	32	41	34	26	26
Callichthyidae	1	0	2	2	3	8	8	7	4	3
Loricariidae	4	4	1	2	0	12	16	10	7	9
Pimelodidae	5	6	5	3	4	7	11	13	11	11
Others	12	5	6	3	5	5	6	4	4	3
Perciformes										
Cichlidae + Nandidae	8	6	11	7	10	13	11	11	10	4
Other freshwater fishes	7	4	1	2	5	1	2	1	2	1
Marine origin	28	22	2	1	1	0	0	0	0	0
TOTAL	92	4	67	42	73	105	124	86	76	75

Cor.low lower Corantijn River
Sur.low lower Suriname River
Kabu Kaburi Creek
Cus Cusewijne River
Par Para River
Kaba Kabalebo River
Cor.mid Corantijn River midstream
Sur.mid Suriname River midstream
Tap Tapanahoni River
Sip Sipaliwini River

Lasiancistrus niger and *Pseudoplatystoma tigrinum* (Pimelodidae).

Several species could not be identified. One or two might prove to be species new to science; most will probably be varieties of species already known.

Table 2 shows enumerations of the species for the larger groups. For most groups the highest diversity is reached in the middle courses of the rivers of the Interior. The Corantijn River is the largest of them and has been thoroughly sampled, yielding 124 species in the middle course. Quite unexpected is the much lower diversity of several groups (Gymnotoidea, Siluriformes) in the upper courses.

The lower sections of the Corantijn and Suriname Rivers are entered by many species that are actually marine, or adapted to estuarine environments. The number of species of most other groups is less than for the interior sections.

The three black-water rivers are inhabited by a moderate number of species. The Characoidea do quite well in this harsh environment, the Silurids much less. Very remarkable is the total absence of Gymnotoidea in the Cusewijne River. Although this river is not thoroughly sampled, a few species would have been found when they had been present.

A further comparison of the fish fauna of these river-sections is very instructive. The similarity of the ichthyofaunas was compared using the coefficient of Jaccard (Krebs 1989)

$$Sj = \frac{a}{a + b + c}$$

a = number of species present in both sample A and B.
b = number of species present in sample B, but not in A.
c = number of species present in sample A, but not in B.

Table 3 shows the resulting values for a comparison of river-sections between different river-systems; Table 4 for a comparison of river-sections within one river-system. The faunal similarity between river-sections of different river-systems is generally low. Even the lower courses, which harbour many marine species originating in a relatively homogeneous environment, are quite dissimilar, the Corantijn River having much more Engraulids and Gobiids. The highest similarity values are observed between the black-water rivers (up to 0.47), the lowest between the Middle Corantijn River and the Tapanahoni River, these river-sections belonging to the most widely separated river-systems (Corantijn and Marowijne). The low similarity coefficients illustrate the differences in the ichthyofauna of Surinamese rivers. For some taxa, e.g. *Corydoras* (Nijssen 1970) and *Hypostomus* (Boeseman 1968), it is known that related species are restricted to a single river-system. However, additional collecting will undoubtedly reduce the differences in the known composition of the fish fauna between different river-systems to some degree.

Because of habitat similarities (inland savannas), another comparison (for the total fish-fauna) was made between the Sipaliwini River and the Rupununi area in Guyana. It is quite remarkable that these faunas have little in common

Table 3. Faunal similarity (coefficient of Jaccard) between sections of different river-systems of Suriname for the Characoidei and Siluriformes (and families of marine origin).

	Marine	Charac.	Silur.
Downstream			
Lower Corantijn and Lower Suriname	0.24	0.29	0.28
Interior midstream			
Middle Corantijn and Middle Suriname		0.31	0.21
Kabalebo and Middle Suriname		0.36	0.22
Interior upstream			
Sipaliwini and Tapanahoni		0.30	0.20
Black water			
Kaburi and Cusewijne		0.33	0.33
Kaburi and Para		0.47	0.37
Cusewijne and Para		0.43	0.47
Midstream and Upstream			
Middle Corantijn and Tapanahoni		0.19	0.14
Middle Suriname and Tapanahoni		0.20	0.22
Middle Suriname and Sipaliwini		0.31	0.33

Table 4. Faunal similarity (coefficient of Jaccard) between sections within one river-system for the Characoidei and Siluriformes.

	Charac.	Silur.
Corantijn		
Lower Corantijn and Middle Corantijn	0.33	0.09
Middle Corantijn and Sipaliwini	0.25	0.22
Lower Corantijn and Sipaliwini	0.08	0.12
Kaburi and Lower Corantijn	0.47	0.29
Kaburi and Middle Corantijn	0.36	0.16
Kabalebo and Lower Corantijn	0.36	0.08
Kabalebo and Middle Corantijn	0.65	0.43
Kabalebo and Sipaliwini	0.30	0.26
Kaburi and Kabalebo	0.42	0.12
Suriname		
Lower Suriname and Middle Suriname	0.16	0.19
Para and Lower Suriname	0.19	0.08
Para and Middle Suriname	0.25	0.12

(Sj= 0.13; between Rupununi and Sipaliwini + Middle Corantijn Sj= 0.19).

The Corantijn is the river-system most extensively sampled, including down-, mid- and upstream parts as well as two major tributaries. Table 4 shows that there is a medium similarity between the down- and midstream parts for the Characoidei, but a very low similarity for the Siluriformes. In all comparisons between different river-sections within one river-system the Siluriformes perform worst. The Kaburi Creek drains into the Lower Corantijn and, not surprisingly, similarity is greatest with this section. The Kabalebo River drains into the Middle Corantijn, and similarity between these parts is the highest recorded (Table 4).

Similarity between river-sections in the Suriname river-system is even lower than in the Corantijn system. Very remarkable is that the Para River, which drains into the Lower Suriname River, shows more similarity with the Middle Suriname than with the Lower.

Although this exercise teaches us a lot, the data are not sufficient to make serious generalizations. These might be possible when the data for more large Guianan rivers (Essequibo, Marowijne) are included as well.

4. Adaptations to the Environment

We do not intend to provide a complete overview of adaptations to certain environments, we will merely touch upon the subject. We will also omit more general adaptations, like the use of electricity for sensing in nocturnal species (Gymnotoidea), etc.

Many fish species of the lower courses and of the black-water streams, remain part of their life in swamps. Swamps require special adaptations. The water is usually deprived of oxygen (the same applies to black water), and desiccation after drying up of the swamp is a real potential danger for aquatic organisms. Many of the fish species living in swamps are able to use oxygen from the air for breathing, by adaptation of one or several organs. In the electric eel (*Electrophorus electricus*) the buccal cavity has a well-vascularized area in which most of its required oxygen is taken up. The gills have degenerated over evolutionary time (Kramer *et al.* 1978). In the Callichthyid genera *Callichthys* and *Hoplosternum* the posterior part of the intestine is specialized for oxygen uptake, an adaptation peculiar to the Amazon region. These fish actually swallow air, while the carbon dioxide is primarily excreted at the gills (Carter and Beadle 1931; Huebner and Chee 1978)). *Synbranchus marmoratus* (Synbranchidae) has a supplementary pair of lunglike sacks of the gill pouch for breathing air (Kramer *et al.* 1978). In the Erythrinids *Hoplerythrinus unitaeniatus* and *Erythrinus erythrinus* the posterior part of the swimbladder is highly vascularized and adapted to air-breathing (Carter and Beadle 1931; Kramer *et al.* 1978). Use of surface water for aquatic respiration is one of the few alternatives to aerial respiration which allow fish to survive extreme hypoxia. Among others *Copella*, *Hemiodopsis*, *Pterophyllum* and *Rivulus* use this very thin zone near the surface in which rapid diffusion raises oxygen concentrations above that of the rest of the water column (Kramer and McClure 1982).

Several fish species that live in swamps or non-permanent ponds are able to move over land. This was observed in *Callichthys callichthys*, *Synbranchus marmoratus*, *Hoplerythrinus unitaeniatus* (Kramer *et al.* 1987) and *Rivulus urophthalmus* (collected in pit traps in the forest some distance from the water).

In the Interior many species are adapted to live on rocks or even in rapids. In order to resist the strong current in rapids, mouth-suckers have developed (esp. in the Loricariidae), and the species living here are usually flattened dorso-ventrally as well. Mouth structures to scrape algae of the rocks are known in the Characidiidae, Hemiodidae and Loricariidae (Power 1984).

The rivers of the Interior have a very poor nutrient content. Fish inhabiting these rivers will not be able to feed on plankton; secondary production will be almost completely based on allochtonous material from the forest. Goulding (1980) has shown that many species are adapted to feed on fruits which fall into the water or on leaves during the rainy season (especially Serrasalmidae and some of the larger Characidae). Other species feed mainly on small invertebrates which fall into the river (Knöppel 1970), their mouth is usually upturned (e.g. Gasteropelecidae, Poeciliidae, *Rivulus* spp., *Pyrrhulina* spp.).

Other adaptations are for instance those of the reproductive cycle being adapted to the water regime. Lowe-McConnell (1987) has presented an overview of the influence of seasonality on the life histories of tropical fish. Here, an example will be given in Chapter 10 on the nesting biology of armoured catfishes.

5. Freshwater Fisheries

As already stated in Chapter 1, hardly any data on freshwater fisheries exist. Consequently, we will restrict ourselves to some general remarks.

Human communities can use fish resources in four ways:

1. As food and protein source for local communities.
2. Commercial exploitation as food fish, resulting in the supply of urban communities or export.
3. Commercial exploitation for the pet trade.
4. Aquaculture, with ultimately the same objectives as for direct commercial exploitation.

All areas of Suriname distinguished here, have sufficient fish resources to supply the local communities (Fig. 3). Many conditions determine the potential of fish species to be commercially exploited as food. Although of the utmost importance, we will not deal here with population dynamic characteristics of the species (hardly any data is available), but will concentrate on numbers and size. Unfortunately, also on fish biomass very few data is available. The only study accomplished is the one by Richter and Nijssen (1980) for the Brokopondo Reservoir. However, the regular sampling trips we made, gave us a good impression of species abundances.

In the rivers of the Interior not a single species is very abundant (see also Eigenmann (1912) for Guyana). This could also be predicted on theoretical grounds: high species diversity goes together with a low number of specimens per species. Commercial exploitation of the interior fish resources is unlikely to be feasible, also because of the inaccessibility of the area.

Figure 3. Inhabitants of the Interior showing the catch of one hour fishing. The large fish is *Phractocephalus hemilioptérus*.

The case study by Richter and Nijssen (1980) for the Brokopondo Reservoir showed that productivity was rather low in 1978, and was even expected to decrease with a further decrease of the nutrient contents of the water. However, productivity would probably be sufficient to supply the local market. In 1978 90% of the biomass was *Serrasalmus rhombeus*. Richter (1980) concluded that the feasibility of commercial exploitation of the Brokopondo Reservoir is questionable due to several biological/technical, as well as socio-economic problems.

Some fish are quite abundant in the black-water rivers. These are the Characoids *Hoplias malabaricus* and *Hoplerythrinus unitaeniatus*, and in some rivers also *Triportheus rotundatus*, the Cichlids *Aequidens tetramerus* or *Krobia guianensis*, the catfish *Ageneiosus brevifilis*, and the tarpon *Megalops atlanticus*. The last mentioned species is actually marine, but enters the black-water rivers far upstream. In the Cusewijne River these species are commercially exploited on a small scale. Large scale exploitation is hindered by the narrow channel of these rivers and the many fallen trees. Large scale exploitation is also undesirable, because it would unbalance the sensitive black-water ecosystem.

Towards the lower courses of the big rivers, fish diversity decreases, but the abundance per species increases. Here, many large and medium-sized species, mostly of marine origin, are of commercial importance (e.g. *Brachyplatystoma* spp., *Plagioscion* spp.), and fishing with trawlers is possible. However, research into the ecology and actual biomass of these species is necessary, before the potential for large scale exploitation on a sustainable basis can be estimated. Great care must be taken not to disturb the nursery function of the estuarine areas.

Although some small-sized fish in the Interior are of value for the pet trade, logistic difficulties make the exploitation of these resources almost impossible. The black-water rivers, however, have a high potential for the exploitation of aquarium fish, since they can be easily reached and hold many interesting small-sized Characids and Cichlids. Of special interest is the successful introduction of the cardinal tetra *Cheirodon axelrodi* in the Para river-system. This species alone accounts for 53% of the commercial value of export of ornamental fish from Amazonia (Junk 1984).

Many fish species from all three areas could have potentials for aquacultural breeding. Most promising are probably some of the catfish, like *Ageneiosus*, *Pimelodus* and *Hoplosternum*. Aquacultural trials already have been started with *Hoplosternum littorale* (Mol in press). Unfortunately, manpower and equipment are insufficient to further research in this direction.

In general, the research efforts in freshwater fisheries need to be increased. At the moment no data whatsoever is available on the number of people occupied in freshwater fisheries, the catch or the impact fisheries have on fish populations, or the ecosystems involved. These gaps in knowledge need to be filled as soon as possible.

Appendix 1A. Characoidea from Suriname. Species marked with * are probably introduced.

Family	Genus	Species	Drainage system									
			KA	SU	SL	CU	CL	PA	TA	SI	CS	KB
1. Anostomidae	Anostomus	anostomus				X						
2. Anostomidae	Leporinus	alternus	X			X						
3. Anostomidae	Leporinus	despaxi							X			
4. Anostomidae	Leporinus	fasciatus	X	X						X		
5. Anostimidae	Leporinus	friderici	X	X		X		X		X	X	X
6. Anostomidae	Leporinus	granti				X						
7. Anostomidae	Leporinus	lebaili							X			
8. Anostomidae	Leporinus	maculatus				X						
9. Anostomidae	Prochilodus	rubrotaeniatus	X			X	X					X
10. Anostomidae	Schizodon	fasciatum		X				X		X		
11. Characidae	Acestrorhynchus	falcatus	X	X		X	X	X			X	X
12. Characidae	Acestrorhynchus	microlepis		X	X	X	X	X			X	
13. Characidae	Astyanax	abramoides	X	X		X	X			X		X
14. Characidae	Astyanax	bimaculatus	X	X		X	X	X				X
15. Characidae	Astyanax	orthodus aff.							X	X		
16. Characidae	Astyanax	polylepis	X	X		X				X		
17. Characidae	Astyanax	spec.1							X			
18. Characidae	Astyanax	spec.2							X			
19. Characidae	Brachychalcinus	quianensis		X		X				X		
20. Characidae	Brycon	falcatus	X			X				X		X
21. Characidae	Bryconamericus	hyphesson				X				X		
22. Characidae	Bryconops	affinis	X		X	X	X		X	X		X
23. Characidae	Bryconops	caudomaculatus			X							
24. Characidae	Bryconops	melanurus	X	X		X		X	X	X	X	X
25. Characidae	Bryconops	spec.								X		
26. Characidae	Chalceus	macrolepidotus	X	X					X	X		

Appendix 1A. (continued)

	Family	Genus	Species	Drainage system									
				KA	SU	SL	CU	CL	PA	TA	SI	CS	KB
27.	Characidae	Charax	gibbosus	X	X		X		X	X	X		X
28.	Characidae	Cheirodon	axelrodi*						X				
29.	Characidae	Cheirodon	gracilis						X				
30.	Characidae	Creagrutus	melanzonus				X						
31.	Characidae	Cynodon	gibbus			X							
32.	Characidae	Cynopotamus	essequibensis			X	X			X	X		
33.	Characidae	Deuterodon	pinnatus	X	X		X			X	X		
35.	Characidae	Deuterodon	spec.								X		
35.	Characidae	Hemigrammus	bellottii						X				
36.	Characidae	Hemigrammus	boesemani						X			X	
37.	Characidae	Hemigrammus	lunatus								X		
38.	Characidae	Hemigrammus	ocellifer						X		X		X
39.	Characidae	Hemigrammus	spec.							X			
40.	Characidae	Hemigrammus	unilineatus	X		X	X	X	X			X	X
41.	Characidae	Holobrycon	pesu							X			
42.	Characidae	Hyphessobrycon	bentosi rosaceus	X			X				X		
43.	Characidae	Hyphessobrycon	georgettae	X							X		
44.	Characidae	Hyphessobrycon	heterorhabus*						X				
45.	Characidae	Hyphessobrycon	minor						X				
46.	Characidae	Hyphessobrycon	spec.	X			X						
47.	Characidae	Megalamphodus	spec.				X						
48.	Characidae	Moenkhausia	bondi				X						
49.	Characidae	Moenkhausia	chrysargyrea	X	X		X	X	X				
50.	Characidae	Moenkhausia	colletti							X	X	X	
51.	Characidae	Moenkhausia	comma						X				
52.	Characidae	Moenkhausia	dichroura	X									
53.	Characidae	Moenkhausia	georgiae							X	X		
54.	Characidae	Moenkhausia	grandisquamis	X	X	X	X	X	X	X	X		X
55.	Characidae	Moenkhausia	intermedia							X			
56.	Characidae	Moenkhausia	lepidura	X			X				X		
57.	Characidae	Moenkhausia	oligolepis	X	X		X			X	X		X
58.	Characidae	Moenkhausia	shidelari							X			
59.	Characidae	Moenkhausia	spec.1							X			
60.	Characidae	Moenkhausia	spec.2							X			
61.	Characidae	Moenkhausia	surinamensis							X	X		
62.	Characidae	Phenacogaster	megalostictus	X			X						
63.	Characidae	Phenagaster	microstictus	X			X						
64.	Characidae	Piabucus	dentatus			X		X					
65.	Characidae	Poptella	longipinnis	X			X						
66.	Characidae	Poptella	orbicularis		X	X			X		X	X	X
67.	Characidae	Pristella	maxillaris						X			X	X
68.	Characidae	Pseudopristella	simulata						X			X	
69.	Characidae	Roeboexodon	guianensis	X			X			X	X		
70.	Characidae	Roeboides	thurni		X								
71.	Characidae	Tetragonopterus	chalceus	X			X						
72.	Characidae	Triportheus	rotundatus			X			X		X	X	
73.	Characidiiae	Characidium	blennoides group				X						
74.	Characidiidae	Characidium	fasciadorsale								X		

Appendix 1A. (continued)

	Family	Genus	Species	Drainage system									
				KA	SU	SL	CU	CL	PA	TA	SI	CS	KB
75.	Characidiidae	Characidium	fasciatum group							X			
76.	Characidiidae	Characidium	spec.	X									
77.	Crenuchidae	Crenuchus	spilurus	X			X	X	X			X	X
78.	Curimatidae	Caenotropus	labyrinthicus				X						
79.	Curimatidae	Caenotropus	maculosus				X						
80.	Curimatidae	Curimata	cyprinoides		X	X	X	X	X				X
81.	Curimatidae	Curimata	helleri		X		X			X			
82.	Curimatidae	Curimata	microcephala	X									X
83.	Curimatidae	Curimata	spilura group	X	X		X			X	X		X
84.	Curimatidae	Curimatella	alburna		X				X				X
85.	Curimatidae	Curimatopsis	crypticus						X			X	X
86.	Curimatidae	Hemicurimata	esperanzae								X		
87.	Erythrinidae	Erythrinus	erythrinus		X								
88.	Erythrinidae	Hoplerythrinus	unitaeniatus	X	X		X		X		X	X	X
89.	Erythrinidae	Hoplias	macrophthalmus	X	X		X					X	
90.	Erythrinidae	Hoplias	malabaricus	X	X	X	X	X	X	X		X	X
91.	Gasteropelecidae	Carnegiella	strigata	X			X		X				X
92.	Gasteropelecidae	Gasteropelecus	sternicla	X	X		X	X	X	X		X	X
93.	Hemiodidae	Bivibranchia	bimaculata	X			X						
94.	Hemiodidae	Bivibranchia	protractila	X			X						
95.	Hemiodidae	Bivibranchia	spec.								X		
96.	Hemiodidae	Hemiodopsis	ocellata	X			X	X					
97.	Hemiodidae	Hemiodopsis	quadrimaculatus	X			X			X	X		
98.	Hemiodidae	Hemiodopsis	semitaeniatus	X	X		X						
99.	Hemiodidae	Hemiodus	unimaculatus		X								
100.	Hemiodidae	Parodon	guyanensis							X	X		
101.	Lebiasinidae	Copella	arnoldi				X	X	X			X	X
102.	Lebiasinidae	Copella	eigenmanni	X				X					X
103.	Lebiasinidae	Nannostomus	beckfordi					X	X			X	X
104.	Lebiasinidae	Nannostomus	bifasciatus			X				X			
105.	Lebiasinidae	Nannostomus	harrisoni*						X				
106.	Lebiasinidae	Nannostomus	marginatus	X			X		X			X	X
107.	Lebiasinidae	Pyrrhulina	filamentosa	X			X	X	X			X	X
108.	Lebiasinidae	Pyrrhulina	stoli	X			X						
109.	Serrasalmidae	Acnodon	oligacanthus		X					X			
110.	Serrasalmidae	Metynnis	altidorsalis						X				
111.	Serrasalmidae	Myleus	pacu		X	X					X		
112.	Serrasalmidae	Myleus	rhomboidalis	X	X						X		
113.	Serrasalmidae	Myleus	rubripinnis	X	X	X	X		X	X			
114.	Serrasalmidae	Myleus	spec.				X						
115.	Serrasalmidae	Myleus	ternetzi goslinei		X					X	X		
116.	Serrasalmidae	Serrasalmus	denticulatus						X			X	
117.	Serrasalmidae	Serrasalmus	rhombeus	X	X	X	X	X		X	X		X
118.	Serrasalmidae	Serrasalmus	striolatus							X			

KA Kabalebo river
SU Upper Suriname river
SL Lower Suriname river
CU Upper Corantijn river
CL Lower Corantijn river
PA Para river
TA Tapanahoni river
SI Sipaliwini river
CS Cusewijne river
KB Kaburi creek

Appendix 1B. Gymnotoidea from Suriname

	Family	Genus	Species	Drainage system									
				KA	SU	SL	CU	CL	PA	TA	SI	CS	KB
1.	Apteronotidae	Apteronotus	albifrons				X						
2.	Electrophoridae	Electrophorus	electricus	X	X		X	X					X
3.	Gymnotidae	Gymnotus	anguillaris	X			X	X	X				
4.	Gymnotidae	Gymnotus	carapo	X	X	X	X	X	X				X
5.	Rhamphichthyidae	Eigenmannia	macrops				X						
6.	Rhamphichthyidae	Eigenmannia	virescens	X			X	X	X	X	X		X
7.	Rhamphichthyidae	Gymnorhamphichthys	rondoni	X	X		X				X		
8.	Rhamphichthyidae	Hypopomus	beebei						X				X
9.	Rhamphichthyidae	Hypopomus	brevirostris	X			X	X					X
10.	Rhamphichthyidae	Hypopomus	spec.		X								X
11.	Rhamphichthyidae	Hypopygus	lepturus	X	X		X	X	X				X
12.	Rhamphichthyidae	Parupygus	savannensis	X			X				X		
13.	Rhamphichthyidae	Parupygus	spec.						X				
14.	Rhamphichthyidae	Steatogenys	elegans				X						
15.	Rhamphichthyidae	Sternopygus	macrurus	X		X	X	X	X	X			X

KA Kabalebo river
SU Upper Suriname river
SL Lower Suriname river
CU Upper Corantijn river
CL Lower Corantijn river
PA Para river
TA Tapanahoni river
SI Sipaliwini river
CS Cusewijne river
KB Kaburi creek

Appendix 1C. Siluriformes of Suriname. Species marked with / are probably new for Suriname.

	Family	Genus	Species	Drainage system									
				KA	SU	SL	CU	CL	PA	TA	SI	CS	KB
1.	Ageneiosidae	Ageneiosus	brevifilis						X		X	X	
2.	Ageneiosidae	Ageneiosus	guianensis					X					
3.	Ageneiosidae	Ageneiosus	marmoratus				X						
4.	Ariidae	Arius	rugispinis			X							
5.	Ariidae	Arius	spixii					X					
6.	Aspredinidae	Agmus	lyriformes					X					
7.	Aspredinidae	Aspredo	aspredo					X					X
8.	Aspredinidae	Bunocephalus	amaurus								X		
9.	Aspredinidae	Bunocephalus	gronovii				X						
10.	Aspredinidae	Platystacus	cotylephorus			X		X					
11.	Auchenipteridae	Auchenipterus	nuchalis			X		X					
12.	Auchenipteridae	Parauchenipterus	galeatus		X			X	X			X	X
13.	Auchenipteridae	Pseudauchenipterus	nodosus			X		X					
14.	Auchenipteridae	Tatia	brunnea		X					X			
15.	Auchenipteridae	Tatia	creutzbergi						X				
16.	Auchenipteridae	Tatia	intermedia		X		X				X		
17.	Auchenipteridae	Tatia	punctata		X					X			
18.	Callichthyidae	Callichthys	callichthys	X	X		X	X		X	X		X
19.	Callichthyidae	Corydoras	aeneus		X								
20.	Callichthyidae	Corydoras	baderi	X			X						
21.	Callichthyidae	Corydoras	bicolor				X				X		
22.	Callichthyidae	Corydoras	boesemani		X								

Appendix 1C. (continued)

	Family	Genus	Species	Drainage system									
				KA	SU	SL	CU	CL	PA	TA	SI	CS	KB
23.	Callichthyidae	Corydoras	bondi	X						X	X		
24.	Callichthyidae	Corydoras	guianensis	X			X			X			
25.	Callichthyidae	Corydoras	melanistius	X			X						
26.	Callichthyidae	Corydoras	nanus		X								
27.	Callichthyidae	Corydoras	octocirrus		X								
28.	Callichthyidae	Corydoras	oxyrhynchus							X			
29.	Callichthyidae	Corydoras	punctatus		X								
30.	Callichthyidae	Corydoras	schwartzi surinamensis	X			X						
31.	Callichthyidae	Corydoras	spec.									X	X
32.	Callichthyidae	Corydoras	wotroi	X			X						
33.	Callichthyidae	Hoplosternum	littorale						X			X	
34.	Callichthyidae	Hoplosternum	thoracatum	X	X		X	X	X				X
35.	Cetopsidae	Hemicetopsis	minutus	X						X			
36.	Doradidae	Acanthodoras	cataphractus					X	X			X	
37.	Doradidae	Anduzedoras	microstomas					X					X
38.	Doradidae	Platydoras	spec.			X			X			X	X
39.	Doradidae	unknown doradid											X
40.	Helogenidae	Helogenes	marmoratus	X			X			X			
41.	Hypophthalmidae	Hypophthalmus	edentatus					X					
42.	Loricariidae	Ancistrus	cirrhosus*							X			
43.	Loricariidae	Ancistrus	hoplogenys				X						
44.	Loricariidae	Ancistrus	temmincki	X									
45.	Loricariidae	Ctenoloricaria	maculata	X	X		X				X	X	X
46.	Loricariidae	Exastilithoxus	spec.*							X			
47.	Loricariidae	Farlowella	spec.	X			X						
48.	Loricariidae	Hartiella	crassicauda								X		
49.	Loricariidae	Harttia	nijsseni				X						
50.	Loricariidae	Harttia	spec.	X			X						
51.	Loricariidae	Harttia	surinamensis		X					X			
52.	Loricariidae	Hypostomus	corantijni	X			X				X		
53.	Loricariidae	Hypostomus	crassicauda								X		
54.	Loricariidae	Hypostomus	gymnorhynchus				X			X			
55.	Loricariidae	Hypostomus	micromaculatus		X								
56.	Loricariidae	Hypostomus	paucimaculatus		X								
57.	Loricariidae	Hypostomus	plecostomus		X	X		X					
58.	Loricariidae	Hypostomus	pseudohemiurus	X			X						
59.	Loricariidae	Hypostomus	pseudohemiurus macrophtha								X		
60.	Loricariidae	Hypostomus	saramacensis	X				X					
61.	Loricariidae	Hypostomus	sipaliwinii								X		
62.	Loricariidae	Hypostomus	spec.	X			X						
63.	Loricariidae	Hypostomus	surinamensis		X								
64.	Loricariidae	Hypostomus	tenuis			X							
65.	Loricariidae	Hypostomus	ventromaculatus			X						X	
66.	Loricariidae	Hypostomus	watwata			X							
67.	Loricariidae	Lasiancistrus	brevispinis		X					X	X		
68.	Loricariidae	Lasiancistrus	niger*							X			
69.	Loricariidae	Lasiancistrus	schomburgkii				X						

Appendix 1C. (continued)

	Family	Genus	Species	Drainage system									
				KA	SU	SL	CU	CL	PA	TA	SI	CS	KB
70.	Loricariidae	Lithoxus	lithoides				X						
71.	Loricariidae	Lithoxus	spec.	X			X						
72.	Loricariidae	Loricaria	cataphracta		X			X					
73.	Loricariidae	Loricariichthys	brunneus	X			X						
74.	Loricariidae	Loricariichthys	maculatus					X					
75.	Loricariidae	Metaloricaria	paucidens nijsseni	X	X						X		
76.	Loricariidae	Metaloricaria	paucidens paucidens							X			
77.	Loricariidae	Pseudancistrus	barbatus				X						
78.	Loricariidae	Rineloricaria	stewarti	X	X		X				X		
79.	Loricariidae	Xenocara	gymnorhynchus				X						
80.	Pimelodidae	Brachyplatystoma	filamentosum			X							
81.	Pimelodidae	Brachyplatystoma	vaillantii			X		X					
82.	Pimelodidae	Hemisorubim	platyrhynchos		X	X							
83.	Pimelodidae	Heptapterus	longior	X	X		X				X		
84.	Pimelodidae	Heptapterus	surinamensis		X								
85.	Pimelodidae	Heptapterus	tapanahoniensis							X			
86.	Pimelodidae	Imparfinis	minutus	X	X		X			X	X		
87.	Pimelodidae	Microglanis	poecilus	X	X		X		X				X
88.	Pimelodidae	Microglanis	secundus		X					X	X		
89.	Pimelodidae	Pimelodella	cristata	X	X		X		X	X	X		
90.	Pimelodidae	Pimelodella	macturki				X						X
91.	Pimelodidae	Pimelodella	procera							X			
92.	Pimelodidae	Pimelodella	spec.1							X			
93.	Pimelodidae	Pimelodella	spec.2							X			
94.	Pimelodidae	Pimelodus	albofasciatus					X			X		
95.	Pimelodidae	Pimelodus	blochii			X	X	X					
96.	Pimelodidae	Pimelodus	heteropleurus				X						
97.	Pimelodidae	Pimelodus	ornatus	X	X		X				X		
98.	Pimelodidae	Pimelodus	spec. (Kaw)						X			X	X
99.	Pimelodidae	Phractocephalus	hemioliopterus								X		
100.	Pimelodidae	Pseudopimelodus	nigricauda	X	X					X	X		
101.	Pimelodidae	Pseudopimelodus	raninus raninus		X		X						
102.	Pimelodidae	Pseudopimelodus	zungaro bufonius		X						X		
103.	Pimelodidae	Pseudopimelodus	fasciatum		X	X	X	X		X	X	X	X
104.	Pimelodidae	Pseudopimelodus	tigrinum							X			
105.	Pimelodidae	Rhamdia	quelen	X	X	X	X	X	X	X	X	X	X
106.	Trichomycteridae	Ochmacanthus	flabelliferus	X			X						
107.	Trichomycteridae	Trichomycterus	guianense	X			X						
108.	Trichomycteridae	Trichomycterus	hasemani	X									

KA Kabalebo river
SU Upper Suriname river
SL Lower Suriname river
CU Upper Corantijn river
CL Lower Corantijn river
PA Para river
TA Tapanahoni river
SI Sipaliwini river
CS Cusewijne river
KB Kaburi creek

Appendix 1D. Cichlidae from Suriname. Species marked with * are probably introduced

	Family	Genus	Species	Drainage system									
				KA	SU	SL	CU	CL	PA	TA	SI	CS	KB
1.	Cichlidae	Aequidens	paloemeuensis							X			
2.	Cichlidae	Aequidens	tetramerus	X	X		X		X	X	X		
3.	Cichlidae	Apistogramma	ortmanni	X			X						X
4.	Cichlidae	Apistogramma	steindachneri	X	X		X		X				
5.	Cichlidae	Chaetobranchus	flavescens						X			X	
6.	Cichlidae	Cichla	ocellaris		X		X	X				X	
7.	Cichlidae	Cichlasoma	bimaculatum	X	X	X							X
8.	Cichlidae	Cleithracara	maronii					X	X	X		X	
9.	Cichlidae	Crenicichla	albopunctata							X			
10.	Cichlidae	Crenicichla	lugubris	X									
11.	Cichlidae	Crenicichla	multispinosa		X					X			
12.	Cichlidae	Crenicichla	nickeriensis	X			X	X					X
13.	Cichlidae	Crenicichla	saxatilis	X	X	X	X	X	X			X	X
14.	Cichlidae	Crenicichla	sipaliwini								X		
15.	Cichlidae	Geophagus	brachybranchus	X			X	X					
16.	Cichlidae	Geophagus	harreri							X			
17.	Cichlidae	Geophagus	surinamensis	X	X	X	X	X					X
18.	Cichlidae	Guianacara	owroewefi		X					X			
19.	Cichlidae	Guianacara	sphenozona	X			X				X		X
20.	Cichlidae	Krobia	guianensis	X	X	X	X		X		X	X	X
21.	Cichlidae	Krobia	itanyi	X						X			X
22.	Cichlidae	Mesonauta	festivus*						X				
23.	Cichlidae	Nannacara	anomala		X	X	X	X	X	X		X	X
24.	Cichlidae	Pterophyllum	scalare*						X				X
25.	Nandidae	Polycentrus	schomburgki	X	X	X		X	X			X	X

KA Kabalebo river
SU Upper Suriname river
SL Lower Suriname river
CU Upper Corantijn river
CL Lower Corantijn river
PA Para river
TA Tapanahoni river
SI Sipaliwini river
CS Cusewijne river
KB Kaburi creek

Appendix 1E. Other groups of freshwater fishes from Suriname

	Family	Genus	Species	Drainage system									
				KA	SU	SL	CU	CL	PA	TA	SI	CS	KB
1.	Belonidae	Potamorrhaphis	guianensis					X				X	
2.	Cyprinodontidae	Rivulus	agilae						X				
3.	Cyprinodontidae	Rivulus	cf. geayi							X			
2.	Cyprinodontidae	Rivulus	holmiae							X			
3.	Cyprinodontidae	Rivulus	urophthalmus	X		X	X	X	X				X
4.	Poeciliidae	Poecilia	parae			X						X	
5.	Poeciliidae	Poecilia	reticulata					X	X				
6.	Poeciliidae	Poecilia	vivipara			X		X	X				
7.	Poeciliidae	Tomeurus	gracilis			X		X					
8.	Potamotrygonidae	Potamotrygon	hystrix				X	X					
9.	Potamotrygonidae	Potamotrygon	spec.								X		
10.	Synbranchidae	Synbranchus	marmoratus		X			X	X				

KA Kabalebo river
SU Upper Suriname river
SL Lower Suriname river
CU Upper Corantijn river
CL Lower Corantijn river
PA Para river
TA Tapanahoni river
SI Sipaliwini river
CS Cusewijne river
KB Kaburi creek

Appendix 1F. Fish families of marine origin from Suriname

	Family	Genus	Species	Drainage system									
				KA	SU	SL	CU	CL	PA	TA	SI	CS	KB
1.	Anablepidae	Anableps	microlepis			X		X					
2.	Batrachoididae	Batrachoides	surinamensis			X							
3.	Carangidae	Caranx	spec.			X							
4.	Centropomidae	Centropomus	ensiferus			X							
5.	Centropomidae	Centropomus	undecimalis			X							
6.	Clupeidae	Odontognathus	mucronatus			X							
7.	Clupeidae	Rhinosardinia	amazonica					X					
8.	Clupeidae	Rhinosardinia	serrata					X					
9.	Echeneidae	Echeneis	naucrates			X							
10.	Engraulidae	Anchoa	spinifer			X		X					
11.	Engraulidae	Anchovia	surinamensis					X					
12.	Engraulidae	Anchoviella	brevirostris					X					
13.	Engraulidae	Anchoviella	guianensis					X					
14.	Engraulidae	Anchoviella	lepidentostole					X					
15.	Engraulidae	Cetengraulis	edentulus			X							
16.	Engraulidae	Lycengraulis	batesi			X		X					
17.	Engraulidae	Lycengraulis	grossidens			X		X					
18.	Engraulidae	Pterengraulis	atherinoides			X		X					
19.	Exocoetidae	Hyporhamphus	roberti					X					
20.	Gobiidae	Awaous	spec.					X					
21.	Gobiidae	Eleotris	amblyopsis					X					X
22.	Gobiidae	Evorthodus	lyricus					X					
23.	Gobiidae	Gobioides	grahamae					X					
24.	Gobiidae	Gobionellus	boleosoma					X					
25.	Megalopidae	Megalops	atlanticus			X		X	X			X	
26.	Mugilidae	Mugil	incilis			X							
27.	Pomadasyidae	Genatremus	luteus					X					
28.	Sciaenidae	Cynoscion	acoupa					X					
29.	Sciaenidae	Lonchiurus	lanceolatus					X					
30.	Sciaenidae	Macrodon	ancylodon			X		X					
31.	Sciaenidae	Pachypops	fourcroi					X					
32.	Sciaenidae	Plagioscion	spec.			X							
33.	Sciaenidae	Plagioscion	surinamensis					X					X
34.	Sciaenidae	Stellifer	microps			X							
35.	Sciaenidae	Stellifer	rastifer			X		X					
36.	Sciaenidae	Stellifer	stellifer			X							
37.	Sciaenidae	Stellifer	venezuelae					X					
38.	Soleidae	Achirus	gronovii			X							
39.	Soleidae	Achirus	paulistanus			X							
40.	Soleidae	Apionichthys	unicolor					X					
41.	Tetraodontidae	Colomesus	psittacus			X		X					

KA Kabalebo river
SU Upper Suriname river
SL Lower Suriname river
CU Upper Corantijn river
CL Lower Corantijn river
PA Para river
TA Tapanahoni river
SI Sipaliwini river
CS Cusewijne river
KB Kaburi creek

References

(incl. literature used to compose the appendices).

Boeseman, M., 1948. On new, little known, or otherwise interesting species of Surinam Fishes. Proc. Kon. Ned. Akad. Wetenschappen 51: 363–371.

Boeseman, M., 1952. A preliminary list of Surinam fishes not included in Eigenmann's enumeration of 1912. Zool. Med., Leiden 31(17): 179–200.

Boeseman, M. 1953. Scientific results of the Surinam Expedition 1948–1949. Part II. Zoology. No.2. The Fishes (1). Zool. Med., Leiden 32(1): 1–24.

Boeseman, M. 1956. On recent accessions of Surinam fishes. Zool. Med., Leiden 34(12): 183–199.

Boeseman, M. 1968. The genus *Hypostomus* Lacép]ede, 1803, and its Surinam representatives. (Silurformes, Loricariidae). Zool. Verh., Leiden 99: 1–89.

Boeseman, M., 1971. The "comb-toothed" Loricarinae of Surinam, with reflections on the phylogenetic tendencies within the family Loricaridae (Siluriformes, Siluroidei). Zool. Verh., Leiden 116: 1–56.

Carter, G.S. and L.C. Beadle, 1931. The fauna of the swamps of the Paraguayan Chaco in relation to its environment. II. Respiratory adaptations in the fishes. J. Linn. Soc., Lond. (Zool.) 37: 327-368.

Eigenmann, C.H., 1912. The freshwater fishes of British Guiana, including a study of the ecological groupings of species and the relation of the fauna of the plateau to that of the lowlands. Mem. Carnegie Mus. 5 (67): 1–578.

Géry, J.R., 1961. Notes on the Ichthyology of Surinam and other Guianas 7. *Hyphessobrycon georgetti* sp. nov., a dwarf species from southern Surinam. Bull. Aq. Biol. 2(22): 121–128.

Géry, J.R., 1966. Notes on characoid fishes collected in Surinam by Mr. H.P. Pijpers, with descriptions of new forms. Bijdr. Dierk. 35: 101–126.

Géry, J.R., 1972. Poissons characoïdes des Guyanes. I. Généralites. II. Famille des Serrasalmidae. Zool. Verh., Leiden 122: 1–250.

Géry, J., 1984. The fishes of Amazonia. In: H. Sioli (ed.), The Amazon. Limnology and landscape ecology of a mighty tropical river and its basin. Dr.W.Junk Publishers, Dordrecht: 353–370.

Goulding, M., 1980. The fishes and the forest. Explorations in Amazonian Natural History. Univ. California Press, Berkeley, 280 pp.

Goulding, M., M.L. Carvalho and E.G. Ferreira, 1988. Rio Negro: rich life in poor water. SPB Academic Publishing, The Hague, 200 pp.

Heitmans, W.R.B., H. Nijssen and I.J.H. Isbrücker, 1983. The mailed catfish genus *Lasiancistrus* Regan, 1904, from French Guiana and Surinam, with descriptions of two new species. (Pisces, Siluriformes, Loricariidae) Bijdr. Dierk. 53(1): 33–48.

Hoedeman, J.J., 1957. Notes on the Ichthyology of Surinam (Dutch Guiana) 4. Additional records of Siluriform Fishes(1). Beaufortia 6(71): 147–160.

Hoedeman, J.J., 1961. Notes on the Ichthyology of Surinam and other Guianas. 8. Additional records of Siluriform Fishes (2). Bull. Aq. Biol. 2(23): 129–139.

Hoedeman, J.J., 1962. Notes on the Ichthyology of Surinam and other Guianas. 9. New records of Gymnotid fishes. Bull. Aq. Biol. 3(26): 53–60.

Huebner, E. and G. Chee, 1978. Histological and ultrastructural specialization of the digestive tract of the intestinal air breather *Hoplosternum thoracatum* (Teleost). J. Morphol. 157: 301–328.

Isbrücker, I.J.H., 1981. Revision of *Loricaria* Linnaeus, 1758 (Pisces, Siluriformes, Loricariidae. Beaufortia 31(3): 51–96.

Isbrücker, I.J.H. and H. Nijssen, 1982. New data on *Metaloricaria paucidens* from French Guiana and Surinam (Pisces, Siluriformes, Loricariidae). Bijdr. Dierk. 52(2): 155–168.

Junk, W.J., 1984. Ecology, fisheries and fish culture in Amazonia. In: H. Sioli (ed.), The Amazon. Limnology and landscape ecology of a mighty tropical river. Dr.W.Junk Publishers, Dordrecht: 443–476.

Kramer, D.L., C.C. Lindsey, G.E. Moodie and E.D. Stevens, 1978. The fishes and the aquatic environment of the Central Amazon basin, with particular reference to respiratory patterns. Can. J. Zool. 56: 717–729.

Kramer, D.L. and M. McClure, 1982. Aquatic surface respiration, a widespread adaptation to hypoxia in tropical freshwater fishes. Env. Biol. Fish. 7: 47–55.
Krebs, C.J., 1989. Ecological Methodology. Harper and Row, New York, 654 pp.
Knöppel, H.A., 1970. Food of central Amazon fishes. Contribution to the nutrient-ecology of the Amazonian rain-forest streams. Amazoniana 2: 257–352.
Kullander, S.O. and H. Nijssen, 1989. The Cichlids of Surinam. E.J. Brill, Leiden, 256 pp.
Le Bail, P.Y., P. Planquette and J. Géry, 1984. Cle de determination des poissons continentaux et cotiers de Guyane. Fasc. IV – Especes dulcaquicoles non Siluriformes. Bull. 9. Inst. Nat. Recherche Agronomique, Kourou, 97 pp.
Lowe-McConnell, R.H., 1964. The fishes of the Rupununi savanna district of British Guiana, South America. Part 1. Ecological groupings of fish species and the effects of the seasonal cycle on the fish. J. Linn. Soc. (Zool.) 45: 103–144.
Lowe-McConnell, R.H., 1987. Ecological studies in tropical fish communities. Cambridge Univ. Press, Cambridge, 382 pp.
Mees, G.F., 1974. The Auchenipteridae and Pimelodidae of Suriname. (Pisces, Nematognathi). Zool. Verh., Leiden 132: 1–256.
Mees, G.F., 1985. Further records of Auchenipteridae and Pimelodidae from Suriname (Pisces: Nematognathi). Zool. Med., Leiden 59(21): 239–249.
Mol, J.H.A., (in press). Sustainable exploitation of neotropical fish resources: aquaculture trials with three callichthyid catfishes, *Hoplosternum littorale*, *Hoplosternum thoracatum* and *Callichthys callichthys*. In: Proceedings of the 2nd Joint Conference of the University of Guyana and the University of Suriname.
Nijssen, H., 1970. Revision of the Surinam catfishes of the genus *Corydoras* Lacepede, 1803 (Pisces, Siluriformes, Callichthyidae). Beaufortia 18(23): 1–75.
Nijssen, H., I.J.H. Isbrücker and J. Géry, J., 1976. On the Species of *Gymnorhamphichthys* Ellis, 1912, Translucent Sand-dwelling Gymnnotid Fishes from South America (Pisces, Cypriniformes, Gymnotoidei). Stud. Neotrop. Fauna Env. 11: 37–63.
Ploeg, A., 1987. Review of the cichlid genus *Crenicichla* Heckel, 1840 from Surinam, with descriptions of three new species (Pisces, Perciformes, Cichlidae). Beaufortia 37(5): 73–98.
Power, M.E., 1984. Grazing responses of tropical freshwater fishes to different scales of variation in their food. In: T.M. Zaret (ed.), Evolutionary ecology of neotropical freshwater fishes. Dr.W.Junk Publishers, The Hague: 25–37.
Richter, C.J.J. and H. Nijssen, 1980. Notes on the fishery potential and fish fauna of the Brokopondo Reservoir (Surinam). Fish. Mgmt. 11(3): 119–130.
Richter, C.J.J., 1980. Exploitatiemogelijkheden van het Brokopondo Reservoir. In: W.L.T. Van Densen and C.J.J. Richter (eds.), Algemene Visserijkunde. LH Algemene visteelt en visserij, Wageningen: 235–250.
Rojas-Beltran, R., 1984. Cle de determination des poissons continentaux et cotiers de Guyane. Fasc. II – Siluriformes. Bull. 7. Inst. Nat. Recherche Agronomique, Kourou, 63 pp.
Stigchel, J.W.B. van der, 1947. The South American Nematognathi of the museums at Leiden and Amsterdam. Zool. Med., Leiden 27(1): 1–204.
Vari, R.P., 1982. *Hemiodopsis ocellata*, a new hemiodontid characoid fish (Pisces: Characoidea) from Western Surinam. Proc. Biol. Soc. Wash. 95(1): 188–193.
Vari, R.P., 1982. Environmental impact of the Kabalebo Project. Final Report. Inventory, biology and ecology of the fishes in the Corantijn River system, Suriname. Ministry of Development, Paramaribo, 129 pp.
Vari, R.P., 1985. A new species of *Bivibranchia* (Pisces: Characiformes) from Surinam, with comments on the genus. Proc. Biol. Soc. Wash. 98(2): 511–522.
Vari, R.P. and H. Ortega, 1986. The Catfishes of the Neotropical Family Helogenidae (Ostariophysi: Siluroidei). Smiths. Contr. Zool. 442: 1–20.

PART II: Case studies

During the sixties and the seventies many ecological and ethological studies of certain animal species were conducted. These research efforts were concentrated on key species for nature conservation, which are mainly birds and mammals (for instance the classical studies by Trial (1983) on the cock-of-the-rock, by Van Roosmalen (1980) on the black spider monkey, and by Duplaix (1980) on the giant otter). Few key species have freshwaters as their main habitat and consequently few studies involved freshwater animals. In this compilation four case studies will be presented, all of which were initiated quite recently.

Although dragonflies spend the adult part of their life out of the water, the aquatic larval phase is sufficiently important to justify their treatise in a book on freshwaters. The taxonomy of Surinamese dragonflies has been studied thoroughly by Belle (e.g. 1970) and Geijskes (e.g. 1970). The microdistribution and microhabitat selection of *dragonflies of a blackwater creek system* will be analyzed in Chapter 9.

Armoured catfish are a group well-known to aquarists with whom the smaller species are very popular. Three species grow to a larger size and are prepared as tasty dishes in certain countries. In order to breed these *armoured catfish*, a preliminary study on their niche-segregation was conducted, yielding interesting differences in *nesting ecology* to be illustrated in Chapter 10.

One of the most remarkable fresh-water ecosystems in Suriname is that of the floating meadows. This ecosystem is well studied in Brazil (e.g. Junk 1973), but hardly so in Suriname. Even its vegetation has not been investigated thoroughly. The fauna of floating meadows is largely unknown. The only study available is on the *herpetofauna* (amphibians, reptiles) of the Para River, which will be presented in Chapter 11. A comparison will be made with *floating meadows* in other parts of South America.

The only recent order of reptiles that is fully aquatic is the Crocodylia. In Suriname there are three species of caimans. For many years they have been studied by several researchers. In Chapter 12 the influence of the water level on *the annual cycle of the spectacled caiman* will be shown. The water level determines the annual and/or life cycle of almost all organisms in the freshwaters of the humid tropics. Unfortunately, very little research has been

Paul E. Ouboter (ed.), Freshwater Ecosystems of Suriname, 155–156.

done on this subject in Suriname, but spectacled caimans clearly demonstrate the principle.

References

Belle, J., 1970. Studies on South American Gomphidae (Odonata). Stud. Fauna Suriname Guyanas 11: 1–158.

Duplaix, N., 1980. Observations on the ecology and behavior of the giant river otter *Pteronura brasiliensis* in Suriname. Rev. Ecol. (Terre Vie) 34: 495–619.

Geijskes, D.C., 1970. Generic characters of the South American Corduliidae, with descriptions of the species found in the Guyanas. Notes on Odonata of Suriname XI. STS 12(44): 1–42.

Junk, W.J., 1973. Faunistisch-ökologische Untersuchungen als Möglichkeit der Definition von Lebensräumen, dargestellt an Uberschwemmungsgebieten. Amazoniana 4(3): 263–271.

Roosmalen, M.G.M. van, 1980. Habitat preferences, diet, feeding strategy and social organization of the black spider monkey (*Ateles paniscus paniscus* Linnaeus 1758) in Surinam. Unpubl. report R.I.N. 80(13): 1–175.

Trial, P.W., 1983. Cock-of-the-rock. Jungle Dandy. Nat. Geogr. 164(6): 830–839.

9. Dragonflies of a black-water creeksystem

MARCEL WASSCHER

1. Introduction

Adult dragonflies are beautiful and colourful and thus an intrinsically attractive group of insects. Since there are not too many problems with the identification of adult dragonflies, it is a group suitable to help to give insight in running water ecosystems in the tropics. Dragonflies are semi-aquatic: the larvae live under water while the adults are aerial creatures. There are approximately 5,000 species in the world. One of these is the spectacular giant damselfly – the largest present-day dragonfly species with a wingspan of 18 cm – which can amaze visitors of the Surinamese jungle by their fairy-like flight. In this chapter a case study is presented on the dragonfly fauna of a black-water creeksystem in Suriname. The main purpose of this study is to define zonation in the distribution of adult dragonflies on a running water ecosystem.

Running waters in Suriname can be roughly divided into white-water, clear-water and black-water streams (see Chapter 4). Black-water creeks are more or less transparent, with olive brown to coffee brown water, and a pH ranging between 3.8 and 4.9 (Sioli 1984). In Suriname these creeks are predominant in the Savanna Belt (Cover Landscape).

2. Study Area and Methods

The study has been conducted at the Sipari Creek and Tibiti River (Fig. 1). The area is situated north of the road from Zanderij to Witagron, 75 km west of Zanderij (5°48′N, 55°85′W), and consists of marsh and swamp forests as well as rather undisturbed rain forests. Some of the drier parts of the forest are used for forestry activities. A small foresters' camp of about 100 man (Kabo-Bruynzeel) was situated in the immediate surroundings of the study area during the period 1985 to 1990, but very few people live permanently in this part of Suriname.

On 14 days between 8 February and 30 March 1989 (short dry season) adult dragonflies were collected on the Sipari Creek and the Tibiti River. The dragon-

Paul E. Ouboter (ed.), Freshwater Ecosystems of Suriname, 157–166.

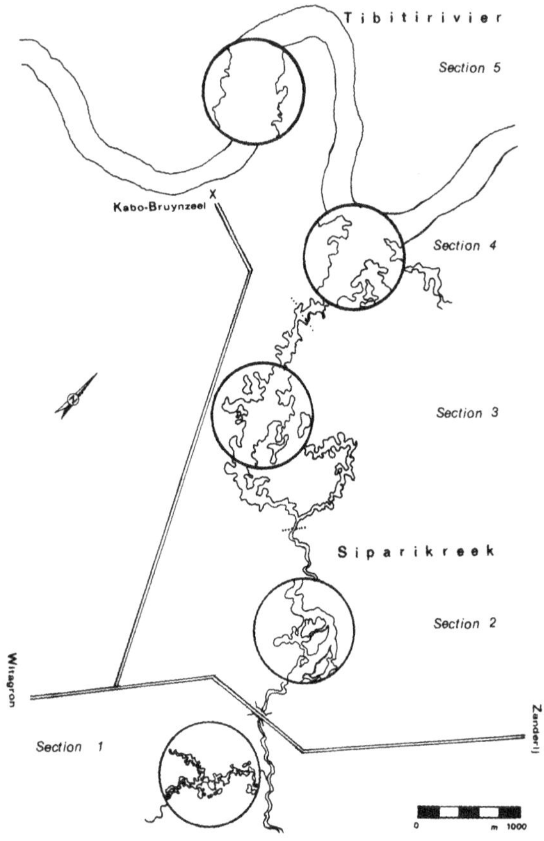

Figure 1. Schematic map of the black-water creeksystem of the Sipari Creek and Tibiti River. Sections: 1: tributary brooklet; 2: central section Sipari Creek; 3: downstream section Sipari Creek; 4: junction section Sipari Creek; 5: Tibiti River. Within the circles a stretch of the section is enlarged three times. The borders between section 2 and 3, and sections 3 and 4 are indicated by a dotted line. Straight lines are the roads in the area.

flies were captured along the stream from the shoreline and out of a boat by means of a sweeping net.

In order to gather information on zonation, the running waters in the area were divided longitudinally into five habitat sections (Fig. 1 and Table 1). The lengths of these sections varied from 1 to 4 km. The time spent in each section varied between 3 and 34 hours (Table 1). From these sections the tributary brooklets (Fig. 2, section 1) and the large Tibiti River (Fig. 3, section 5) were the most distinct zones. The three zones of the Sipari Creek more or less merge into one another.

The name black-water creeksystem is related to the reddish brown colour of the water in the brooklets and central section of the Sipari Creek. The water of the creek and the river is acidic (a pH value of approximately 5). There were very

Table 1. Physical characters of 5 habitat sections in the Sipari Creek and Tibiti River, and the time spent collecting in each section. Sections: 1: tributary brooklet; 2: central section Sipari Creek; 3: downstream section Sipari Creek; 4: junction section Sipari Creek; 5: Tibiti River. Legend: temp. = temporary, perm. = permanent. (Most values are estimated).

	1.	2.	3.	4.	5.
min. width (m)	0.5	5	15	30	60
max. width (m)	2	20	40	50	80
min. flow rate	0.1	0.1	0.1	0	0
max. flow rate	0.2	1.0	0.4	0.2	0.6
permanence	temp.	temp.	perm.	perm.	perm.
bottom	sand	sand	clay	clay	clay
influence sea	none	none	none	some	some
% shadow (at noon)	95	75	25	15	5
time spent collecting (h)	12	34	16	3	17

few water plants, except for some water-hyacinths (*Eichhornia crassipes*), which formed at certain locations a dense vegetation along the banks of the Tibiti River. Although the study area is situated 80 km from the sea, low tide and high tide affect flow rates and water level of the Tibiti River, and of the Sipari Creek where it confluences with the Tibiti River.

The numbers per section were estimates of dragonflies which were easy to catch and/or identify (see Table 2). The method employed is based on a method generally used for vegetation surveys. Abundances were based on rough estimations of the number of adult dragonflies present per stretch of water. For zygopteran species a stretch of water was 25 metres long and for small anisopteran species a stretch of water was 50 metres. Abundant was defined as more than 11 specimens per stretch, frequent 4–10, and occasionally 1–3. Rare being only one or two specimens for the whole section.

Figure 2. The tributary brooklet (section 1).

Figure 3. The Tibiti River (section 5)

The definitive identification of the collected dragonflies took place in the Netherlands. The main source used for this identification was the reference collection present at the National Museum of Natural History (RMNH) in Leyden. The most important part of the reference collection for the purpose of this study was the Geijskes collection for Suriname. Moreover, some dragonflies could be identified by means of literature on neotropical Odonata (for instance, *Epipleoneura pereirai*).

Most of the collected specimens are in the National Zoological Collection at the University of Suriname (Paramaribo).

3. Abundance

In the study area 236 specimens (155 Zygoptera and 81 Anisoptera) were collected. The number of reliably identified species recorded for the study area is 40. Seven species collected in the study area are still unidentified.

In Table 2 only those species are listed for which the abundance could be estimated. This is the case for nearly all Zygoptera. The one exception is the giant damselfly *Mecistogaster ornata* which is hard to catch. Numbers could also not be estimated for three red coloured *Neoneura* species (*rubriventris*, A and B) separately, and the abundances of these are lumped together in Table 2. The only anisopteran species for which the abundance could be properly estimated are three small, rather easy to catch species (*Perythemis cornelia, Perythemis lais*, and *Oligoclada abbreviata*), and three easy to recognize species: two dark winged species (*Zenithoptera fasciata* and *Diastatops pullata*), and one species with dark wingtips (*Uracis ovipositrix*). The species listed in Table 2 give an impression of a selection of dragonflies useful for answering ecological

Table 2. Dragonflies collected on a black-water creeksystem (Sipari Creek and Tibiti River) in February – March 1989: the species from which the abundance could be estimated. Sections: 1: tributary brooklet; 2: central section Sipari Creek; 3: downstream section Sipari Creek; 4: junction section Sipari Creek; 5: Tibiti River; A = abundant, F = frequent, O = occasionally, R = rare, + = recorded, abundance of separate species unknown. (Anisopteran species are marked with a, after their names).

Species	1.	2.	3.	4.	5.
AA	R				
Metaleptobasis brysonima	R				
Metaleptobasis A	R				
Hetaerina moribunda	O				
Argia A	O				
Argia no 13	O				
Argia no 15	F				
Oxystigma williamsoni	R	O			
Metaleptobasis fernandezi	O	O			
Epipleoneura lamina	R	O	O		
Acanthagrion apicale		R			
Hetaerina laesa		R			
Argia no 14		O			
Perythemis cornelia'		R			
Zenithoptera fasciata'		R	R		
Uracis ovipositrix'		O	F	O	
Hetaerina caja dominula		F	O	R	
Neoneura (rubriventris + A + B)		F	F	O	O
Neoneura rubriventris		+	+	+	
Neoneura A			+	+	
Neoneura B					+
Neoneura joana		O	F	O	A
Acanthagrion indefensum		R	O	F	F
BA				R	
Aeolagrion flammeum				O	
Diastatops pullata'				O	F
Perythemis lais'				F	O
Oligoclada abbreviata'					O
Epipleoneura pereirai					O
Neoneura bilinearis					O
Ischnura fluviatilis					F

questions like zonation. The other dragonflies collected (or seen) in the study area, of which the abundance could not be estimated are listed in Table 3.

The small numbers of all dragonfly species present in the study area, are striking. The low densities found here are not always the case for the species occurring in the study area. This can be illustrated by the fact that *Acanthagrion indefensum* can be very abundant on woodland creeks in French Guiana (Machet 1989).

In the surroundings beyond the study area seven species more could be collected on stagnant waters. These species are not included in the numbers mentioned above and in the tables. On an oxbow lake (old river arm) along the

Table 3. Dragonflies collected (or seen) on a black-water creeksystem (Sipari Creek and Tibiti River) in February – March 1989: additional recorded species. Sections: 1: tributary brooklet; 2: central section Sipari Creek; 3: downstream section Sipari Creek; 4: junction section Sipari Creek; 5: Tibiti River; c = collected, s = seen. (Anisopteran species are marked with a, after their names).

Species	1.	2.	3.	4.	5.
Mecistogaster ornata		s			
Aphylla dentata'		c			
Progomphus brachycnemis'		c			
Dythemis multipunctata'		c			
Elga leptostylla'		c			
Misagria parana'		c			
Oligoclada walkeri'		c			
Orthemis attenuata'		c			
Uracis A'		c			
Erythrodiplax famula'		c			
Uracis imbuta'		c	c		
Anatya guttata'		c	c		c
Tricanthagyna septima'		c			c
Gynacantha nervosa		c			c
Staurophlebia r.reticulata'		c	c		
Staurophlebia spec.		s	s	s	s
Orthemis ferruginea'				c	
Micrathyria spinifera'				c	
Oligoclada pachystigma'				c	

Tibiti River: *Aeolagrion dorsale*. On a waste land used for transshipment of logs at Kabo-Bruynzeel, near the Tibiti River: *Ischnura capreola*, *Erythrodiplax umbrata*, *Erythrodiplax basalis*, *Erythrodiplax fusca*, *Pantala flavescens*, and *Uracis fastigiata*.

4. Zonation

In the black-water creeksystem zonation in the distribution of adult dragonflies is obvious (Table 2). Most species (18) are found exclusively in one habitat section while none of the species occurs in all five sections. Only two species have been recorded in four sections, four species in three sections, and five species in two sections. The two species recorded in four sections, *Neoneura joana* and *Acanthagrion indefensum*, do both only occur in a small part of South America: Venezuela and the Guianas. Although they are rather widely distributed in the creeksystem of the Sipari Creek, they are not widely distributed in the Neotropics as a whole.

Zonation of fishes and aquatic insects is a well-known phenomenon in running water ecosystems. This study implies the phenomenon also holds for the distribution of dragonflies on a black-water creeksystem in the tropical rainforest of Suriname. Although some studies on the distribution of dragonflies do suggest that zonation may occur in these systems (for instance in Liberia (Lempert 1988), and in Indonesia (van Tol 1987)) no study on zonation of dragonflies in the tropics has been found to compare the findings of this study with.

5. Habitat Selection

Little is known on the ecology of dragonflies (as well as other aquatic invertebrates) in the tropics. This emphasizes the importance of discussing some ecological features of the dragonflies occurring in this black-water creeksystem. It must be said, however, that the distribution of dragonflies is not only defined by the local aquatic environment (important for the larvae, and oviposition by the adult females), but also by the terrestrial surroundings (important for foraging, resting and finding a partner for the adults).

In Table 1 the minimum and maximum width is given of the water in the various sections. These values are only valid for the main stream. In most sections there are several tributaries (Fig. 1). As the bank of the first two sections is sandy, the tributaries of these sections are generally separated from the main stream by ridges of sand, forming small backwaters. In these shallow backwaters species are found of the genus *Metaleptobasis*. The species of this genus have been found more than once in syntopy. Further downstream no specimens of this genus have been collected, although shallow backwaters are present there, too.

Other species are restricted to sections 1 and 2 as well, for instance *Oxystigma williamsoni*, and the species of the genus *Argia*. Unlike the species of the genus *Metaleptobasis*, these species in general can be found in the direct surroundings of the main stream.

Of the – rheophilous (larvae live only in running waters) – genus *Hetaerina*, *H. caja dominula* was the most common species. Although this species was not recorded on the Tibiti River within the study area, it has been collected in considerable numbers on this river upstream from the study area at Tibitisoela (a rapid in the river). This species did not occur on the tributary brooklet (Fig. 2, section 1). Instead its congener *Hetaerina moribunda* was found here. Judging from other records concerning the latter species in Suriname, brooklets are its most important habitat. From the third *Hetaerina* in the study area (*H. laesa*) only one male was collected on a tributary of the central section (Fig. 4, section 2) of the Sipari Creek.

The presence of shaded and unshaded locations can influence the (micro)distribution of the adults within a section. The first two sections are mainly shaded. Adult specimens of *Epipleoneura lamina*, and *E. pereirai* prefer shaded spots. The first is most abundant in section 2, but occurs in section 3 (Fig. 5) mainly in shaded localities. The distribution of *E. pereirai* in the study area is confined to the Tibiti River, where it only occurs in the shadow of large trees near the river bank.

Figure 4. The central section of the Sipari Creek (section 2)

Figure 5. The downstream section of the Sipari Creek (section 3)

The spatial separation of *Epipleoneura lamina* and *E. pereirai* in general appears to be similar in other parts of the Interior of Suriname. Spatial separation, however, does not occur in the Savanna Belt, where both species regularly can be found together (Wasscher 1991).

Two anisopteran species (*Oligoclada abbreviata*, and the dark-winged *Diastatops pullata*) prefer the open, unshaded places, for instance parts of the Tibiti River overgrown with water-hyacinths.

On the shaded creek in section 2, only locally sunny spots are present in places were trees have fallen down. In only one locality, in the direct surroundings of the bridge, a large clearing is present in this section. This locality near the bridge differs in dragonfly composition compared to the rest of section 2. For both *Acanthagrion apicale* and *Erythrodiplax famula* this is the only place where these species were recorded near the creek. Individuals of the third species (*Acanthagrion indefensum*) were only recorded in section 2 near the bridge. Further downstream, in open parts of sections 3, 4, and 5 they occur in fairly large numbers. The implication that the dragonfly population composition near bridges differs from other (shaded) parts of the creek can be important in the interpretation of habitat requirements for dragonflies from collected material.

The habitat preferences of the dragonflies that emerge from this study do on one hand correspond with those mentioned in literature, while on the other hand they add to our knowledge of dragonfly ecology. For instance, in the case of *Oxystigma williamsoni* the habitat in the study area corresponds with the findings of Geijskes (1976), namely in lowland creeks and slowly running creeks in the Interior, where it preferred shaded parts. For the two gomphids *Aphylla dentata* and *Progomphus brachycnemis* the bottom structure is of importance, because the larvae of these species burrow in sandy soils of creeks (J. Belle pers. comm.). This corresponds with the occurrence of these species in section 2.

In the case of *Oligoclada abbreviata*, this species has a broader habitat preference than suggested by Geijskes (1984), who found the species "inhabiting rocky places in and along rivers in the interior". In the study area this species was only found on the Tibiti River, where no rocky places are found. The limitation "near rocky places" is in Suriname as a whole not true.

Neoneura bilinearis, a species only found on the river during this study, is known from creeks in the coastal area of neighbouring countries (Machet 1989), while Williamson (1917) even found the species on a small muddy brooklet. The species seems to have a preference for shaded parts of water influenced by the tides.

At least 260 species of dragonflies are known from Suriname (Geijskes 1967). In this study one species (*Aeolagrion flammeum*) could be recorded new for the fauna of Suriname. Yet, in view of the number of unidentified species present in the RMNH collection in Leyden (for instance, those found in this study), much taxonomical work on the dragonflies of Suriname still remains to be done. This will not, however, cross ecological studies which in the future will prove their importance in understanding ecological relationships in the tropical rainforests, one of the most complex ecosystem on earth.

Summary

The distribution of adult dragonflies was studied in 1989 in a black-water creeksystem in the Interior of Suriname. During 14 days in the field, 47 species of dragonflies were collected in the study area. In the distribution of these species, zonation is obvious. The restricted distribution of some dragonfly species in the creeksystem is discussed in the view of some selected physical factors.

Acknowledgements

This study would have been impossible without the large amount of work on the dragonflies of Suriname carried out by the late Dr. D.C. Geijskes. I would like to thank the following persons in Suriname: Jos Beerlink who was my host and made the field work possible, Andre Jomi and Maclaen Sabayo, my local guides; in Paramaribo, Paul Ouboter, who arranged working accommodation for me in the Zoological Collection of the Anton de Kom University of Suriname, and Pieter Teunissen for his enthusiastic help and discussions. In the Netherlands, I would like to thank the Uyttenboogaart-Eliasen Foundation which gave me financial support to elaborate the gathered data in Leyden and Jan van Tol, curator of the dragonfly collection of the RMNH in Leyden, for giving me the opportunity to work in the collection, and for critically reading the manuscript of this chapter.

References

Geijskes, D.C., 1967. De insektenfauna van Suriname, ook vergeleken met die van de Antillen, speciaal wat betreft de Odonata. Ent. Ber. Amsterdam, 27: 69–72.

Geijskes, D.C., 1976. The genus *Oxystigma* Selys, 1862 (Zygoptera: Megapodagrionidae). Odonatologica 5(3): 213–230.

Geijskes, D.C., 1984. What is *Oligoclada abbreviata* (Rambur, 1842)? (Odonata: Libellulidae). Zool. Meded., Leiden 58 (12): 175–185.

Lempert, J., 1988. Untersüchungen zur Fauna Ökologie und zum Fortpflazungsverhalten von Libellen (Odonata) an Gewassern des tropischen Regenwaldes in Liberia, Westafrika. Diplomarbeit. Bonn, Rheinischen Friedrich Wilhelms Universitat, 238 pp.

Machet, P., 1989. Contribution à l'étude des odonates de GuyanaFrançaise. 1. Zygoptera. Opuscula Zoologica Fluminensia 40: 1–16.

Sioli, H. (ed.), 1984. The Amazon: limnology and landscape ecology of a mighty tropical river and its basin. Dordrecht, Dr. W. Junk Publicers, 763 pp.

Tol, J. van, 1987. The Odonata of Sulawesi (Celebes), Indonesia: an introduction. Adv. Odonatology 3: 147–155.

Wasscher, M., 1991. Ecological notes on the genus *Epipleoneura* in Suriname. Abstr. Papers XI Int. Symp. Odonatol.: 32–33.

Williamson, E.B., 1917. The genus *Neoneura* (Odonata). Trans. Am. Entom. Soc. 18: 211–246.

10. Structure and function of floating bubble nests of three armoured catfishes (Callichthyidae) in relation to the aquatic environment

JAN H.A. MOL

1. Introduction

Armoured catfishes of the family Callichthyidae are represented in Suriname by 17 small-sized fishes of the well-known genus *Corydoras* (Nijssen 1970) and three larger food fishes of the genera *Hoplosternum* and *Callichthys* (Hoedeman 1952). The latter three species, *Hoplosternum littorale* (Hancock 1828), *Hoplosternum thoracatum* (Cuvier and Valenciennes 1840) and *Callichthys callichthys* (Linnaeus 1758), are the most preferred and expensive food fishes of Suriname. They are the only fish species in the country protected by law in their main reproductive season (Rondeel 1965). Still, overfishing, human-induced habitat modifications and the introduction of the exotic Tilapia *Oreochromis mossambicus* present serious threats to the survival of these popular fishes. Especially *H. littorale* is in danger of local extinction, its distribution being confined to the more densely populated Coastal Plain (Mol in prep.). In 1987 the University of Suriname started a research programme on the three larger callichthyids with the ultimate aim of producing the fishes on a large scale, thus removing some of the pressure that now rests on the natural stocks (Mol in press). *H. littorale* also attracted considerable regional attention as food fish with an aquacultural potential (Lowe-McConnell 1984; Luquet *et al.* 1989).

Hoplosternum littorale, *Hoplosternum thoracatum* and *Callichthys callichthys*, together with two Amazonian species of the genus *Dianema*, make up the subfamily Callichthyinae (Gosline 1940; Hoedeman 1952; Burgess 1989). Contrary to the species belonging to the second callichthyid subfamily Corydoradinae, the Callichthyinae are aphrophilic nest-guarding fishes (Balon 1975). The aphrophilic reproductive guild is characterized by the building and guarding of floating bubble nests. Carter and Beadle (1931) consider the structure of the nests of *H. littorale* of importance in providing an oxygen-rich environment for the developing eggs and early larvae, but alternative advantages of the foam-nesting habit have also been suggested (Balon 1975). As Carter and Beadle (1931) point out, *Hoplosternum* and *Callichthys* often live in tropical standing waters deprived of oxygen. Their description of the swamps of

Paul E. Ouboter (ed.), Freshwater Ecosystems of Suriname, 167–197.

the Paraguayan Chaco (Carter and Beadle 1930) closely corresponds with the water conditions of the standing waters of the Surinamese Coastal Plain (see Mol, Chapter 7). Another well established adaptation of *Hoplosternum* and *Callichthys* to the oxygen-depleted aquatic environment is their obligatory air-breathing habit and intestinal respiration (Carter and Beadle 1931; Gee and Graham 1978; Huebner and Chee 1978; Kramer *et al.* 1978).

Floating bubble nests of *H. littorale* are associated with the annual flooding of shallow swamps (Carter and Beadle 1931; Singh 1978; Machado-Allison and Zaret 1984). Detailed descriptions of the nests of *H. littorale* are provided by Carter and Beadle (1931) and Machado-Allison and Zaret (1984). The nests of *H. thoracatum* and *C. callichthys* are only known from observations in aquariums (see Breder and Rosen 1966). Contrary to *H. littorale*, which in Suriname is collected only in Coastal Plain standing waters, *H. thoracatum* and *C. callichthys* also inhabit the lotic environment of small creeks in the interior of Suriname (Mol in prep.). The aquatic habitat of these headwater streams, which drain the very poor soils of the Precambrium Guiana Shield, is very different from the Coastal Plain swamps and swamp forests. Leentvaar (1975), following the terminology of Sioli (1950), characterizes these creeks as clear water, very low in dissolved nutrients and minerals (e.c.) and often saturated with oxygen (also see Haripersad and Ouboter, Chapter 4).

The nests of *H. littorale* were studied during a five-year period (1987–92) in a *Typha* dominated swamp at Weg Naar Zee, only 2 km south of the Atlantic Ocean, and in an *Eleocharis* dominated swamp at Lelydorp, 20 km south of Paramaribo. The nests of *H. thoracatum* and *C. callichthys* were studied in 1990 and 1991 in the Lelydorp swamp. Confirmation of the nest building species by rearing the newly hatched larvae for one week was often necessary in order to distinguish nests of *H. thoracatum* from nests of *C. callichthys*. Fecundity of the three species was determined by counting ripening eggs present in the ovaries, preserved with Gilson's fluid (Bagenal and Braum 1978). This study aims at linking the differences in structure of the nests of the three species with differences in their environment.

2. The Nests

2.1. *The Nests of* Hoplosternum littorale

2.1.1. Nest Construction, Oviposition and Incubation of the Eggs

Floating bubble nests of *Hoplosternum littorale* (Sur. soke or hé-ede kwikwi) are built by the male in shallow, herbaceous swamps during the rainy season (Fig. 1). Nests are also abundant in flooded rice fields. In standing waters of the Coastal Plain of Suriname the first nests are usually observed in late December or January when the water level in the swamps reaches about 25 cm (Mol in prep). The reproductive season lasts until July/August and is considerably longer than reported for Venezuela (Machado-Allison and Zaret 1984) and

Trinidad (Singh 1978) but shorter than reported for ponds in French Guiana (Hostache *et al.* 1990). The length of the reproductive season changes with increasing rainfall from Venezuela towards French Guiana (Snow 1976). Within the main reproductive period, nesting of *H. littorale* in the Surinamese Coastal Plain was concentrated in the two rainy seasons: December–February and April–August (Mol in prep.).

Figure 1. Nest of *Hoplosternum littorale* in a *Typha* dominated swamp at Weg Naar Zee, Young Coastal Plain, Suriname. The guarding male is caught with the scoop net.

Although the actual construction of the nest of *H. littorale* was not observed, daily surveys in the swamps early in the morning suggest that nest-building is concentrated in the late evening or during the night. Completed nests without eggs were often observed between 07.00 and 09.00 hours at localities in the swamp where in the afternoon of the day before no nest was present. Gautier *et al.* (1988) describe in detail the construction of the nest of *H. littorale* in tanks, but they do not mention the time of nest-building activities. Luquet *et al.* (1989) state that nest-building of *H. littorale* in fish ponds of French Guiana occurs during the night.

In the swamps of Suriname, oviposition is usually between 10.00 and 14.00 hours on the first day after nest construction and is indicated by more or less violent movements of the nest. I have observed the process of oviposition three times and have indirect confirmation of oviposition between 10.00 and 14.00 hours for another four nests. Local fishermen are well acquainted with the timing of oviposition and catch both male and females by positioning a bottomless basket over the nest as soon as they notice the nest shaking. All females spawn in the nest simultaneously, so the eggs in the nest are always of the same age and developmental stage (Table 1).

The eggs hatch on an average three days after oviposition, the exact time of development depending on nest temperature during incubation (Hostache *et al.* in press). Although small variations in incubation time occur in the swamps, probably due to weather conditions and microclimate in the immediate

Table 1. Selected data (mean ± SE) on the floating bubble nests of three callychthyid catfish *Hoplosternum littorale, Hoplosternum thoracatum* and *Callichthys callichthys.* Number of observations in parentheses

	Nests of *H. littorale*	Nests of *H. thoracatum*	Nests of *C. callichthys*
Nest construction material	grass	dicotyledonous leaves	dicotyledonous leaves
Number of leaves	–	5.2 ± 0.21 (225)	2.8 ± 0.21 (77)
Dry weight nest material (g)	33.1 ± 3.1 (48)	4.5 ± 0.5 (20)	1.5 ± 0.2 (16)
Nest surface area (cm^2)	706 ± 20 (109)	413 ± 35 (20)	181 ± 30 (16)
Nest height (cm)	6.0 ± 0.4 (109)	0.5 – 1.0	0.5 – 1.0
Number of eggs	23,339 ± 4,473 (11)	3,467 ± 342* (20)	2,953 ± 762* (16)
Fresh weight egg mass (g)	73.02 ± 12.89 (11)	11.82 ± 1.05 (18)	11.03 ± 2.66 (16)
Dry weight egg mass (g)	13.05 ± 2.40 (11)	2.14 ± 0.21 (18)	2.29 ± 0.57 (16)
Volume egg mass (ml)	73.73 ± 13.23 (11)	11.24 ± 1.00 (18)	10.29 ± 2.52 (16)
Egg developmental stages in nest	1	1 – 3	1 – 3
Egg hatching time (days)	3.3 ± 0.1 (59)	2 – 3	2 – 3
Life span of nest (days)	3.9 ± 0.1 (89)	5.7 ± 0.2 (190)	4.6 ± 0.3 (56)
Standard length guarding male (cm)	15.9 ± 0.2 (76)	11.0 ± 0.3 (6)	14.3 ± 0.5 (5)
Attack distance guarding male (m)	0.57 ± 0.07 (71)	0	0

* average number of eggs/nest present at one point in time

surroundings of the nest, the overall hatching time is fairly constant (Table 1; Fig. 2). It is not clear whether the nest of *H. littorale* deteriorates and flattens on the water surface shortly before hatching or immediately after hatching. The newborn larvae stay for one or two days among the floating nest materials before they move to the bottom of the swamp and start living on exogenous food (Singh 1978; Machado-Allison 1986; Mol in press).

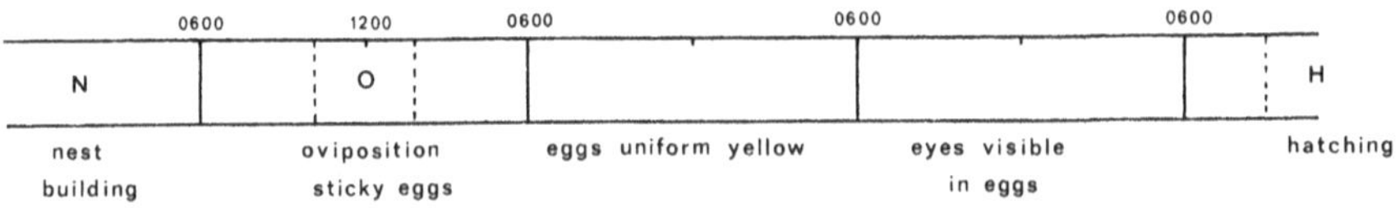

Figure 2. Time schedule of nest construction, oviposition and hatching for *Hoplosternum littorale* in the swamps of northern Suriname.

2.1.2. Nest Structure

Figure 3A shows a well developed nest of *Hoplosternum littorale* floating on the surface of a grass swamp in the Coastal Plain of Suriname. The internal structure of the nest is revealed by a cross-section through a frozen nest illustrated diagrammatically in Fig. 3B. The nest consists of a dome of dead or dying plant material, particularly grasses (e.g. *Hymenachne amplexicaulis*), *Typha angustifolia*, *Thalia geniculata*, *Hydrocotyle umbellata*, *Nymphoides indica*, *Nymphaea* spec. and floating aquatics like *Pistia*, *Salvinia*, *Azolla* and Lemnaceae. Rarely are nests built of leaves of dicotyledonous trees. The thickness of the plant cover is 2 to 3 cm. The inner layer of about 1 cm consists of tightly packed grass and other plant material, contrary to the outer surface of the nest, where the plant material is more loosely interwoven. The dry weight (70 °C, 48 hours) of the plant material averages 33.1 ± 3.1 g (Table 1). The fresh weight of the plant material is more difficult to measure in a uniform manner since part of the nest material is always soaked with swamp water, but a rough estimate is 240 ± 22 g (mean ± SE). Nests of *H. littorale* average 30.0 ± 5.0 cm in diameter and are 6.0 ± 0.4 cm in height above the water surface (Table 1). Large nests can measure up to 42 cm in diameter and are 10 cm in height. The eggs are glued together and to the underside of the dome of plant material. Below the eggs and covering the whole of the underside of the nest is a mass of foam. The eggs are not in contact with the oxygen-depleted swamp water.

The description of the nests of *H. littorale* in the Coastal Plain of Suriname closely corresponds with the drawings of Carter and Beadle (1931) for nests in the Paraguayan Chaco and with the description of Machado-Allison and Zaret (1984) for nests in the Venezuelan Llanos.

2.1.3. Nest Fauna

The huge and complex construction making up the nest of *Hoplosternum littorale* provides a habitat for a rich and diverse fauna of aquatic, semi-aquatic

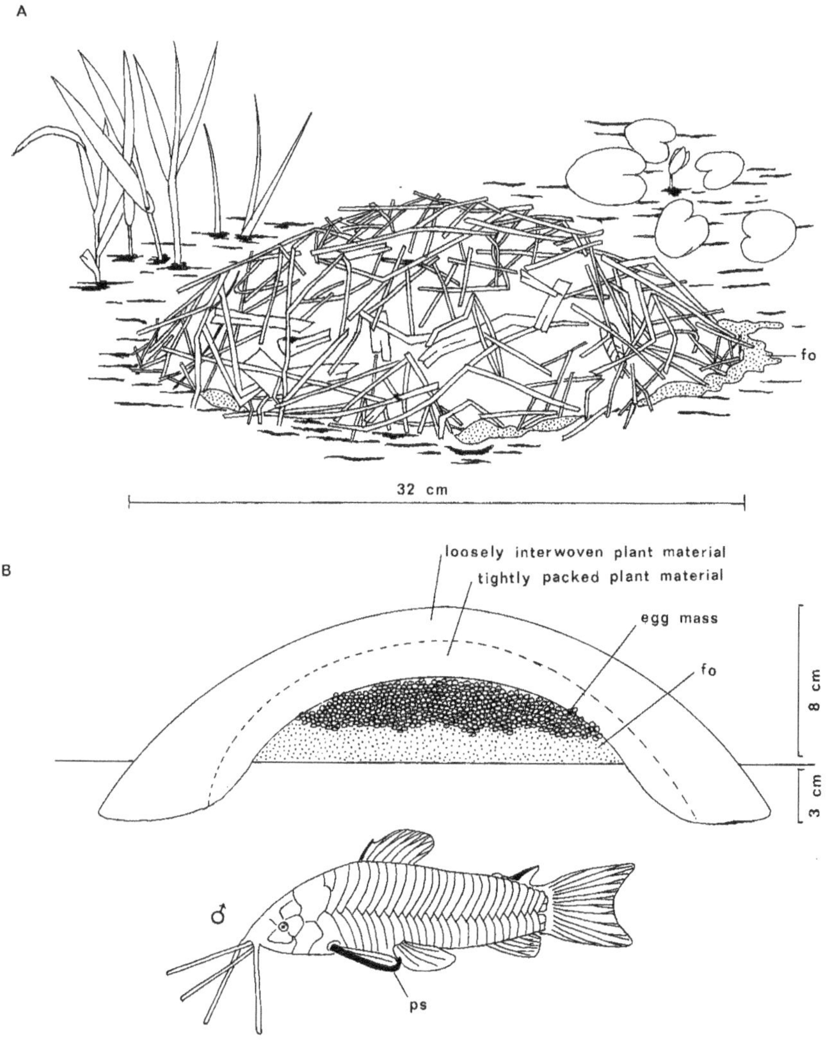

Figure 3. The floating bubble nest of *Hoplosternum littorale*. A. The huge and complex construction of dead and dying grass leaves floats on the surface of the swamp. The foam (fo) which keeps the nest in shape can be seen at its edge. B. Cross-section of a frozen nest illustrating in the center of the nest the mass of sticky eggs glued to the underside of the plant cover, out of the oxygen-depleted swamp water and embedded in the foam. Also shown is the guarding male, in the breeding season easily distinguished from the female by the large, modified, ski-like pectoral spine (ps).

and even terrestrial animals (Table 2). In 48 nests examined an average of 764 individuals/nest or 11,012 ind/m² nest surface were counted. This number is still low compared to the 26,665 ind/m² recorded for the root zone of the floating vegetation in the Coastal Plain of Suriname (Table 2; see also Mol, Chapter 7). The composition of the fauna of these nests is quite different from that of the root zone of the floating vegetation, however. Relatively well represented in the nests are Gastropoda, aquatic Coleoptera, Collembola, Isopoda, terrestrial Arthropoda and especially Formicidae. In 12.5% of the nests examined, ant colonies, including larvae, have been established. A negative effect of the presence of the ant colonies on the development of the eggs of *H. littorale* could not be detected. The ants may even provide some protection for the nests against egg predators like snakes or birds. In two nests a juvenile of the swamp eel *Synbranchus marmoratus* was detected. Examination of the stomach contents of these two specimens did not reveal the presence of eggs of *H. littorale*.

Table 2. Comparison of the fauna of floating bubble nests of *Hoplosternum littorale* (ind/nest and ind/m^2) with the fauna in the root zone of the floating vegetation (ind/m^2) in the Coastal Plain of Suriname

Taxon	Nests of *H. littorale* mean ± SE (n = 48) ind/nest	Nests of *H. littorale* mean ± SE (n = 48) ind/m^2	Floating vegetation mean ± SE (n = 25) ind/m^2
Oligochaeta, Naididae	186 ± 50	2,679 ± 727	6,000 ± 1,658
Hirudinea	0.6 ± 0.2	8 ± 3	40 ± 27
Nematoda	150 ± 66	2,159 ± 952	1,376 ± 657
Gastropoda	48 ± 26	692 ± 374	67 ± 24
Diptera larvae	78 ± 21	1,127 ± 304	5,536 ± 1,173
Coleoptera larvae	8.5 ± 2.5	123 ± 36	167 ± 36
Odonata larvae	0.7 ± 0.2	10 ± 3	151 ± 56
Ephemeroptera & Trichoptera larvae	0.2 ± 0.1	3 ± 1	31 ± 16
aquatic Coleoptera	13 ± 2.6	191 ± 37	33 ± 10
Hemiptera	2.7 ± 0.7	39 ± 10	196 ± 60
Ostracoda	153 ± 109	2,209 ± 1,568	2,394 ± 1,044
Copepoda	19 ± 9.1	279 ± 131	6,070 ± 1,672
Cladocera	5.7 ± 1.9	82 ± 28	3,906 ± 1,140
Conchostraca	0.7 ± 0.3	10 ± 4	468 ± 338
Hydracarina	7.6 ± 1.4	109 ± 20	72 ± 14
Hydra spec.	3.5 ± 1.6	50 ± 22	128 ± 72
Collembola	4.5 ± 1.9	65 ± 27	8 ± 3
terrestrial Arthropoda	2.6 ± 0.6	38 ± 9	21 ± 14
Formicidae	79 ± 44	1,135 ± 634	–
Isopoda	0.3 ± 0.1	4 ± 1	–

2.1.4. Number of Eggs, Spawnings per Nest and Parental Care

The mean number of eggs counted in eleven nests of *Hoplosternum littorale* in the Coastal Plain of Suriname is 23,340 (Table 1). Minimum and maximum numbers recorded are respectively 3,888 and 51,981 eggs/nest. Machado-Allison and Zaret (1984) report an average of 17,100 eggs/nest and a range of 5,600–55,300 eggs/nest for nests of *H. littorale* in the Venezuelan Llanos. Singh (1978), working in the swamps of Trinidad and Guyana, recorded only 10,200 (2,000–22,800) eggs/nest. In swamps and fish ponds in French Guiana, Luquet *et al.* (1989) report an average of 14,700 eggs/nest and a minimum and maximum of respectively 3,100 and 51,500 eggs/nest.

Data on the fecundity of *H. littorale* in Suriname are summarized in Table 3.

Table 3. Fecundity of three callychthyid catfish *Hoplosternum littorale, Hoplosternum thoracatum* and *Callichthys callichthys* (mean ± SE)

Weight class (g)	Frequency	Weight (g)	Standard length (cm)	Gonad weight (g)	Number of ripe eggs (≥ 1 mm)	Number of immature eggs (< 1 mm)
Hoplosternum littorale						
30.0 – 60.0	4	52.5 ± 3.69	10.9 ± 0.2	5.49 ± 1.44	2,898 ± 663	894 ± 273
60.0 – 90.0	5	81.5 ± 2.32	12.6 ± 0.3	4.70 ± 1.34	3,598 ± 956	431 ± 235
90.0 – 120.0	7	99.5 ± 3.06	13.7 ± 0.2	10.05 ± 1.21	6,914 ± 772	1,141 ± 467
120.0 – 150.0	2	133.1 ± 7.05	15.0 ± 0.2	19.60 ± 6.00	10,596 ± 5,944	1,432 ± 561
	18	87.8 ± 5.99	12.9 ± 0.3	8.76 ± 1.47	5,509 ± 868	922 ± 213
Hoplosternum thoracatum						
0.0 ± 16.0	1	12.2	6.8	2.02	926	515
16.0 ± 32.0	3	29.2 ± 0.70	9.2 ± 0.2	2.67 ± 0.48	1,667 ± 847	527 ± 331
32.0 ± 48.0	11	43.1 ± 1.04	10.5 ± 0.2	5.00 ± 0.28	2,944 ± 296	1,246 ± 415
48.0 ± 64.0	6	53.3 ± 1.80	11.2 ± 0.3	5.68 ± 1.18	2,857 ± 678	2,011 ± 616
64.0 ± 80.0	1	68.6	12.0	3.20	1,007	2,350
	22	43.8 ± 2.53	10.4 ± 0.2	4.65 ± 1.72	2,566 ± 284	1,374 ± 285
Callichthys callichthys						
23.0 ± 33.0	1	25.2	9.5	1.30	426	237
33.0 ± 43.0	4	38.6 ± 1.85	10.8 ± 0.3	2.05 ± 0.29	1,636 ± 294	3,231 ± 2,336
43.0 ± 53.0	6	49.8 ± 0.73	12.3 ± 0.1	4.00 ± 0.53	1,330 ± 148	3,429 ± 528
53.0 ± 63.0	4	57.5 ± 1.93	12.7 ± 0.1	4.93 ± 1.16	2,230 ± 549	3,504 ± 1,289
63.0 ± 73.0	1	63.5	12.9	5.90	1,708	2,194
	16	48.2 ± 2.57	11.9 ± 0.3	3.61 ± 0.46	1,599 ± 191	3,122 ± 659

An average of 5,509 ripe eggs (diameter > 1 mm) are present in females ready to spawn. Machado-Allison and Zaret (1984) counted an average of 4,500 mature eggs/female and Luquet *et al.* (1989) report separate spawnings in aquariums of 600 to 10,000 eggs for 100 g females. Table 3 also shows immature eggs (< 1 mm) present in most of the ovaries. Their number is on average 922

$\pm$ 213. Machado-Allison and Zaret (1984) conclude that the females spawn only once in the breeding season, but Luquet *et al.* (1989) report multiple, successive spawnings for females in tanks and in fish ponds. In the prolonged reproductive season in Suriname females probably spawn at least in both the short and the long rainy season (Mol in prep.).

Following Machado-Allison and Zaret (1984) the number of females that contributed to the egg mass in a particular nest of *H. littorale* may be estimated by dividing the number of eggs/nest by the mean number of mature eggs/female. Estimates of the mean and maximum number of spawnings/nest are, respectively, four and nine, so most nests receive their eggs from more than one female. This corresponds well with reports of local fishermen catching many specimen of *H. littorale* during oviposition at one nest. The fairly constant life-span of the nests of *H. littorale* (three to four days) and the observation that the eggs in the nests are always of the same developmental stage, indicate that multiple spawnings are all performed at the same time.

Figure 4A shows that there is a positive correlation between the size of the nest of *H. littorale* and the number of eggs contained in the nest.

Adult males of *H. littorale* are of larger average size than the females. In the reproductive season they are easily distinguished from the females by their elongate and modified pectoral spines (Fig. 3B) and fat deposits at the base of the pectoral fin (Winemiller 1987). The male defends the nest site, and vigorously attacks every moving object in the direct surroundings of the nest (Winemiller 1987). Gautier *et al.* (1988) describe the aggressive behaviour of the male towards the female after oviposition in tanks. This aggressive guarding behaviour makes it possible in 74.5% of the nests containing eggs to catch the male with a 50 cm diameter scoop-net by gently beating the water surface above the net with a small stick. The large nests are usually built and guarded by large males (Fig. 4B). When defence is measured, however, by the maximum distance from the nest where an attack on the stick can still be provoked, smaller males are more aggressive (Fig. 4C). The territory established by the male is defined as the defended area around the nest: a circle with an average radius of 57 cm (Table 1). The minimum distance recorded between two nests of *H. littorale* in rice fields is 3 m. In swamps the nests are generally built more than 10 m from one other.

Mark-recapture experiments with nest-guarding males of *H. littorale* in the swamps have so far not been very successful, possibly due to the crude marking method of clipping parts of the pectoral fin. At one nest site, however, a marked male was replaced by a second one that took over the nest site and nest material within a week after the hatching of the eggs. This second male constructed, after hatching of his first batch of eggs, a second nest at the same nest site, using the same nest material, thus demonstrating that males can built and guard more than one nest in a season. This experiment also shows that nests constructed at one nest site are not necessarily built by the same male.

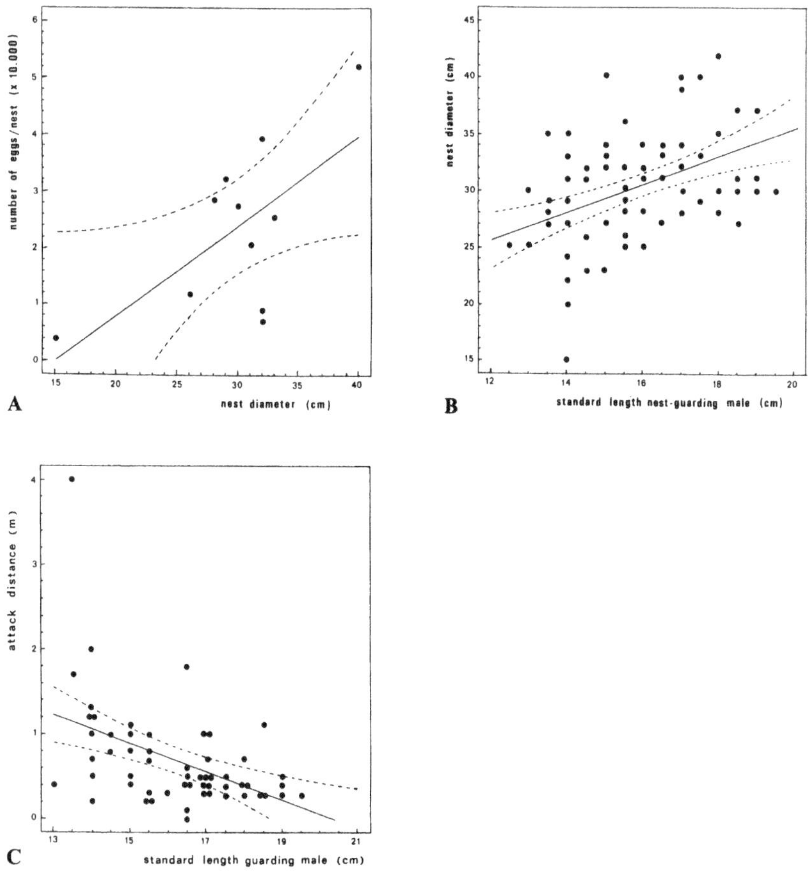

Figure 4. A. Relationship between nest diameter of floating bubble nests of *Hoplosternum littorale* and the number of eggs/nest. The linear regression equation (± 95% confidence limits) is: Y = 23,600 + 1,574 X; r = 0.64 (p = 0.0327).
B. Relationship between the size of nest-guarding males of *Hoplosternum littorale* and the diameter of the floating bubble nest. The linear regression equation (± 95% confidence limits) is: Y = 10.8 + 1.23 X; r = 0.45 (p = 0.00006).
C. Relationship between the size of nest-guarding males of *Hoplosternum littorale* and the distance from the center of the nest at which the male attacks a small stick gently beating the water surface. The linear regression equation (± 95% confidence limits) is: Y = 3.37–0.16 X; r = 0.45 (p = 0.00052).

2.2. *The Nests of* Hoplosternum thoracatum

2.2.1. *Nest Construction, Oviposition and Incubation of the Eggs*

Nests of *Hoplosternum thoracatum* (Sur. catharina kwikwi) are built by the male in shallow swamps or swamp forests during the rainy season (Fig. 5). The nests are always located in the direct environment of a dicotyledonous tree or shrub

Figure 5. Nest of *Hoplosternum thoracatum* in a swamp at Lelydorp, Old Coastal Plain, Suriname.

(Fig. 6A). Nests were observed in the same periods as those of *H. littorale* (December – August), but *H. thoracatum* started nesting one or two weeks earlier then *H. littorale* (Mol in prep.).

Actual construction of the nest of *H. thoracatum* by the male was not observed in the swamps, but detailed descriptions of observations in aquariums are reported (Schubert 1948; Zukal 1981; Burgess 1989). The nests do not have

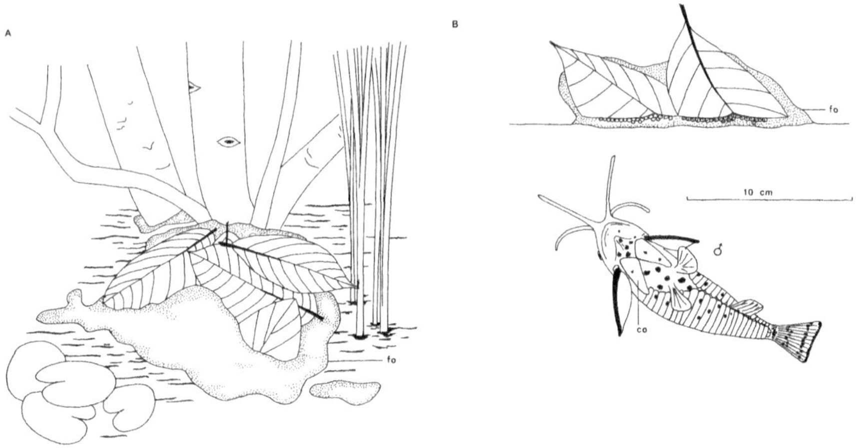

Figure 6. The floating bubble nest of *Hoplosternum thoracatum*. A. The nest consisting of up to nineteen dicotyledonous leaves floating on top of a thin layer of foam (fo) on the water surface is built at the foot of a swamp tree. B. A cross-section of a frozen nest reveals the mono-layer of eggs glued to the underside of the leaves embedded in foam and not in touch with the oxygen-poor swamp water. Also note the two different developmental stadia of the eggs in the nest. The guarding male, distinguished from the female by the ventrally meeting coracoids (co), is shown on its way to the surface to supply new foam to the nest.

the complex structure of nests of *H. littorale* and nest building probably does not take much time. I suspect that nest building takes place shortly before oviposition because nests without eggs were rarely observed in the swamps.

I have confirmation of the time of oviposition for only four nests. In three nests oviposition was between 09.30–12.00 hours and in one nest eggs were deposited between 11.00 and 18.00 hours. It is quite possible, however, that oviposition is concentrated in the night or early morning because many nests with sticky eggs were observed between 07.00 and 08.00 hours.

Oviposition in nests of *H. thoracatum* is not limited to a single spawning session. Multiple successive spawnings are indicated by the many observations of the presence of eggs of more than one developmental stage in one nest at one point in time and by the large variation in the life-span of the nests (2–21 days). Nests with eggs of only one developmental stage have a life-span of two or three days and this seems to be a reasonable estimate of the incubation time of the eggs in the swamps. Only 17% of the 239 nests observed in the swamps did not acquire more than one spawning.

2.2.2. Nest Structure

An illustration of a nest of *Hoplosternum thoracatum*, situated at the base of a swamp tree, is presented in Fig. 6A. The nests consist of 1–19 dicotyledonous leaves floating on top of a thin layer of foam on the water surface. The structure of the nest is revealed in Fig. 6B, representing a diagrammatic illustration of a cross-section through a frozen nest. The leaves of most nests originate from dicotyledonous trees. In the Coastal Plain of Suriname these are e.g. *Pterocarpus officinalis, Syzygium cumini, Triplaris surinamensis* and *Vochysia tetraphylla.* Rarely is a nest built under floating leaves of *Nymphaea* spec. The average thickness of the leaves is 0.5 ± 0.06 mm (mean ± SE).

The eggs are glued firmly to the underside of the leaves of the nest. They are deposited almost in a mono-layer, embedded in foam and not in contact with the oxygen-depleted swamp water. Table 1 summarizes some selected data on the nests of *H. thoracatum.*

2.2.3. Number of Eggs, Spawnings per Nest and Parental Care

The reproductive strategy of *Hoplosternum thoracatum* makes it very difficult to arrive at a reliable estimate of the total number of eggs deposited in one nest. The male, distinguished from the female by the coracoids meeting ventrally (Fig. 6B), maintains the floating bubble nest for quite a long time. The mean life-span of a nest is 5.7 ± 0.2 days (Table 1), but nests of up to 12 days old are not uncommon. One nest even lasted for 21 days. During this period females spawn in the nest and one or more batches of eggs are always present. A nest without eggs is rarely observed.

The average number of eggs counted in a nest of *H. thoracatum* at one point in time is 3,467 ± 342 (Table 1). Minimum and maximum numbers are respectively 1,159 and 7,715 eggs/nest. In 60% of the 20 nests examined, eggs of more than one developmental stage were present, indicating batches of successive spawnings. A nest which reveals at one point in time both recently deposited sticky eggs and eggs about to hatch, must have a life-span of at least five or six days. This happens to be the average life-span of the nests of *H. thoracatum* in the swamp (Table 1) and it means that the average number of eggs present in a nest at one point of time is probably representative for the average total number of eggs deposited in the nest. A conservative estimate of the maximum number of eggs/nest for the nest which lasted 21 days is 3,467 + the eggs of 5 spawnings. The mean fecundity of *H. thoracatum* is 1,374 eggs/female (Table 3) and thus an estimate of the maximum number of eggs/nest must be 10,337 eggs/nest. The estimated mean and maximum number of spawnings/nest are respectively two and seven spawnings/nest. In general it is very difficult to compare these data for nests of *H. thoracatum* in the swamps with data on nest building in aquariums because of spatial and social limitations (number of available females) in captivity.

Figure 7 shows that the number of eggs present in a nest at one point in time

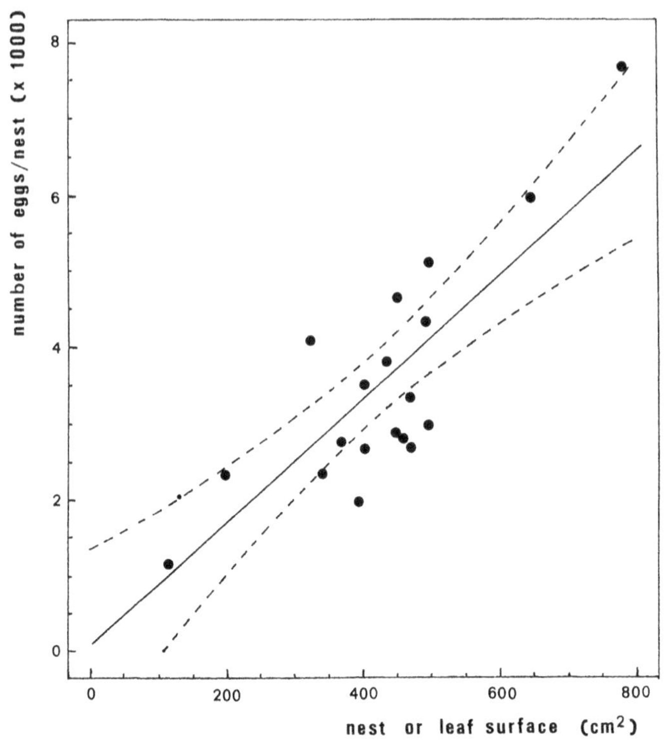

Figure 7. Relationship between total leaf surface of floating bubble nests of *Hoplosternum thoracatum* and the number of eggs/nest. The linear regression equation (± 95% confidence limits) is: Y = 99.4 + 8.2 X; r = 0.83 (*p* = 0.00001).

correlates with the total leaf surface available for oviposition. As the eggs are deposited almost in a mono-layer (Fig. 6B) and on average 91% of the available leaf surface is covered with eggs, this is not very surprising.

The nest-guarding male of *H. thoracatum* is not very aggressive. Compared to *H. littorale* only 9 males out of 209 nests tested (4%) displayed attack behavior. Males of *H. thoracatum* probably do not establish a territory of the size realized by *H. littorale* because nests in the swamps are often built within 50 cm of each other. Figure 6B shows the guarding male on its way to the nest in order to supplement the nest with fresh foam.

2.3. The Nests of Callichthys callichthys

2.3.1. Nest Construction, Oviposition and Incubation of the Eggs

The nests of *Callichthys callichthys* (Sur. platkop or plata-ede kwikwi) are built by the male near the very edge of shallow swamps and swamp forests. Most nests of *C. callichthys* are located in very shallow water, in cavities of earth and tree roots (Fig. 8). Taking into account the size of *C. callichthys* (Table 1) it is hard to believe one's eyes when a nest of this species is detected in a shallow and small cavity of tree roots (Fig. 8A and 8C) or in an isolated pool just 30 cm in diameter and only accessible through a narrow underwater corridor (Fig. 8B). In fact, characteristics of the nest site microhabitat are the most reliable field marks in distinguishing the nests of *C. callichthys* from those of *H. thoracatum* (Mol in prep.). Confirmation of the nest-building species was obtained for 31 of the 85 nests of *C. callichthys* observed, mostly by rearing the newborn larvae in the laboratory. Nests were observed from mid-December to August when the swamps were flooded (Mol in prep.).

Construction of the nests of *C. callichthys* by the male was not observed in the swamps, but observations of nest building and spawning in aquariums are reported by Breder and Rosen (1966). The structure of the nest is very simple and nest building is probably completed in a short period preceding oviposition.

Twice oviposition of *C. callichthys* was actually observed in the swamps of Lelydorp, both times between 08.15 and 08.45 hours in the morning. In two nests oviposition was between 18.00 and 07.00 hours and once oviposition was recorded between 08.30 and 10.00 hours. Oviposition is not strictly limited to a single spawning session as was observed for *H. littorale*. Multiple successive spawnings are indicated by the variation in life-span of the nests (2–17 days), but most nests examined displayed eggs of only one developmental stage. It seems that in nests of *C. callichthys* spawning only takes place after a preceding batch of eggs has hatched. The incubation time of the eggs is probably between two and three days (Table 1), which means that 30% of the 85 nests observed in the swamps acquired more than one spawning.

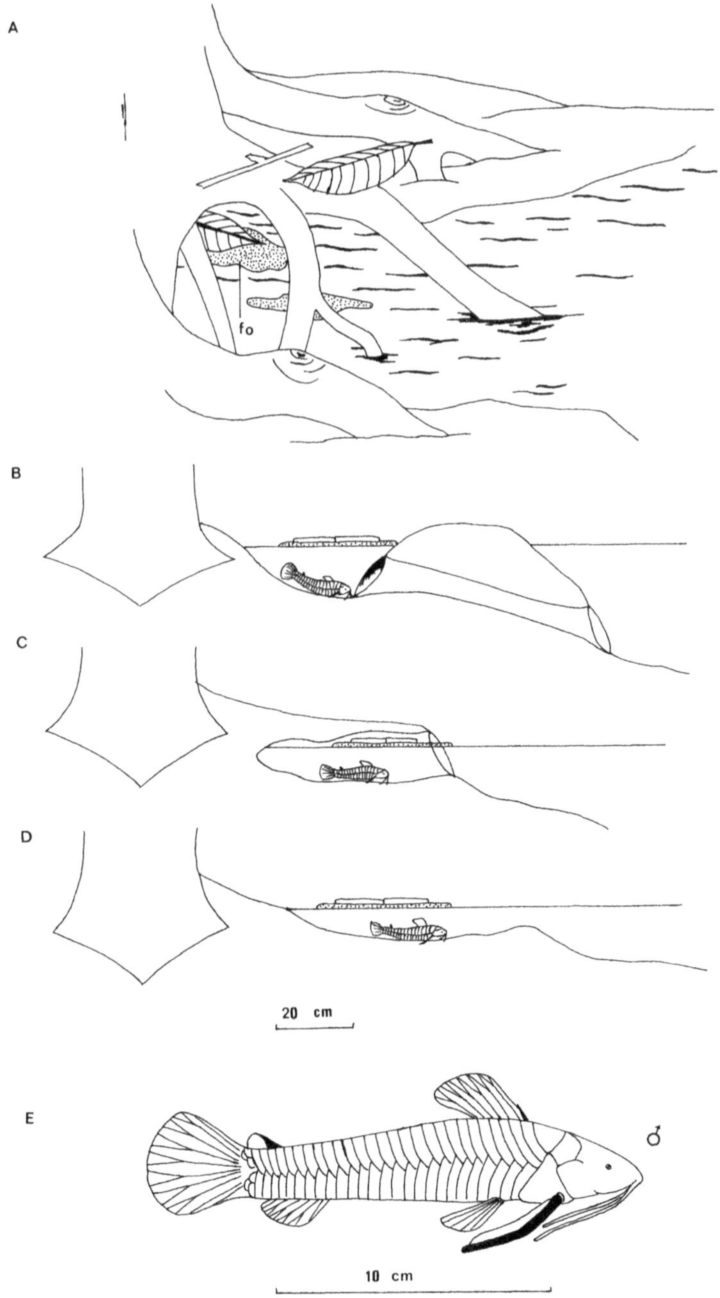

Figure 8. The floating bubble nest of *Callichthys callichthys*. A. The very simple nest consists of a few dicotyledonous leaves floating on top of a thin layer of foam (fo) on the water surface. The nest is often constructed in very shallow water near the edge of the swamp and in cavities of tree roots. B, C, D. Some examples of habitats where nests of *C.callichthys* were found. E. Guarding male.

2.3.2. Nest Structure

Figure 8A shows a nest of *Callichthys callichthys* floating on the water surface and partly hidden in a cavity of tree roots. The structure of the nest, as revealed by cross-sections through a frozen nest, is very similar to the structure of nests of *H. thoracatum* (Fig. 6B). The only major difference I could detect is the irregular clustering of eggs at the underside of the leaves of nests of *C. callichthys* in contrast with the almost perfect mono-layer of eggs in nests of *H. thoracatum* (Fig. 9). The slipshod deposition of eggs in nests of *C. callichthys* is probably a consequence of the lack of freedom of movement for the spawning female(s) at the nest site (see Fig. 8B, 8C and 8D) and the lack of available leaf surface in the nest. The average number of eggs/cm² of leaf surface is 9.8 ± 0.75 (mean ± SE) for nests of *H. thoracatum* and 22.5 ± 6.06 for nests of *C. callichthys*, the difference being significant (*t*-test, $p < 0.05$). Selected characteristics of the nests of *C. callichthys* are summarized in Table 1.

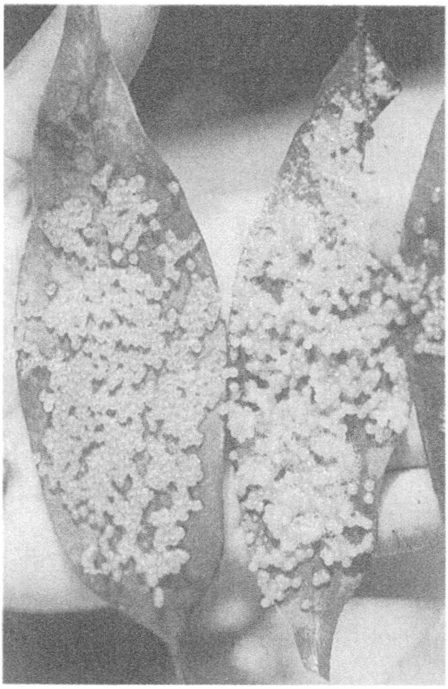

Figure 9. Irregular clustering of eggs at the underside of a leaf from a nest of *Callichthys callichthys* (right) and eggs deposited in a mono-layer in a nest of *Hoplosternum thoracatum* (left).

2.3.3. Number of Eggs, Spawnings per Nest and Parental Care

The average number of eggs present at one point in time in the nests of *Callichthys callichthys* is 2,953 ± 762 (Table 1). Minimum and maximum numbers recorded are 237 and 8,474 eggs/nest respectively.

Data on the fecundity of *C. callichthys* are summarized in Table 3. The average number of ripe eggs/female is 1,599 ± 191.

As already noted for nests of *H. thoracatum*, these figures are not representative for the total number of eggs spawned in a nest. Contrary to nests of *H. thoracatum*, the 16 nests of *C. callichthys* examined in the laboratory only contained eggs of one developmental stage. Surveys in the swamp confirmed that nests with eggs of more than one developmental stage are very rare, although a few of such nests were observed. Nests without eggs are also uncommon and it seems that following hatching of a batch of eggs, the nests are either abandoned or obtain a new spawning shortly afterwards. With knowledge of the mean hatching time of the eggs in the swamp and the mean life-span of the nests, it is possible to calculate the total number of eggs spawned in one nest from the number of eggs present at one point in time as 5,433 eggs/nest. An estimate of the maximum number of eggs/nest obtained from the nest with a life-span of 17 days is 20,080. Mean and maximum number of spawnings/nest are, respectively, three and twelve.

With nests of *C. callichthys* a significant linear relationship between the number of eggs/nest present at one point in time and the leaf surface of the nest was not present (ANOVA, $p = 0.26$) and this is not very surprising taking into account the irregular deposition of the eggs in these nests.

The male of *C. callichthys* guards the nest after finishing its construction. In the swamps guarding males displayed the attack response with only 7% of the nests examined.

3. Function of the Foam Nest

3.1. Foam Nests of Insects, Frogs and Fishes

The construction of a foam nest is reported for insects (Richards and Davies 1977), frogs (Duellman and Treub 1986) and fishes (Balon 1975). Table 4 summarizes the data on foam nest construction in the animal kingdom.

Insect foam nests produced by the female during oviposition (oothecae) are attached to twigs, bark, walls and other objects by mantids or buried in the ground by grasshoppers (Richards and Davies 1977). The frothy excretion of the female is viscid and very rapidly hardens to the consistency of a spongy substance. A second type of foam nest is produced by the nymphs of the families Cercopidae and Aphrophoridae. The nymphs of these "cuckoo spit" insects envelop themselves in froth on plants (Richards and Davies 1977). Richards and Davies (1977) suggest that the foam may protect eggs and nymphs from predation by birds or lizards and from desiccation in the sun. The nests did not offer protection against parasitic Hymenoptera.

Within the Anura Duellman and Treub (1986) distinguish five modes of reproduction which include the construction of a foam nest. Nests are constructed on the water surface of permanent water bodies, in small temporal pools next to a stream, in burrows next to ponds or streams (aquatic tadpoles), in burrows with tadpoles completing their development in the nest and in trees

Table 4. Construction of foam nests in the animal kingdom. In parentheses the number of species that built foam nests

Classis Ordo Familia Genus	Producer of foam	Parental care after construction of foam nest	Geographical distribution	References
INSECTA				
Dictyoptera				
Empusidae	♀	–	Old world	Richard & Davies 1977
Mantidae	♀	–	Cosmopolitan	Richard & Davies 1977
Orthoptera				
Acrididae	♀	–	Cosmopolitan	Richard & Davies 1977
Pygomorphidae	♀	–	Old world	Richard & Davies 1977
Homoptera				
Aphrophoridae	nymph	–	Cosmopolitan	Richard & Davies 1977
Cercopidae	nymph	–	Cosmopolitan	Richard & Davies 1977
OSTEICHTHYES				Breder & Rosen 1966
Siluriformes				
Callichthyidae (5)	♂	♂	S.America	Burgess 1989
Callichthys (1)	♂	♂	S.America	Breder & Rosen 1966
Dianema (2)	♂	♂	S.America	Petrovicky 1979
Hoplosternum (2)	♂	♂	S.America	Carter & Beadle 1931
Cypriniformes				
Characidae (1)				
Hepsetus odoe	♂ + ♀	♂ + ♀	Africa	Merron et al. 1990
Synbranchiformes				
Synbranchidae (1)				
Monopterus (Fluta)	?	♀(?)	S.Asia	Vogt 1970
Perciformes				
Belontidae (23)			S.Asia	Richter 1979
Belontia (2)	♂	♂ + ♀	S.Asia	Richter 1979
Macropodus (2)	♂	♂(+ ♀)	S.Asia	Breder & Rosen 1966
Pseudosphromenus (2)	♂	♂	S.Asia	Richter 1979
Malpulutta (1)	♂	♂	S.Asia	Richter 1979
Betta (6)	♂	♂	S.Asia	Breder & Rosen 1966
Trichopsis (3)	♂	♂	S.Asia	Breder & Rosen 1966
Trichogaster (4)	♂	♂	S.Asia	Breder & Rosen 1966
Colisa (3)	♂	♂	S.Asia	Breder & Rosen 1966
Anabantidae (4)				
Ctenopoma (4)	♂	♂	Africa	Richter 1979
Nandidae (1)				
Polycentropsis (1)	♂	♂	Africa	Scherpe 1981
AMPHIBIA, ANURA				Duellman & Treub 1986
Leptodactylidae	♂		S.Amercia	Heyer 1969
Adenomera	♂	–	S.America	
Edalorhina	♂	?	S.America	
Leptodactylus	♂	?	S.America	
Lithodytes	♂	?	S.America	

Table 4. (continued)

Classis Ordo Familia Genus	Producer of foam	Parental care after construction of foam nest	Geographical distribution	References
Physalaemus	♂	?	S.America	
Pleurodema	♂	?	S.America	
Myobatrachidae	♀		Australia	Martin 1970
Adelotus	♀	♂	Australia	
Heleioporus	♀	?	Australia	
Kyarranus	♀	–	Australia	
Lechriodus	♀	?	Australia	
Limnodynastes	♀	?	Australia	
Megistolotis	♀	?	Australia	
Philoria	♀	?	Australia	
Rhacophoridae				Duellman & Treub 1986
Chirixalus	?	?	Africa/Asia	
Chiromantis	♂ + ♀	♀	Africa	
Philautus	?	?	Asia	
Polypedates	?	?	Asia	
Rhacophorus	?	?	Asia	
Hyperoliidae				Duellman & Treub 1986
Opisthothylax	?	?	Africa	

with hatching tadpoles dropping into the water. The overall evolutionary trend within these five modes of reproduction toward terrestriality, associated with avoidance of aquatic predation on eggs and larvae, is beautifully demonstrated in the families of Leptodactylidae (Heyer 1969) and Myobatrachidae (Martin 1970). Gorzula (1977) stresses the thermal properties of the foam in preventing extreme temperatures during development. A third function of the foam might be protection of eggs and larvae from desiccation (Heyer 1969). Duellman and Treub (1986) mention experimental evidence of growth-inhibiting properties of the foam. Heyer (1969) describes the viscosity of leptodactylid foam as having the consistency of beaten egg white with some hardening of the outer layer.

Balon (1975) defines the aphrophilic reproductive guild of fishes and reports the building of foam nests for the African pike *Hepsetus odoe*, the callichthyids *Hoplosternum* and *Callichthys* and a number of anabantoid species. To this list the nandid *Polycentropsis abbreviata* (Scherpe 1981), the synbranchyid *Monopterus alba* (Vogt 1970) and two species of the callichthyid genus *Dianema* (Petrovicky 1979) may be added. A number of functions of the foam nest have been proposed (Breder and Rosen 1966; Machado-Allison and Zaret 1984; Hostache *et al.* in press):

1. oxygen reservoir for eggs and newborn larvae
2. protection against predation (fishes and aquatic insects)

3. thermal properties in relation with incubation of the eggs
4. regulation of microbial growth
5. concealment of previous spawns to courting females
6. protection of eggs from sunlight (desiccation)
7. depot for deposition of the eggs
8. anchor of eggs amid plants.

Protection of eggs from desiccation by the foam (6) is mentioned for insects, anurans and fishes. This function of the foam seems very important for aquatic organisms depositing their eggs out of the water. Most nests do function as a depot for deposition of eggs (7). It is possible that properties of the foam, e.g. its stickiness, help in keeping the eggs together (Heyer 1969). The floating bubble nest of fishes may anchor the eggs amid plants (8). However, the eggs of foam nesting anabantoids and callichthyids have a higher density than water (Richter 1979; Burgess 1989), making drift of eggs unlikely. In absence of the foam, these eggs would sink to the bottom.

3.2. Oxygen Supply to the Developing Eggs

Bagenal and Braum (1978) consider oxygen, temperature and waterflow ecological features of fundamental importance for the development and survival of fish eggs. There is a close relationship among these three parameters. In mountain streams with a strong water current, oxygen is continuously supplied to developing eggs (e.g. of salmonids). If waterflow is not present, fishes can create a current over their eggs by fanning their fins (e.g. cichlids). However, if high temperatures and decaying organic materials cause oxygen to be absent, as in tropical standing waters, developing eggs are in trouble. Carter and Beadle (1931) point out that the structure of the nest of *Hoplosternum littorale* may assist in the supply of oxygen to the developing eggs and larvae in an aquatic environment without dissolved oxygen.

Cross-sections through frozen nests of *H. littorale* (Fig. 3B), *H. thoracatum* (Fig. 6B) and *C. callichthys* reveal the eggs embedded in foam above the water surface. Figure 10 illustrates the results of measurements of oxygen (Ag/Pt electrode), temperature and pH during a 24 hour period in the foam of a nest of *H. littorale*, in the swamp water next to the nest and in the air above the nest. In close agreement with measurements by Carter and Beadle (1930) in the Paraguayan Chaco, Fig. 10A shows the little oxygen present in the water during the day (photosynthesis) absent at night. The foam in this period is 86% oxygen-saturated and a significant difference in oxygen content between day and night was not present (t-test, $p = 0.53$). Analyses of the foam at the Academic Hospital of Paramaribo confirm the measurements in the field (partial pressure of O_2 is 179.9 mm Hg/cm² at 37 °C, pH is 7.82). Observations during this experiment revealed the guarding male regularly supplying fresh foam to the nest.

In a second experiment the removal of the guarding male from the nest resulted in the immediate deterioration of both the nest and its eggs (Fig. 11).

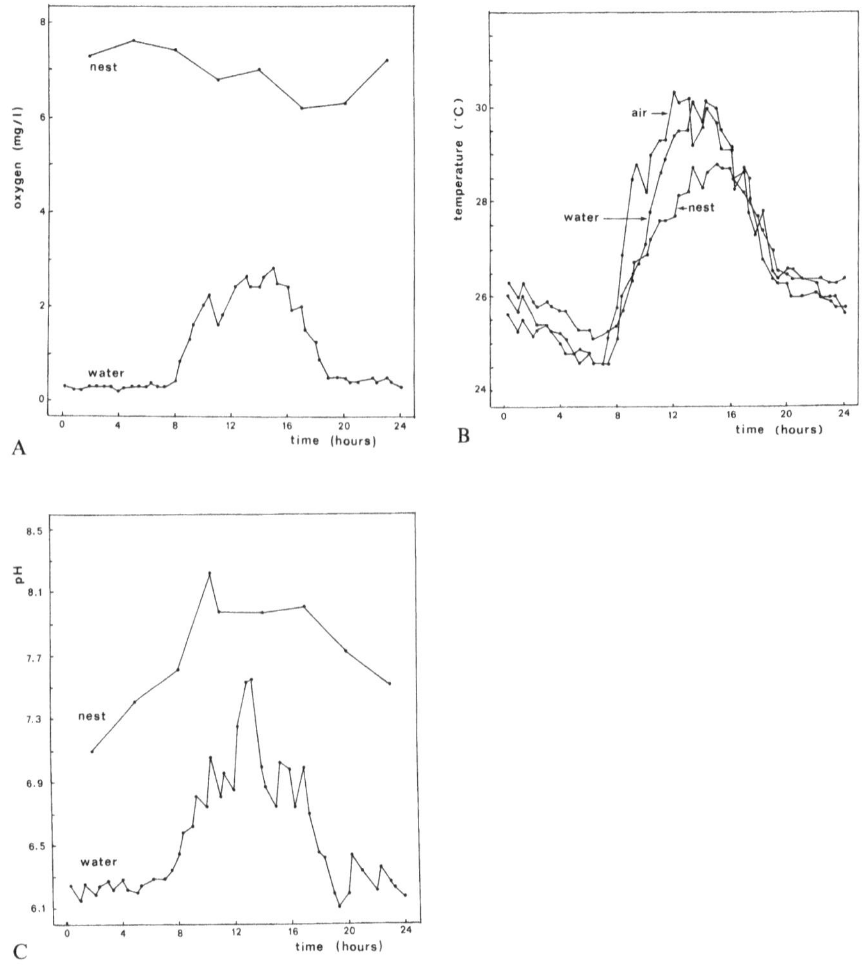

Figure 10. Diurnal fluctuations of oxygen (A), temperature (B) and pH (C) during incubation of eggs of *Hoplosternum littorale*. Recordings are 5 cm under the water surface, in the foam next to the incubating eggs and 50 cm above the water surface in the shade (temperature only) on a warm and cloudless day without rain (27th January 1990).

Hostache *et al.* (in press) demonstrate hatching of eggs of *H. littorale* in conical incubators with a continuous flow of oxygen-saturated water. The deterioration of the eggs after contact with swamp water is probably caused by oxygen deficiency (see Fig. 10A).

The male produces foam rich in oxygen, which keeps the nest in shape and the eggs out of the oxygen-depleted swamp water. The question remains whether the foam just lifts the eggs out of the swamp water in the oxygen-rich air or if it has a more direct function in the oxygen supply to the eggs. In nests of *H. thoracatum* atmospheric oxygen must diffuse through a 0.5 mm thick leaf

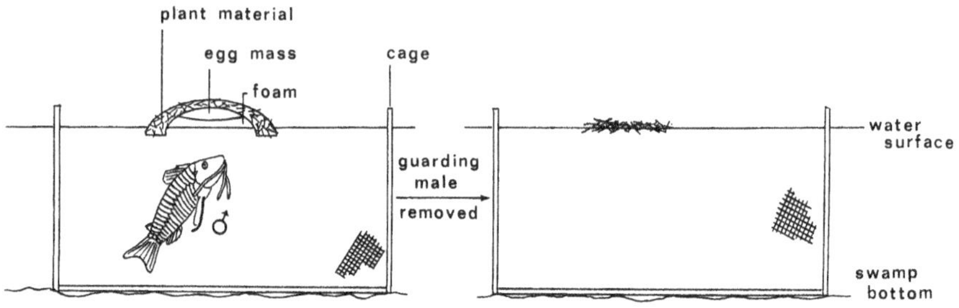

Figure 11. Experiment demonstrating the effect of removing the nest-guarding male on the development of the eggs of *Hoplosternum littorale*. A cage of mosquito mesh (1.0 × 0.5 × 0.5 m) prevents disturbance from intruders. One day after removing the guarding male the nest, out of shape, flattens on the water surface. The eggs deteriorate after contact with the oxygen-depleted swampwater.

with closed stomata in order to reach the eggs. In rainy weather oxygen must probably cross a 1 cm thick layer of water-soaked plant material in nests of *H. littorale* on its way to the egg mass. On the other hand, callichthyids are well-known for their aerial respiration (Carter and Beadle 1931; Johansen 1970; Gee and Graham 1978) and they are probably able to produce foam rich in oxygen by gulping air at the surface.

In order to detect a possible function of direct oxygen supply to the eggs through the foam, a nest of *H. littorale* was sealed from atmospheric oxygen by positioning a plastic basin upside down over the nest at 07.00 hours on the first day after oviposition (Fig. 12). Development of the eggs proceeded normally during the 48 hours the eggs were excluded from atmospheric oxygen. Experiments of Braum (1978) demonstrate that oxygen consumption of fish eggs increases as development proceeds. Following Braum (1978) it is estimated that the eggs of *H. littorale* use 80% of the oxygen needed for their development in the last 48 hours before hatching. Using a Q_{O2} of 1.5 μg O_2/mg dry weight/hour (Blaxter 1969) the total oxygen consumption of the egg mass with a mean dry weight of 13.05 g (Table 1) during the experiment is calculated as 1.13 liter of oxygen. A second estimate based on an average oxygen consumption during development of 100 μl/egg (Blaxter 1969) results in a total oxygen consumption of the egg mass of 1.87 liter of oxygen. If the foam does not take part in the oxygen supply to the eggs, at least the conservative estimate of 1.13 liter O_2 must be supplied by the air enclosed between the nest and the basin. The amount of oxygen present in the air enclosed in the basin is between 0.637–0.768 liter, depending on the percentage of water vapor in the air and the amount of air present in the outer layer of the nest. The experiment demonstrates that the oxygen available in the air enclosed by the basin was not sufficient for the development of the eggs and that additional oxygen must have been provided, presumably by the regular supply of fresh foam by the guarding male. The oxygen consumed by the nest fauna (Table 2) and by the micro-organisms in the decaying nest material is not taken into account.

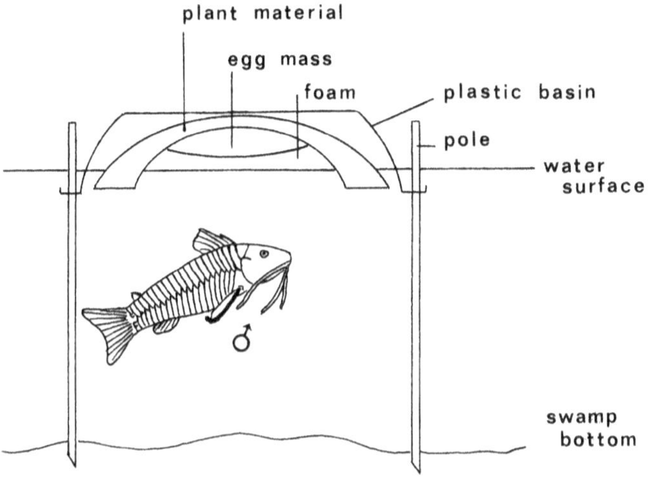

Figure 12. Experiment with nest of *Hoplosternum littorale* sealed from atmospheric oxygen by means of a plastic basin positioned upside down above the nest. The eggs hatch as usual on the third day after oviposition, demonstrating that oxygen in the foam produced by the guarding male, is provided to the developing eggs (see text).

It seems that the primary function of the foam nest of fishes is associated with the supply of oxygen to the eggs. A close relation exists between the foam nesting habit and the habitat of the fishes that build these nests. The African pike *Hepsetus odoe* is known from swamps and slow flowing rivers (Merron *et al.* 1990). The five callichthyinids are also able to survive oxygen-depleted standing waters (Carter and Beadle 1931; Burgess 1989). Of the 25 foam nest building anabantoids for which a description of the habitat is available (Richter 1979), 84% lives in standing or slow flowing water with dense aquatic vegetation. A good description of a habitat of foam nest building anabantoids (where *Monopterus alba* (Synbranchidae) was collected as well) is provided by Furtado and Mori (1982). In addition, callichthyinids, anabantoids and synbranchyids are adapted to an oxygen-depleted aquatic environment by supplementary respiratory organs (Carter and Beadle 1931; Johansen 1970).

3.3. Protection Against Predators

Avoidance of aquatic predation by small fishes and aquatic insects is considered the primary driving force behind the evolutionary trend towards terrestrial reproduction in frogs (Duellman and Treub 1986). Heyer (1969) and Martin (1970) discuss the trend toward terrestriality and the function of the foam nest in the families of Leptodactylidae and Myobatrachidae with respect to protection against aquatic predators.

Balon (1975) states that two factors play a leading role during embryonic development of fish eggs: predation and the availability of oxygen. In many families of freshwater fishes the construction and defence of a nest is quite common (Breder and Rosen 1966). All species belonging to the aphrophilic reproductive guild of Balon (1975) continue parental care after construction of the foam nest and oviposition (Table 4). This additional parental care always includes the maintenance and defence of the nest and sometimes removal of eggs infected with micro-organisms and defence of newborn larvae (Vogt 1970; Richter 1979; Scherpe 1981; Burgess 1989; Merron *et al.* 1990).

All nests of fish guarded by one or both parents probably offer some protection against predation, but the structure of the floating foam nest does not necessarily improve this protective function. The rich fauna of the nests of *H. littorale* collected in the swamp (Table 2) indicates that at least small potential predators are not effectively expelled from the nest. Nymphs of Odonata (Hostache *et al.* 1990), Hemiptera (pers. obs.) and possibly Coleoptera larvae predate on larvae of *H. littorale*. Two nests even yielded a juvenile of the swamp eel *Synbranchus marmoratus* also known to predate on larvae of *H. littorale* (Singh 1978). Moreover, the floating foam nest places terrestrial predators like egrets and snakes in a favourable position to prey upon the eggs.

Foam nests of leptodactylid frogs were sometimes observed floating on the swamp near nests of *H. thoracatum*, *C. callichthys* and *H. littorale*. However, the stickiness and consistency of the foam of the leptodactylid nests probably offers excellent protection against both aquatic and terrestrial predation (Heyer 1969).

3.4. Thermal Properties of the Foam Nest

Hostache *et al.* (in press) use conical incubators and oxygen saturated water to determine the relation between incubation temperature and hatching time of eggs of *Hoplosternum littorale*. The minimum hatching time of 36.4 hours was associated with an incubation temperature of 33.6 °C, but incubation temperatures above 35 °C already proved lethal to the developing eggs. Eggs and larvae survived without problems at incubation temperatures as low as 24 °C, the lowest temperature tested, and this is not surprising if the geographical distribution of the species is considered (south to Buenos Aires in northern Argentina (Burgess 1989; Mol in prep.)).

Figure 10B shows temperatures measured during a 24 hour period in the nest of *H. littorale*, in the swamp water around the nest and the air above the nest. The diurnal fluctuation of the temperature in the water and the air is, to some degree, damped in the nest. Compared with water and air temperatures, nest temperatures are low in daytime and high at night. The results differ in some respects from measurements by Hostache *et al.* (in press) in and around nests of *H. littorale* in fish ponds in French Guiana. In these ponds, emptied of aquatic vegetation, very high nest temperatures are recorded in daytime resulting in diurnal fluctuations up to 11 °C (Hostache *et al.* in press). The average hatching time of 51 hours recorded in the ponds is also relatively short compared with the approximately 72 hours incubation in the swamps of Suriname (Fig. 2). It is my impression that the nests of *H. littorale* in the fish ponds of Socoumou, French Guiana, are not of the same size and nest materials as the nest in the Surinamese swamps (pers. obs.). I think it probable that in the swamps the aquatic vegetation surrounding the nest and the large quantities of nest materials used in the construction of the nest (see Table 1) account for the damping of extreme temperatures as shown in Fig. 10B.

Hostache *et al.* (in press) propose a brooding effect of the structure of the nests of *H. littorale* based on the high nest temperatures measured in the ponds. However, if it is lethal to the eggs if they are exposed for length to temperatures above 35 °C, prevention of such extreme temperatures, not rare in the swamps, seems to me of more importance. Gorzula (1977) demonstrated a cooling effect of the foam of nests of leptodactylid frogs and stressed the function of the foam in the damping of extreme temperatures in the environment of foam nests exposed to intense sunlight. In addition to cooling the nest in daytime, the foam also keeps the nest temperature from falling very low at night, speeding up the development of the eggs. The nests of *H. thoracatum* and *C. callichthys* are often built in the shade of a dicotyledonous tree and probably do not experience the extreme temperatures of the open grass swamp.

3.5. Properties of the Foam in Relation to Microbial Growth

Balon (1975) suggests that the foam produced by fish of the aphrophilic reproductive guild may attract micro-organisms on which the larvae can feed. Braum (1978) discusses the detrimental effect of fungi and bacteria on fish eggs. Duellman and Treub (1986) mention experimental evidence of growth-inhibiting properties of foam of frogs' nests.

The observation that eggs of *H. littorale* deteriorate after the male was removed from the nest (Fig. 11) did raise the question whether an attack by micro-organisms made possible by the absence of protective foam caused the decay of the eggs. To detect antimicrobial properties of the foam the sensitivity of nine species of bacteria and two species of fungi (Table 5) to foam of nests of *H. littorale* sterilized with a Seitz filter was tested with a modified agar-diffusion method (Fig. 13). As shown in Table 5 the experiments failed to detect any antimicrobial action of the foam.

It is possible that the foam attracts micro-organisms in the way the decaying nest material certainly does, but I do not think that the newly hatched larvae feed on this microflora. The larvae start feeding two days after hatching when their yolksac is consumed (Machado-Allison 1986; Mol in press), but at this point in time they have already moved to the bottom of the swamp. One day after hatching of the eggs of *H. littorale* the larvae are rarely seen at the nest site and foam is never present.

Table 5. Results of agar-diffusion sensitivity tests of sterilized foam of nests of *Hoplosternum littorale*

Test micro-organism	Culture conditions			sensitive to foam of *H. littorale*
	substrate	temperature (°C)	pH	
Laboratory stock-culture bacteria				
Escherichia coli	BBL-Mueller/Hinton	37	7.0	no
Proteus spec.	BBL-Mueleer/Hinton	37	7.0	no
Haemophilus influenzae	Chocolate agar	37	7.0	no
Staphylococcus spec.	Trypticase soy agar	37	7.0	no
Streptococcus pyogenes	Trypticase soy agar	37	7.0	no
Bacteria isolated from swampwater				
Pseudomonas spec.		35	7.0	no
species A (gram negative rod)		35	7.0	no
species B (gram negative rod)		35	7.0	no
species C (gram negative rod)		35	7.0	no
Fungi				
Aspergillus niger	PDA	30	5.6	no
Penicillium spec.	PDA	30	5.6	no

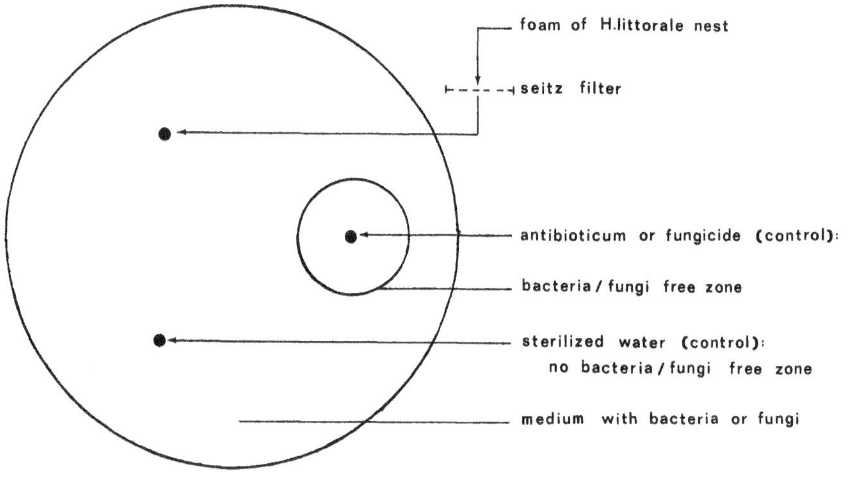

Figure 13. Agar-diffusion sensitivity test for antimicrobial action of foam of floating bubble nests of the callichthyid catfish *Hoplosternum littorale.* The test is considered positive if a bacteria/fungi free zone is detected around the pit with foam.

3.6. Concealment of Previous Spawns

Machado-Allison and Zaret (1984) point out that the structure of the foam nest of *H. littorale* allows the male to conceal the presence of previously deposited eggs from courting females. While it is certainly possible that the foam hides the eggs in a nest from the view of courting females, this property of the foam is probably irrelevant to multiple spawnings of *H. littorale* which all take place at the same time. *H. thoracatum*, and to a lesser extent *C. callichthys*, do spawn at successive intervals in one nest as is revealed by the presence at one point in time of eggs of more then one developmental stage. Consequently, concealment of previous spawns by the foam might have some importance to the guarding male of these species.

4. The Nests of *Hoplosternum littorale, Hoplosternum thoracatum* and *Callichthys callichthys* Compared

The foam nests of the three related armoured catfishes of the subfamily of Callichthyinae discussed in this text share many important characteristics, but also differ in a number of striking features.

All three species deposit their eggs in foam nests out of the oxygen-poor water under a cover of plant materials. Multiple spawns are deposited in the nests. The nests are constructed and guarded by the male, the most common form of parental care in bony fishes (Blumer 1979; Baylis 1981). For all three species nesting starts with the flooding of the swamps and reproductive activities continue until the swamps dry up (Mol in prep.). The eggs are more or less of the same size, and probably hatching times are quite similar (Table 1).

The main difference is between the nests of *Hoplosternum littorale* on the one hand and those of *Hoplosternum thoracatum* and *Callichthys callichthys* on the other hand. The nest of *H. littorale* is constructed from larger quantities of plant material than the nests of *H. thoracatum* and *C. callichthys* (ANOVA, $p < 0.0001$). The latter two species use dicotyledonous leaves as construction material, while *H. littorale* prefers grass, *Typha*, twigs and other filamentous materials. In selecting a suitable nest site *H. littorale* is not dependent on the presence of dicotyledonous trees. *H. thoracatum* and *C. callichthys* rarely build their nests in open herbaceous swamps (but a few nests under leaves of a water-lily were observed). The number of eggs deposited in nests of *H. littorale* is larger than the numbers of eggs/nest recorded for *H. thoracatum* and *C. callichthys* (ANOVA, $p < 0.0001$). It is also striking that in nests of *H. littorale* all eggs are spawned at the same time, resulting in a relatively shorter life-span of the nest than recorded for nests of *H. thoracatum* and *C. callichthys*.

Nests of *H. thoracatum* and *C. callichthys* differ in the microhabitat at the nest site. Although the number of eggs/nest is very similar for both species, nests of *C. callichthys* are smaller with respect to the nest material used in the construction of the nest. *T*-tests show dry weight of nest material ($p < 0.0001$), the number of leaves/nest ($p < 0.0001$) and leaf surface area ($p < 0.0001$) to be significantly different. Overlap of successive spawnings, measured as the percentage of nests with eggs of more than one developmental stage, is characteristic of nests of *H. thoracatum* (60%) but is almost absent in nests of *C. callichthys*.

The foam nests of the three callichthyid catfishes represent an adaptation to oxygen-depleted water. In Suriname, deoxygenation is generally associated with standing waters of the Coastal Plain (see Mol Chapter 7). The distribution of *H. littorale* is limited to this type of water (Mol in prep.). *H. thoracatum* and *C. callichthys* inhabit Coastal Plain standing waters too, but are restricted to habitats where trees are present, e.g. swamp forests.

Although the foam nesting habit probably evolved in oxygen-poor tropical standing waters, *H. thoracatum* and *C. callichthys* are also well represented in the fish fauna of rainforest creeks in the Interior of Suriname (Mol in prep.). The rugged landscape of the Interior is characterized by steep slopes and rocky or stony soils. Surface runoff after heavy local rains creates an unpredictable and violent increase of water velocity in these headwater streams (Gray and Wigham 1970; Noordam, Chapter 2). Small streams often rise suddenly overnight but cease to flow shortly after the rains and are then reduced to interconnected pools. Swift currents may easily sweep away the floating bubble nests of *H. thoracatum* and *C. callichthys*. However, even in these headwater streams quiet backwaters and obstacles that block the current provide some lentic habitats. *C. callichthys* probably avoids nesting in turbulent water by building its nests near the shore in very shallow water where drag reduces velocities (Gray and Wigham 1970) or in relatively calm water in cavities of tree roots. *C. callichthys* is even collected in brooks high in the Peruvian Andes (Lowe-McConnell 1987).

In the extreme biotope of headwater creeks it must be of adaptive advantage to produce several batches of eggs where the first may be endangered by unpredictable fluctuations in water velocity during peak flow events (Horn and Rubenstein 1984). Both *H. thoracatum* and *C. callichthys* reduce the risk of losing large quantities of eggs by building flexible nests and spreading successive spawnings in time. If a nest of *H. thoracatum* or *C. callichthys* is swept away by the current, the loss is not as serious as might have been if the strategy of simultaneous spawning of *H. littorale* was adopted. As long as conditions are favourable the nest is prolonged in time.

In the extensive swamps of the flat Coastal Plain swift water currents and sudden changes in water level are out of question (see Amatali, Chapter 3). Seasonal flooding is generally non-catastrophic and under these more stable conditions an aggressive defence of a large reproductive investment becomes profitable. The elaborate nests of *H. littorale* and the establishment of a well-defended territory are probably associated with the predictability of the aquatic environment in the Coastal Plain.

Summary

For five years, the floating bubble nests of three armoured catfishes, *Hoplosternum littorale*, *Hoplosternum thoracatum* and *Callichthys callichthys*, were studied at two localities in the Coastal Plain of Suriname. The structure of the nests is described in some detail mainly from cross-sections through frozen nests. Multiple spawnings and nest guarding of the males are described. Based on the structure of the nest, field experiments and laboratory tests, the function of the foam nest is discussed in relation to the habitat of the fish. The nests of *H. littorale*, *H. thoracatum* and *C. callichthys* differ in many striking features. The geographical distribution of the three species is also distinct, *H. littorale* being limited to the swamps of the Coastal Plain. *H. thoracatum* and *C. callichthys* also inhabit the lotic environment of small forest creeks in the Interior. An attempt is made to relate differences in the reproductive strategy of the three species to the environment in which they live.

Acknowledgements

I thank Ir. J.H. Mol and Ir. A.P. Mol for their generous support and Professor J.W. Osse for the stimulus he has given to me. Mr. Stolk (Weg Naar Zee) and Mr. J. Bissumbhar (Lelydorp) provided facilities in the swamps. I also thank Mr. S. Ramanand for help in field work at Lelydorp. Mr. A. Lieuw of the Academic Hospital of Paramaribo and Mrs. S. Van Dijk (CELOS) assisted with the agar-diffusion experiments on, respectively, bacteria and fungi.

References

Bagenal, T. and E. Braum, 1978. Eggs and early life history. In: T. Bagenal (ed.), Methods for assessment of fish production in freshwaters (3rd ed.). Blackwell, Oxford: 165–201.

Balon, E.K., 1975. Reproductive guilds of fishes: a proposal and definition. J. Fish. Res. Bd. Can. 32: 821–864.

Baylis, J.R., 1981. The evolution of parental care in fishes, with reference to Darwin's rule of male sexual selection. Env. Biol. Fish. 6: 223–251.

Blaxter, J.H., 1969. Development: eggs and larvae. In: W.S. Hoar and D.J. Randall (eds.), Fish Physiology, Vol.III. Academic Press, London: 177–252.

Blumer, L.S., 1979. Male parental care in the bony fishes. Quarterly Review Biology 54: 149–161.

Braum, E., 1978. Ecological aspects of the survival of fish eggs, embryos and larvae. In: S.D. Gerking (ed.), Ecology of freshwater fish production. Blackwell, Oxford: 102–131.

Burgess, W.E., 1989. Family Callichthyidae (armored catfishes). An atlas of freshwater and marine catfishes. A preliminary survey of the Siluriformes. T.F.H. Publications, Neptune City, 784 pp.

Breder, C.M. and D.E. Rosen, 1966. Modes of reproduction in fishes. Natural History Press, Garden City N.Y., 941 pp.

Carter, G.S. and L.C. Beadle, 1930. The fauna of the swamps of the Paraguayan Chaco in relation to its environment. I. Physico-Chemical nature of the environment. J. Linn. Soc., Lond. (Zool.) 37: 205–258.

Carter, G.S. and L.C. Beadle, 1931. The fauna of the swamps of the Paraguayan Chaco in relation to its environment. II. Respiratory adaptations in the fishes. J. Linn. Soc., Lond. (Zool.) 37: 327–368.

Duellman, W.E. and L. Treub, 1986. Biology of amphibians. McGraw-Hill, New York, 670 pp.

Furtado, J.L. and S. Mori (eds.), 1982. Tasek Bera: the ecology of a freshwater swamp. Dr.W.Junk, The Hague.

Gautier, J.Y., P. Planquette and Y. Rouger, 1988. Etude éthologique de la relation mâle-femelle au cours du cycle de reproduction chez *Hoplosternum littorale*. Rev. Ecol. (Terre Vie) 43: 389–398.

Gee, J.H. and J.B. Graham, 1978. Respiratory and hydrostatic functions of the intestine of the catfishes *Hoplosternum thoracatum* and *Brochis splendens* (Callichthyidae). J. Exp. Biol. 74: 1–16.

Gorzula, S.J., 1977. Foam nesting in leptodactylids: a possible function. British J. Herpetol. 5: 657–659.

Gosline, W.A., 1940. A revision of the neotropical catfishes of the family Callichthyidae. Stanford Ichthyological Bulletin 2: 1–29.

Gray, D.M. and J.M. Wigham, 1970. Peak flow: rainfall events. In: D.M. Gray (ed.), Handbook on the principles of hydrology. Water Information Center, Huntington: 8.1–8.102.

Heyer, W.R., 1969. The adaptive ecology of species groups of the genus *Leptodactylus* (Amphibia, Leptodactylidae). Evolution 23: 421–428.

Hoedeman, J.J., 1952. Notes on the ichthyology of Surinam (Dutch Guiana). The catfish genera *Hoplosternum* and *Callichthys*, with key to the genera and groups of the family Callichthyidae. Beaufortia 1: 1–11.

Horn, H.S. and D.I. Rubenstein, 1984. Behavioural adaptations and life history. In: J.R. Krebs and N.B. Davies (eds.), Behavioural ecology: an evolutionary approach (2nd ed.). Blackwell, London: 279–298.

Hostache, G., M. Pascal, M. Kernen and C. Tessier, in press. Température et incubation chez l'Atipa, *Hoplosternum littorale* (Teleostei, Siluriforme). Aquat. Living Resour.

Hostache, G., P. Vallat and C. Tessier, 1990. Maîtrise du cycle d'élevage de l'Atipa. Nature Guyanaise 4: 33–41.

Huebner, E. and G. Chee, 1978. Histological and ultrastructural specialization of the digestive tract of the intestinal air breather *Hoplosternum thoracatum* (Teleost). J. Morphol. 157: 301–328.

Johansen, K., 1970. Air breathing in fishes. In: W.S. Hoar and D.J. Randall (eds.), Fish Physiology, Vol.IV. Academic Press, London: 361–411.

Kramer, D.L., C.C. Lindsey, G.E. Moodie and E.D. Stevens, 1978. The fishes and the aquatic environment of the Central Amazon basin, with particular reference to respiratory patterns. Can. J. Zool. 56: 717–729.

Leentvaar, P., 1975. Hydrobiological observations in Surinam, with special references to the man-made Brokopondo Lake in 1964. Stud. Fauna Suriname Guyanas 15: 1–175.

Lowe-McConnell, R.H., 1984. The status of studies on South American freshwater food fishes. In: T.M. Zaret (ed.), Evolutionary ecology of neotropical freshwater fishes. Dr.W.Junk, The Hague: 139–156.

Lowe-McConnell, R.H., 1987. Ecological studies in tropical fish communities. Cambridge University Press, Cambridge, 382 pp.

Luquet, P., T. Boujard, P. Planquette, Y. Moreau and G. Hostache, 1989. The culture of *Hoplosternum littorale*: state of the art and perspectives. In: Advances in tropical aquaculture, AQUACOP-IFREMER, Paris: 511–516.

Machado-Allison, A., 1986. Aspectos sobre la historia naturel del "Curito" *Hoplosternum littorale* (Hancock, 1828) (Siluriformes-Callichthyidae) en el bajo llano de Venezuela: desarrollo, alimentacion y distribucion espacial. Acta Cient. Venez. 37: 72–78.

Machado-Allison, A. and T.M. Zaret, 1984. Datos sobre la biologia reproductiva de *Hoplosternum littorale* (Siluriformes-Callichthyidae) de Venezuela. Acta Cient. Venez. 35: 142–146.

Martin, A.A., 1970. Parallel evolution in the adaptive ecology of leptodactylid frogs of South America and Australia. Evolution 24: 643–644.

Merron, G.S., K.K. Holden and M.N. Bruton, 1990. The reproductive biology and early development of the African pike, *Hepsetus odoe*, in the Okavango Delta, Botswana. Env. Biol. Fish. 28: 215–235.

Mol, J.H.A., in press. Sustainable exploitation of neotropical fish resources: aquaculture trials with three callichthyid catfishes, *Hoplosternum littorale*, *Hoplosternum thoracatum* and *Callichthys callichthys*. In: Proceedings of the 2nd Joint Conference of the University of Guyana and the University of Suriname.

Nijssen, H., 1970. Revision of the surinam catfishes of the genus *Corydoras* Lacepede, 1803 (Pisces, Siluriformes, Callichthyidae). Beaufortia 18: 1–75.

Petrovicky, I., 1979. Die Zucht der Panzerwelse *Hoplosternum thoracatum* (Cuvier and Valenciennes, 1840) und *Dianema longibarbis* Cope, 1972 ist fast gleich. Aquarien Terrarien Zeitschrift 22: 73–78.

Richards, O.W. and R.G. Davies, 1977. Imm's general textbook of entomology (10th ed.). Chapman & Hall, London, 1354 pp.

Richter, H.J., 1979. Das Buch der Labyrinthfische. Neumann Verlag, Leipzig (Dutch edition, 168 pp.).

Rondeel, A.J., 1965. The Surinam fish protection legislation. De Surinaamse Landbouw 4: 63–64.

Scherpe, U., 1981. Nicht ganz einfach, aber interessant sind Nandiden im Aquarium. Das Aquarium 139: 18–24.

Schubert, T.J., 1948. Breeding the spotted *Callichthys*. Aquarium 7: 80–81.

Singh, T.B., 1978. The biology of the cascadura *Hoplosternum littorale* Hancock 1834 with reference to its reproductive biology and population dynamics. Ph.D. thesis, University of the West Indies, Trinidad, 298 pp.

Sioli, H., 1950. Das Wasser im Amazonasgebiet. Forsch. Fortschr. 26: 274–280.

Snow, J.W., 1976. The climate of Northern South America. In: W. Schwerdtfeger (ed.), Climates of Central and South America. Elsevier, Amsterdam: 295–379.

Winemiller, K.O., 1987. Feeding and reproductive biology of the currito, *Hoplosternum littorale*, in the Venezuelan llanos with comments on the possible function of the enlarged male pectoral spines. Env. Biol. Fish. 20: 219–227.

Zukal, R., 1981. *Hoplosternum thoracatum*, an unusual bubblenest breeder. Trop. Fish Hobbyist 30: 87–92.

Vogt, D., 1970. Synbranchiformes. In: B. Grzimek, Enzyklopädie des Tierreiches. Vol.V. Kindler Verlag, Zurich (Dutch edition: 51–52).

11. The herpetofauna of floating meadows[1]

MARINUS S. HOOGMOED

1. Introduction

The Amazon area, which for convenience's sake is here considered to include the Guianas, and therefore could better be called the Central cis-Andean tropical lowlands (Hoogmoed 1979b; Lynch 1979), is renowned for the many rivers and the large volume of water (20% of the Earth's freshwater) flowing through it. The water in this area can be divided into three classes (Junk 1983; Sioli 1984), viz.:

(1) Black water has a dark brown colour, has a very low concentration of dissolved minerals, has no suspended particles and has a high acidity (pH 3.8–4.9), due to dissolved tannines leached out of leaflitter. Notable examples of black-water rivers are the Rio Negro in northern Brazil, draining the westernmost part of the Guiana Shield, and the creeks in the white sand Savanna Belt of the Guianas, of which one in Suriname is known as Cola Creek.

(2) Clear water is greenish to transparent, with no suspended particles and a concentration of dissolved minerals that may range from very low to relatively high, and a pH that ranges between very acid and neutral (pH 4.5–7.8). Good examples are the Rio Tapajos and the Rio Xingu, and forest creeks in Suriname (witiewatra) south of the Savanna Belt.

(3) White water has a milkish white to grey colour due to the suspended inorganic material, it is turbid and contains a relatively high concentration of dissolved minerals and has a pH of about 7 (6.2–7.2). Examples are the Rio Solimoes, the Rio Madeira, and in Suriname the lower courses of the larger rivers.

Generally main river channels are unsuitable for the growth of macrophytes, because of the water current and the depth, but when the level of dissolved nutrients is sufficient (e.g. in white water rivers), large quantities of free floating

[1] This is Notes on the herpetofauna of Suriname XII. Notes on the herpetofauna of Suriname XI appeared in Studies in honour of Dr. Pieter Wagenaar Hummelinck, Found. Sc. Res. Surinam and the Netherlands Antilles 123: 65–73.

Paul E. Ouboter (ed.), Freshwater Ecosystems of Suriname, 199–213.

plants may drift to the main river channels from the shores and from flood plain lakes (Junk and Howard-Williams 1984). The quantity of floating plants and herbaceous shoreline vegetation decreases with decreasing levels of nutrients. In extremely nutrient-poor and acid rivers aquatic macrophytes may only be represented by small patchy stands of *Montrichardia arborescens* near the shores (Junk and Howard-Williams 1984). When in nutrient-poor black water rivers the nutrient level and the pH rise, because of human agricultural activities, the aquatic vegetation may increase, as was observed in the Anavilhanas archipelago in the Rio Negro, Brazil (Junk and Howard-Williams 1984).

The floating vegetation (popularly known as "floating meadows") is independent of fluctuations in water level, and the species of plants encountered in these floating meadows "are more typical of swamp vegetation than real flood plain vegetation" (Junk and Howard-Williams 1984). Floating meadows may appear in many different shapes. In the large river systems of the Amazon proper, they may be attached to the shore on one side, bordered by open water on the other side. Initially, the floating meadows are formed by free floating plants, which secondarily may be invaded by terrestrial plants that cause the entire structure to become more firm. They may be severed from the shore during high water, and turn into islands of free floating vegetation (called matupá in Brazil), serving as rafts and an important means of long-range dispersal for all plants and animals living in them. In small rivers they may form a vegetation cover from shore to shore, thus effectively blocking any boat traffic, as happens on the Cusewijne and Para Rivers in Suriname. Another form of this vegetation can be observed in large, open, swampy areas, the so-called "zwiebel zwampen" (=swaying swamps), which are characterised by a more or less firm superficial layer, with water underneath (Lindeman 1953). Floating meadows can be bordered by open formations (savanna, agricultural land) or by forests (gallery, terra firme or varzea forest), and as such form a distinct habitat by themselves.

Studies on the herpetofauna of floating meadows are scarce (Hödl 1977, 1990; Hill *et al.* 1987), and in the case of Hödl restricted to frogs. Hödl's (1977) study, which was mainly intended to study frog calls in the surroundings of Manaus, shows that the majority of the frogs found there are treefrogs (Hylidae); one species belongs to the aquatic family Pseudidae and one to the (generally terrestrial) family Leptodactylidae. Hödl (1990) gives an overview of the modes of reproduction used by the species of frogs living in floating meadows. Hill *et al.* (1987) give a detailed description of floating meadow vegetations in N. Australia, both the botanical composition and the structure of the mats. Observations on herpetofauna were scarce, and apparently were only made during daytime. Only two frog species were observed, but if observations had also been made at night, the number of species might have been higher. Schmidt (1928), when discussing *Caiman yacare* from Paraguay, mentions that "The vast mats of floating grass and water hyacinth, the 'camelotte', constitute a biocaenosis of unusual interest. The camelotte affords shelter to the juvenile caimans, as is shown by the capture of two small specimens in this situation. .... The importance of the camelotte as an agent of dispersal has frequently been commented upon. Floating islands of this vegetation of varying extent are of frequent occurrence in the lower Paraná, and often reach Buenos Aires and even Montevideo. During exceptionally high water, the camelotte forms the last available refuge for many land forms, and the rafts sent down river at such times carry tropical lizards and snakes, as well as both small and large caimans, to the mouth of the La Plata". Lang (1989) says that the breeding cycle of *Caiman crocodilus* appears to be closely linked to fluctuations in water level, and that "in the Amazon basin this species prefers to build nests in the grass mats that cover the shallow parts of large lakes".

2. Methods

For details about the botanical aspects of floating meadows in Suriname, I may refer to the chapter by Teunissen (Chapter 5).

Since 1966 I have been studying the herpetofauna of Suriname and adjacent countries, and since 1968 field surveys have been conducted in a somewhat regular manner in most major habitats in Suriname. From my own observations I could give data on the open swamps in the coastal area with floating vegetation mats (1969–1990), and on floating meadows in the Para River (October 23 – November 5, 1990). Other data are available from the notebooks of, and the collection made by P.H.Creutzberg, one of the members of the Suriname Expedition 1948–49 (Bakker and Lanjouw 1949; Lindeman 1953), who paid special attention to reptiles and amphibians. In order to place the Suriname data in perspective, comparisons are made with a locality in Bolivia (Perseverancia), where I obtained data from 5 to 17 September 1990, with a locality in Brasil (Manaus area) from the literature, and with another Brazilian locality (Caxiuana) where I collected data from October 21 to November 19, 1992.

All areas were sampled at night from boats or on foot, with headlamps as a light source. Frogs collected were anaesthesized with MS 222 (methane-sulphonate of meta-aminobenzoic acid ethyl-ester), reptiles were killed with an injection of Nembutal (Pentobarbitol sodium) and subsequently specimens were fixed in formalin 4% and preserved in ethyl-alcohol. The material referred to here is in the collections of the Nationaal Natuurhistorisch Museum (formerly Rijksmuseum van Natuurlijke History), Leiden, Netherlands, in the Nationale Zoölogische Collectie, Paramaribo, Suriname, in the Museu Paraense Emilio Goeldi, Belém, Brasil, and in the Museo de Historia Natural "Noel Kempff Mercado", UAGRM, Santa Cruz de la Sierra, Bolivia.

3. Description of Areas Discussed

3.1. Suriname: Para River Near Confluence with Coropina Creek

This is a narrow (width 10–30 m) black water river (average pH 4.3; secchi depth (visibility) 80 cm) running through the belt of white sand that is situated more or less parallel to the coast. Up to Hanover approximately, it is influenced by the tides. It is bordered by swamp forest, mixed terra firme forest (= mesophytic dryland forest) and marsh forest, and savannas (Teunissen 1978) with areas influenced by agricultural activities on a small scale. At the time of my visit the floating meadows extended along both banks of the river, where they blended into well-developed stands of *Montrichardia arborescens*, but the mats from each side were not in contact. In the free water some loose *Salvinia* and small *Eichhornia* occurred, and in some places large stands of *Cabomba aquatica* (mainly submerged). In one place part of the vegetation mat that had been present the year before was still evident, containing some toppled, medium trees. The thickness of the mat was estimated to be 5 m. Approximately one year before my visit the Para was entirely closed by this vegetation and at present the river apparently is closed again. In some places near the bank these floating mats of vegetation, consisting mainly of grasses and sedges, but also supporting bushes and small trees, were strong enough to carry a person and effectively prevented boats from reaching the shore (Fig. 1A).

3.2. Bolivia: Perseverancia, Rio Negro

Frey (1990), gives the following detailed description:

> Open water is bordered in some places by steep, bare clayey banks that sometimes show stony layers, but generally the banks are gently sloping to nearly flat. These flat parts are covered by large mats of floating vegetation that often extend across the whole river. the greatest part of the river has open water, bordering either on gallery forest, on floating meadows, or on open, marshy vegetation.
>
> Large parts of the river, especially shallow ones and the lagunas are covered by a special type of vegetation that extends from the riverbank into the open water for varying distances. It forms extremely dense floating mats of partly rooting, partly floating species together with free-floating plants. They are composed mainly of *Eichhornia crassipes*, *Eichhornia azurea*, *Ipomoea tuba*, *Cyperus* sp., *Utricularia* sp., *Salvinia* sp., *Azolla* sp., *Hydrocotyle* sp., *Merremia* sp., *Cissus* sp., Gramineae spp., *Polygonum* sp., *Ludwigia helminthorrhiza* and two other *Ludwigia* spp. In some places a person could walk on these meadows, but for the most part they were too thin and could not support any large weight.

The water of the Rio Negro (Bolivia) is opaque brown, with a visibility of about 1 m and a pH of 5.5–6. The current at the time of my stay was slow, and

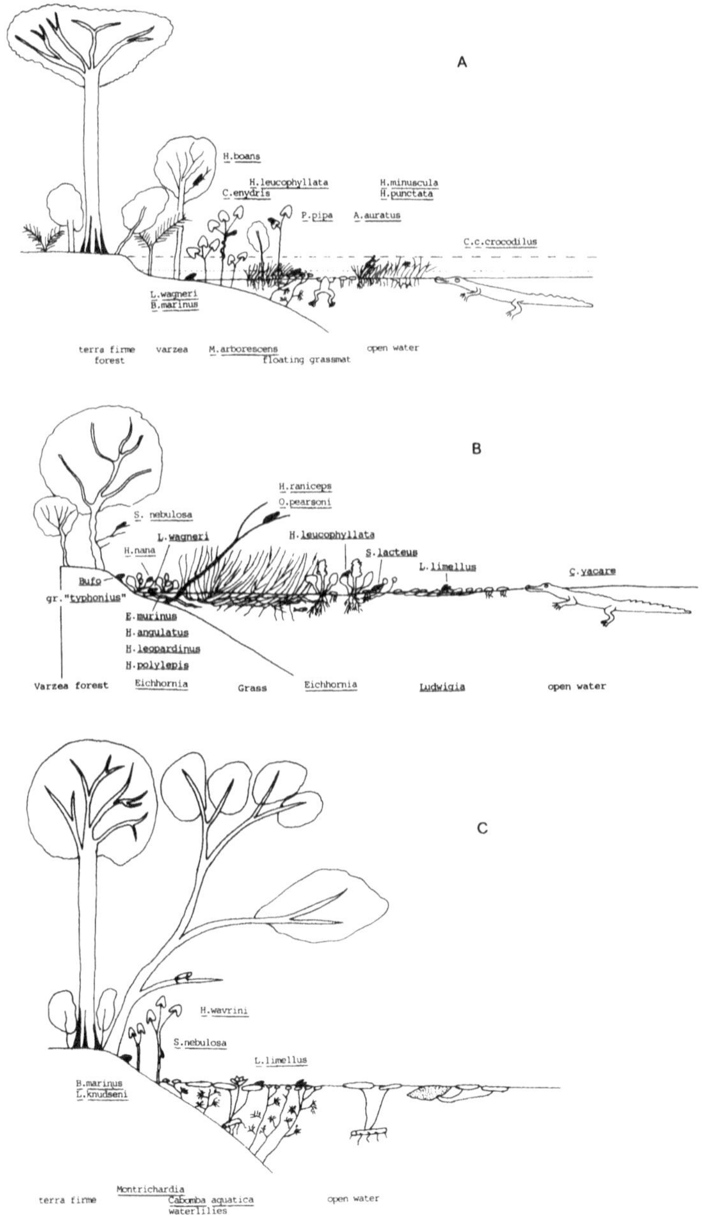

Figure 1. Schematic cross-section of floating vegetation in the Para River, Suriname (A), in Perseverancia, Bolivia (B) and in Caxiuana, Brazil (C). The line at the bottom of each drawing indicates the type of vegetation above it. Names above the water surface, or just below it indicate species of herpetofauna present in that specific area. The broken lines indicate the water level during the rainy season, the dotted lines indicate high tide in the dry season in areas influenced by the tides. In the bottom drawing the stippled area presents a mass of uprooted waterlily rhizome with mud attached. Vegetation and animals are not to scale.

the temperature was generally 26 °C. Vegetation along the banks consisted of varzea (temporarily flooded) forest (Fig. 1B).

3.3. Brazil: Caxiuana, Estaçao Cientifica "Ferreira Penna", Rio Curuá

The water of the Rio Curua belongs to the clear water type, and although it had a slightly brownish colour the visibility was at least 2.5 m. The water level was influenced by the tides. In its lower course the river is wide, with large stretches of open water in which lay several islands covered by terra firme forest. Near its confluence with the Rio Caxiuana, there are floating mats mainly consisting of *Eichhornia* sp. and grasses, which are attached to the shore. The upper course of the river narrows considerably over a short distance and turns into a narrow forest creek. In the area just before the river turns into a forest creek, extensive fields of floating vegetation were present along the shores, consisting of water lillies, *Cabomba aquatica*, and near the shore *Montrichardia* and grasses. Also some still waters branching off from the main river channel were covered with this vegetation. Nowhere was this vegetation thick enough for a person to stand on. Generally, the vegetation consisted of individual plants not forming a mat, although uprooted water lillies sometimes were provided with big masses of mud around their rhizome (Fig. 1C).

Because no systematic survey of herpetofauna was done in the coastal swamps of Suriname no data are presented here on the vegetation or on the herpetofauna in the table, although ecological data on species obtained there have been incorporated in the discussion of the species.

4. Potential Herpetofauna of Floating Vegetation in the Para River

It is not easy to delimit exactly the species of amphibians and reptiles belonging to the association of floating meadows, as some of them were not (yet) found in the floating meadows proper, but on the riverbank or in trees and bushes in them or close to them, or in the water close to the floating meadows, although it seems likely they enter them on occasion. I therefore include them in this overview, but in Table 1 mark their presence by (x).

4.1. Reptiles

4.1.1. Snakes

Corallus enydris (Linnaeus) (Boidae) is mainly present in branches of trees, bushes or on stems of *Montrichardia arborescens* at the edge of the water, or overhanging the floating vegetation. It has also been observed in bushes in sedge swamps. In the Para River and Coropina Creek area it was quite abundant in

Table 1. Species known to occur in or near floating meadows (resp. x and (x)). Based on own fieldwork and on literature data. Numbers for localities: 1. Brazil, Manaus, Lago Janauari (Hödl, 1977); 2. Suriname, Para River; 3. Bolivia, Perseverancia; 4. Brazil, Caxiuana. The question-marks in column 1 indicate that no data on reptiles are available from that locality.

	1	2	3	4
AMPHIBIA				
Hylidae				
Hyla nana-like	x		(x)	
Hyla minuscula		x		
Hyla haraldschulzei	x			
Hyla rossalleni	x			
Hyla triangulum	x			
Hyla leucophyllata		x	x	
Hyla punctata	x	x	x	
Hyla raniceps	x		(x)	
Hyla lanciformis	x			
Hyla geographica		(x)	x	
Hyla wavrini	x			x
Hyla boans		(x)		
Osteocephalus pearsoni			x	
Scinax boesemani	x			
Scinax nebulosa	x	x	x	x
Scinax rubra		x		
Sphaenorhynchus carneus	x			
Sphaenorhynchus dorisae	x			
Sphaenorhynchus lacteus	x		x	
Pseudidae				
Lysapsus limellus bolivianus			x	
Lysapsus limellus caraya				x
Lysapsus limellus laevis	x			
Leptodactylidae				
Leptodactylus knudseni				(x)
Leptodactylus wagneri	x	(x)	x	
Bufonidae				
Bufo marinus		x		(x)
Pipidae				
Pipa pipa		x		
REPTILIA				
Sauria, Iguanidae				
Anolis auratus	?	x		
Teiidae				
Kentropyx striata	?	(x)		
Ophidia, Boidae				
Corallus enydris	?	(x)		(x)
Eunectes murinus	?		x	
Ophidia, Colubridae				
Helicops angulatus	?		x	
Helicops leopardinus	?		x	
Helicops polylepis	?		x	
Oxyrhopus petola	?		x	

Table 1. (continued)

	1	2	3	4
REPTILIA (continued)				
Viperidae				
Bothrops atrox	?	(x)		
Crocodylia, Alligatoridae				
Caiman crocodilus crocodilus	?	x		
Caiman (crocodilus) yacare	?		x	
Paleosuchus palpebrosus	?	x		

stands of *M. arborescens*. The population in this area showed a great variety in colour, with grey, beige, brown, yellow, orange and reddish specimens present.

Bothrops atrox (Linnaeus) (Viperidae) according to Campbell and Lamar (1989) prefers damp locales and is found in close association with creeks, lakes, or river concourses. In Suriname it is generally found in low vegetation close to creeks and swamps, but also near human habitations. In the Coropina Creek one specimen was found on the muddy ground at the edge of the water among a stand of *M. arborescens*.

Neither *B. atrox*, nor *C. enydris* were found in the floating meadows proper.

4.1.2. Lizards

Anolis auratus Daudin (Iguanidae) is found in open habitats throughout the country, e.g. on beaches, in gardens, in meadows, in roadsides, on savannas (Hoogmoed 1973; 1975). Vanzolini (1972) reported the species from a single-log walk across a stretch of grass-covered swampy ground. In the Para River several specimens were found at night, sleeping on grass stalks 100 cm above the water, in mats of floating vegetation that could support the weight of a person.

Kentropyx striata (Daudin) has been reported from wet habitat by several authors (Boos and Quesnel 1971; Müller 1971; Vanzolini 1972; Hoogmoed 1973, 1975). There is a clear relationship with swamps and inundated areas. In Suriname several specimens were found in *Cyperus giganteus* swamps by the Suriname Expeditie 1948–49, in coastal Suriname, while I collected several specimens on the airstrip Kaysergebergte in inundated savannas. In the Para River area this species was not found on or near the floating mats of vegetation, but it was present on partly inundated savannas along the bank of the river near Zanderij Airport.

4.1.3. Turtles

In grassmats of the Upper Marataka River P.E. Ouboter (pers. comm.) observed *Rhinoclemmys punctularia* (Daudin) (Emydidae). The species was not observed in the Para River, but most probably occurs there as well.

4.1.4. Caimans

Two species of caimans have been found associated with floating mats of vegetation in the Para River. More data on caimans and their habitats will be published by Ouboter in forthcoming publications.

Caiman crocodilus is the most abundant species. It was not found in or on the mats of floating vegetation during the period of observation, but specimens were observed in open water at the edge of the floating mats. They were also observed in open water, and near the riverbank among stands of *Montrichardia arborescens*. In other seasons *C. crocodilus* was regularly seen on the mats (P.E. Ouboter, pers. comm.). It is not clear what the animals do when the river is completely blocked by vegetation. Presumably open areas remain, and possibly the caimans concentrate there. The species is known to construct its nest on small elevations remaining dry during the rainy season, often among stems of *M. arborescens*, in other areas (Ouboter and Nanhoe 1985, 1987). Ir. F.C. Bubberman sent me some pictures of a nest of (presumably) this species, constructed in the extensive swamps with grassy vegetation of the Barbakoeba Creek in northeastern coastal Suriname. Nowhere near the nest was high vegetation evident. For more details on that area see Versteeg and Bubberman (1992). Hill *et al.* (1987) cite several authors who report nests of this species built on floating meadows.

Paleosuchus palpebrosus (Cuvier) seems to be less abundant than *C. crocodilus*. In some places they occur close to each other, in other places there seems to be a predominance of one of the species. It was not observed near floating vegetation, but either in open water close to the bank or on the bank. However, P.E. Ouboter (pers. comm.) found it during the rainy season on the mats of floating vegetation in inundated areas. In contrast to the opinion generally held that this is a species of creeks and rivers in the interior, with fast flowing water, one specimen (in RMNH) was collected by the Suriname Expedition 1948–49 in the area between Wia Wia Beach and the Wane Creek, in a swamp (fieldnr. 3099).

Paleosuchus trigonatus only occurs in the open reaches of the Lower Para, where floating vegetation was not present.

4.2. Amphibians

Treefrogs are most evident in the floating mats of vegetation. No quantitative data could be collected, but my impression was that *Hyla punctata*, *Hyla misera* and *Scinax nebulosa* were the most abundant species.

Hyla boans (L.) is the largest species of the genus (snout-vent length (svl) 11 cm). It is present in forested areas which are not inundated (terra firme), and males call from both the ground (where they construct circular nests) and from trees up to a height of several meters. In the Para River area in Suriname the species was not noted in the floating vegetation itself, but was calling from trees on the riverbanks. It occurs from eastern Panama throughout the Guianas, the

Orinoco and Amazon River basins and the Pacific lowlands of Colombia and in Trinidad (Frost 1985; Hoogmoed 1990). In Suriname it occurs throughout the country south of the coastal Savanna Belt. Hoogmoed (1990) discussed the distribution of this taxon and had doubts about the conspecificity of specimens from eastern Panama and Colombia on the one hand, and specimens from lowland cis-Andean South America on the other hand. Hödl (1977) reported "*Hyla boans*" from trees at the edge of the floating meadows near Manaus, but as pointed out by Hoogmoed (1990), the material he studied belonged to *Hyla wavrini* Parker, a species restricted to varzea along large rivers in Amazonia. *H. wavrini* was also observed in Caxiuana where specimens were calling from trees projecting over the river and on stems of *Montrichardia* growing near the riverbank and surrounded by floating vegetation.

Hyla geographica Spix is a relatively large treefrog (svl 6–8 cm) which can be found in areas where open vegetation (swamps or pools) borders on forests. In the interior of Suriname it is concentrated in places where trees have fallen along creeks ("chablis"). It calls from rushes and low vegetation at low elevations 30–100 cm above the water. In the Para River area in Suriname it was not observed in floating vegetation, but was heard calling from the riverbank. It is distributed throughout tropical South America east of the Andes and in Trinidad (Frost 1985). In Suriname it occurs throughout the country.

Hyla leucophyllata (Beireis) is a medium-sized (svl 4 cm) frog which can be found throughout Suriname where forest borders on open vegetation. Its call is very characteristic and easily recognised at a distance. Males call from leaves 2–5 m above the ground/water surface. It occurs throughout the Amazon and the Guianas (Frost 1985), and also along the Brazilian coast (Duellman 1977).

Hyla minuscula Rivero is a small (svl < 2 cm) species that can be found mostly in the same places as *H. punctata*, but also in meadows. Males call sitting on grass around pools or on grass growing in water, but also on leaves 5 –200 cm above ground level. It occurs from Venezuela to Belém, Brazil (Frost 1985). In Suriname its distribution seems to be restricted to the coastal area.

Hyla punctata (Schneider) is a medium-sized (svl 4 cm) species present in coastal Suriname in wet places like swamps, roadside ditches, rice paddies, and floating meadows. It has a wide distribution in South America (Hoogmoed 1979a; Frost 1985). It calls from low vegetation between 20 and 200 cm above the water surface, and in the floating mats of vegetation it was found calling from grasses.

Scinax nebulosa (Spix) can be found in most wet vegetations (see *H. punctata*) throughout Suriname (mostly coastal area, and Sipaliwini). It is the smallest member of the *Scinax rostrata* group, reaching a snout-vent length of about 2.5 cm. It calls from low vegetation, where it sits on leaves with its head pointing down at 30–60 cm above the water. In all localities from where it was reported the species not only occurred in the floating vegetation, but also in the vegetation immediately bordering the water. It was known from the Guianas, the Lower Amazon and northeastern Brasil (Duellman 1972; Frost 1985), and is here reported for the first time from Bolivia. This species was reported upon by Hödl (1977) as *Hyla egleri*.

Scinax rubra (Laurenti) is a very common frog (svl 4 cm) in the coastal area of Suriname where it is very abundant in perianthropic situations. It freely enters houses, but can also be found in semi-natural and natural situations. Males call from within low vegetation (grasses) at about 10 cm above the ground or the water surface. It also can be found on stems of *Montrichardia arborescens*, and on twigs and leaves of bushes up to 100 cm above ground level. It has an extensive distribution in lowland cis-Andean South America (Frost 1985).

Leptodactylus wagneri (W. Peters) is a medium-sized frog (svl 6–7 cm) that lives throughout Suriname in wet habitats in forested areas. Males are found calling from the ground close to water, in low vegetation or among leaflitter. No specimens of this species were found in the floating vegetation, though one specimen was collected on a muddy riverbank close to floating vegetation. It was present in floating meadows in Manaus and in Perseverancia (Table 1). According to Frost (1985) it occurs on several Antilles and in South America south to the Tropic of Capricorn. W.R. Heyer (Washington), is studying this group, in which several species are involved, at the moment.

Bufo marinus (L.) is one of the largest species of anurans known. In the collection of the Nationaal Natuurhistorisch Museum, Leiden is a specimen (RMNH 4981) from the Gonini River, Suriname that has a snout-vent length of 23 cm and measures 23.5 cm across the paratoids. Such large specimens are only known from the interior of Suriname, whereas in the coastal area specimens stay much smaller. In the Para River area one specimen was observed at night sitting on floating vegetation at the open water edge. Males call sitting in shallow water. The original distribution of the species extended from southern Texas to Central Brazil. It has been introduced on many islands of the Antilles and in the Pacific, often with disastrous results for the local fauna.

Pipa pipa (L.) is a large (svl 14 cm), flat, exotic looking aquatic frog which occurs in swamps, rivers and creeks throughout Suriname. It was present in the Para River, but was especially abundant in the Coropina Creek, where it was observed floating obliquely in the water, either freely, or between stems of *Montrichardia arborescens*, between roots, or between dead branches fallen into the water. Specimens were mainly observed near the riverbanks, and were also present in areas of open water within floating vegetation. It occurs in the entire Amazon basin and in Trinidad (Frost 1985).

5. Discussion

In the literature, data on the herpetofauna of floating meadows of South America are only presented by Hödl (1977). He only considered frogs, and did not provide data on reptiles.

Only one species (*Scinax nebulosa*) was present in all four localities (Table 1) of which data are available. One species (*Hyla punctata*) was present in three localities, and because of the terra firme vegetation bordering on the river was not expected to occur (and was not found) in Caxiuana. *Leptodactylus wagneri* also was present in three localities. Most species occur in two or one locality only. Most of these (*Hyla leucophyllata*, *Scinax boesemani*, *S. rubra*, *Sphaenorhynchus lacteus*, *Bufo marinus*, *Pipa pipa*, *Anolis auratus*, *Eunectes murinus*, *Corallus enydris*, *Helicops angulatus*, *H. leopardinus*, *Caiman crocodilus*, *Paleosuchus palpebrosus*) on the basis of their general distribution and their habitat requirements could be expected to occur in other areas as well. The remaining species have more restricted distributions and are not expected to occur in the other areas. In some instances related species pairs seem to replace each other in different localities (*H. boans* – *H. wavrini*; *H. leucophyllata* – *H. triangulum*), although both species of the last pair mentioned are known to occur sympatrically in Santa Cecilia, Ecuador (Duellman 1978).

Only one locality (Manaus: Hödl 1977) of the four mentioned here was studied systematically (for frogs only) for an extended period, and yielded the highest number of frogs. In the Para River in Suriname seven species of frogs and three species of reptiles (species with (x) in Table 1 not counted) were found associated with floating meadows, after a short fieldwork period of thirteen days only. Several species occurring in the area probably will be found associated with the floating meadows in the future after more protracted studies. In Perseverancia eight species of frogs and six species of reptiles were found. Relatively little is known about the general herpetofauna of the area (Frey 1990), so it is difficult to predict if and how much these numbers could increase. In Caxiuana, during a field period of one month, only three species of frogs and no reptiles were found associated with floating vegetation. In the area studied in Caxiuana no real floating meadows were present like in the other three localities. Here we only could speak of floating patches of aquatic vegetation (no grasses present), and consequently for many species the proper habitat may have been lacking.

The presence of two terrestrial lizards (*Anolis auratus*, *Kentropyx striata*) in floating meadows or in other very wet habitats merits some comments. As described under the species these heliotropic lizards occur in open habitats (wet or dry) throughout Suriname and elsewhere. This leads to the conclusion that one of the most important, if not *the* most important, factor for these lizards determining their occurrence in habitats is the amount of sunshine entering it, and that it is irrespective of the wetness of the habitat. In Amazonia *Kentropyx altamazonica* Cope seems to be associated with floating meadows (Avila-Pires, pers. comm.), and other open vegetation (Gallagher *et al.* 1986). Hill *et al.* (1987) mention that in N. Australia they found one lizard to be more abundant on the floating meadows than on the adjacent bank. No data are available to check whether something similar happens in Suriname.

Although in the Rio Negro, Bolivia, turtles of several species abound, none were ever found associated with the floating meadows. In Caxiuana several specimens of *Podocnemis unifilis* (Troschel) were observed, a juvenile eating *Eichhornia* leaves of flotsam plants. Here they were neither observed in or around the floating vegetation.

During fieldwork in Perseverancia temperatures suddenly dropped considerably over a very short period (a phenomenon known as surazo in Bolivia, friagem in Amazonian Brazil, with cold winds blowing from the south) to a low of 9 °C early in the morning of September 15, 1990. During this surazo it was observed that *Lysapsus limellus bolivianus* and *Hyla punctata* were still calling at night at air temperatures of 16 °C (as contrasted to > 20 °C normally). The temperature of standing water was down to 21 °C, from a normal of 26 °C. Such cold spells occasionally occur in the southern Amazon area, but are unknown in the Lower Amazon and in Suriname. Thus, it can be concluded that both species involved physiologically are quite adaptable, with different populations in different areas being confronted with widely diverging environmental conditions.

6. Conclusions

From the above it can be concluded that there is no such thing as *the* herpetofauna of floating meadows in the Amazon basin. Some species are present in all areas studied, other species join the assemblage in the areas where they occur. No species is exclusively limited to this vegetation, although *Lysapsus limellus* comes pretty close to this. Apparently there is no special adaptation to this habitat, and floating meadows are invaded by species (mainly Hylidae) generally present in other open herbaceous vegetations in wet places (often bordering on riverbanks), or by aquatic frogs (Pseudidae, Pipidae). Terrestrial frogs (Bufonidae, Leptodactylidae) only rarely enter this type of vegetation. Reptiles associated with this vegetation can generally be considered as aquatic species (caimans, most snakes), and only in a few cases as terrestrial (*Anolis auratus*). Some terrestrial (*Oxyrhopus petola*, *Bothrops atrox*) and arboreal snakes (*C. enydris*) may on occasion enter floating meadows.

Judging from the species of reptiles and amphibians found in several "swampy" areas the number of species that potentially could inhabit floating meadows in Suriname could increase by about ten (not including the ones reported as (x) in Table 1).

Summary

An overview is given of the structure of floating meadows in general and is described in more detail for three localities in Suriname, Bolivia and Brazil where fieldwork was done. The herpetofauna occurring in the floating meadows in these sites is described briefly. Comparisons are made with the only article known dealing with the frogs of a floating meadow in Manaus, Brazil. The general conclusion is that a unique herpetofauna of floating meadows does not exist. All species present in them can be found in other open habitats and sometimes forest species enter floating meadows from adjacent forest.

Acknowledgements

Fieldwork in Suriname in 1968 was funded by grant W 956–2, in 1974–75 by grant W 87–78, both of WOTRO (Netherlands Foundation for the Advancement of Tropical Research); in 1979 in Suriname by WOTRO grant WR 87–127; in 1980 and 1981 in Suriname by WOTRO grant W 84–191; in Suriname in 1990 by funds of the Nationaal Natuurhistorisch Museum (acronym RMNH), Leiden (export permits 01171–75 and CITES 01892); that in Bolivia in 1990 by a grant from the Van Tienhoven Foundation and personnal funds, and in Brazil in 1992 (portaria SCT no. 583,23-ix-1992, licença especial 077/92-DEVIS) by funds of the RMNH, Leiden.

J.J.P. Paats, D.G. Reeder, W.N. Polder and P.E. Ouboter were pleasant field companions in Suriname, Teresa C.S. de Avila Pires and Reginaldo Rocha in Brazil, and Nelson Vaca and W.N. Polder in Bolivia.

References

Bakker, J.P. and J. Lanjouw, 1949. Indrukken van de Natuurwetenschappelijke Expeditie naar Suriname 1948–1949. Tijdschr. Ned. Aardr. Gen. 66: 538–557.

Boos, J.O. and V.C. Quesnel, 1971. Notes on *Kentropyx striatus* in Trinidad, West Indies. Herpetologica 27(4): 477–481.

Campbell, J.A. and W.W. Lamar, 1989. The venomous reptiles of Latin America. Comstock Publishing Associates, Cornell University Press, Ithaca, 425 pp.

Duellman, W.E. 1972. South American frogs of the *Hyla rostrata* group (Amphibia, Anura, Hylidae). Zool. Med., Leiden 47: 177–192.

Duellman, W.E. 1977. Liste der rezenten Amphibien und Reptilien. Hylidae, Centrolenidae, Pseudidae. Das Tierreich 95: 1–225.

Duellman, W.E. 1978. The biology of an equatorial herpetofauna in Amazonian Ecuador. Misc. Publ. Univ. Kansas 65: 1–352.

Frey, R. (ed.) 1990. Report Expedition Perseverancia 1990. Inst. Syst. Bot. Univ. Zürich, 23 pp.

Frost, D.R. (ed.) 1985. Amphibian species of the world. A taxonomic and geographical reference. Assoc. Syst. Coll., Lawrence, 732 pp.

Gallagher, D.S., J.R. Dixon, and D.J. Schmidly, 1986. Geographic variation in the *Kentropyx calcarata* species group (Sauria: Teiidae): a possible example of morphological character displacement. J.Herp. 20(2): 179–189.

Hill, R., G.J.W. Webb, and A.M.A. Smith, 1987. Floating vegetation mats on a floodplain billabong in the Northern territory of Australia. Hydrobiologia 150(2): 153–164.

Hödl, W. 1977. Call differences and calling site segregation in Anuran species from Central Amazonian floating meadows. Oecologia (Berl.) 28: 351–363.

Hödl, W. 1990. Reproductive diversity in Amazonian lowland frogs. In: W. Hanke (ed.), Biology and physiology of the Amphibians. Forts. Zool. 38: 41–60.

Hoogmoed, M.S. 1973. Notes on the herpetofauna of Surinam IV. The lizards and amphisbaenians of Surinam. Biogeographica 4: 1–419.

Hoogmoed, M.S. 1975. De hagedissen en wormhagedissen van Suriname III. de teju's. Lacerta 33(9/10): 143–160.

Hoogmoed, M.S. 1979a. Resurrection of *Hyla ornatissima* Noble (Amphibia, Hylidae) and remarks on related species of green tree frogs from the Guiana area. Notes on the herpetofauna of Surinam VI. Zool. Verh., Leiden 172: 1–46.

Hoogmoed, M.S. 1979b. The herpetofauna of the Guiana region. In: W.E. Duellman (ed.), The

South American herpetofauna: Its origin, evolution, and dispersal. Mus. Nat. Hist. Univ. Kansas Monogr. 7: 241–279.

Hoogmoed, M.S. 1990. Resurrection of *Hyla wavrini* Parker (Amphibia: Anura: Hylidae), a gladiator frog from northern South America. Zool. Med., Leiden 64(6): 71–93.

Junk, W.J. 1983. Aquatic habitats in Amazonia. In: Ecological structures and problems of Amazonia. IUCN, Commission on Ecology Papers 5: 24–34.

Junk, W.J. and C. Howard-Williams, 1984. Ecology of aquatic macrophytes in Amazonia. In: H. Sioli (ed.), The Amazon. Limnology and landscape ecology of a mighty tropical river and its basin. Junk Publishers, Dordrecht: 269–293.

Lang, J.W. 1989. Sex determination. In: C.A. Ross (ed.), Crocodiles and alligators. Facts on file, New York: 120–121.

Lindeman, J. 1953. The vegetation of the coastal region of Suriname. The Vegetation of Suriname 1(1): 1–135.

Lynch, J. 1979. The amphibians of the lowland tropical forests. In: W.E. Duellman (ed.), The South American herpetofauna: Its origin, evolution, and dispersal. Mus. Nat. Hist. Univ. Kansas Monogr. 7: 189–215.

Müller, P. 1971. Herpetologische Reiseeindrücke aus Brasilien. Salamandra 7(1): 9–30.

Ouboter, P.E. and L.M.R. Nanhoe, 1985. An ecological study of *Caiman crocodilus crocodilus* in northern Surinam. Unpublished report Zoöl. Lab. Dieroec., Univ. Nijmegen 233: 1–65.

Ouboter, P.E. and L.M.R.Nanhoe, 1987. Notes on nesting and parental care in *Caiman crocodilus crocodilus* in northern Suriname and an analysis of crocodilian nesting habitats. Amphibia-Reptilia 8: 331–348.

Schmidt, K.P. 1928. Notes on South American caimans. Field Mus. Nat. Hist. Zool. Series 12(17), Publ. 252: 205–231.

Sioli, H. 1984. The Amazon and its main affluents: Hydrography, morphology of the river courses, and river types. In: H. Sioli (ed.), The Amazon. Limnology and landscape ecology of a mighty tropical river and its basin. Junk Publishers, Dordrecht: 127–165.

Teunissen, P.A. 1978. Overzichtskaart Surinaamse Laagland Ecosystemen (Kustvlakte en savannegordel): sheet 3. - Stinasu, Paramaribo.

Vanzolini, P.E. 1972. Miscellaneous notes on the ecology of some Brasilian lizards (Sauria). Pap. Av. Zool. 26(8): 83–115.

Versteeg, A.H. and F.C. Bubberman, 1992. Suriname before Columbus. Meded. Sur. Museum 49A: 1–64.

12. The annual cycle of the spectacled caiman

PAUL E. OUBOTER and GEERT M.T. PEETERS

1. Introduction

Freshwater animals in the humid tropics are generally well adapted to water cycles. Most aquatic insects, fishes and amphibians show a life cycle strongly correlated to rain and water levels (e.g. Lowe-McConnell 1987). Changes in the water level determine food availability and the opportunities for breeding (often related to food availability as well). The principle of freshwater animals showing a life cycle strongly linked to the rain and water regime will be illustrated by this case study on the spectacled caiman.

The spectacled caiman (*Caiman crocodilus*) is one of three species of caimans that occur in Suriname (Fig. 1). The other two species are Cuvier's smooth-fronted caiman (*Paleosuchus palpebrosus*), and Schneider's smooth-fronted caiman (*Paleosuchus trigonatus*). The distribution and ecology of all three species have been the subject of intensive studies during the past 10 years (see e.g. Ouboter and Nanhoe 1989, and Ouboter and Hoogmoed in press). It seems that *C. crocodilus* is most abundant in the Coastal Plain, where it is an inhabitant of ditches, swamps, creeks and rivers. It is the most aquatic of the three species, usually seen in the water with only part of the head emerging. Spectacled caimans are only occasionally seen ashore in Suriname. Considering its aquatic habits, it is evident that the life history of these caimans will be much affected by the regime of the water. Here, we will describe how general activity and breeding are related to the season.

2. Study Area and Methods

Most of the data referred to in this chapter was collected on the Cusewijne River. This is a rather narrow river with a length of 190 km, draining a vast savanna, swamp forest and swamp area in the coastal lowlands of Suriname (Fig. 2). Observations were made in two areas. In 1982 and 1983 the area surrounding Zeekoe Creek, about 118 km upstream of the mouth of the Cusewijne River, was the main study area, whereas in 1991 and 1992 the area

Paul E. Ouboter (ed.), Freshwater Ecosystems of Suriname, 215–222.

Figure 1. Head of spectacled caiman *Caiman crocodilus.*

around Kaaiman Creek, about 34 km further downstream, was studied. The habitats of both areas are swamps, swamp forests and, more limited, pieces of high dryland forest.

Suriname has four seasons (see also Amatali, Chapter 3): a long rainy season (May to July/August), a long dry season (July/August to November), a short rainy season (December and January) and a short dry season (February to April). During the long dry season, most tributaries of the Cusewijne River become completely dry and in the largests of them the water level drops to between 20 and 70 cm. In the short rainy season, some areas may become flooded; in the long rainy season, nearly the whole area is inundated.

Various methods were employed to collect the data presented here. At the start of all surveys the caimans in the study area were counted during the night, and often size estimates were made. During the following days (or weeks), caimans were captured, marked and released. In this way, data on population structure and movement were obtained. In 1982/1983 radio-transmitters were attached to several caimans in order to follow their movements. During the rainy seasons the study areas were searched for nests. Due to the inaccessability of the habitats concerned, only a few nests were discovered. Consequently, most conclusions on the breeding season were extrapolated from the observation of newly hatched young. A more detailed description of the methods employed is presented by Ouboter and Nanhoe (1989).

3. Reproduction

In general, crocodilians exhibit two nesting stategies (Ouboter and Nanhoe 1987). The nesting strategy followed by many Crocodylids is to dig hole-type

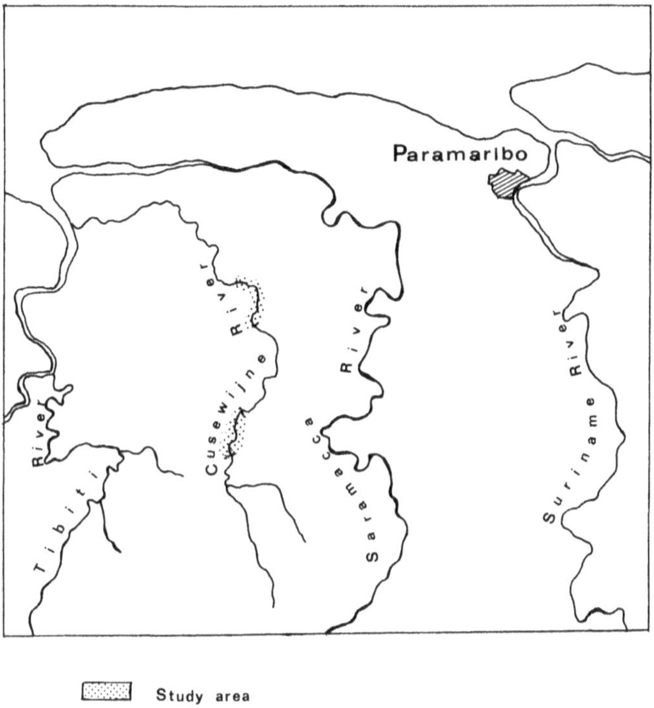

Figure 2. Study areas along the Cusewijne River.

nests in the sand during the period of receding or low water levels (e.g. Cott 1961). Most Alligatorids, however, construct mound-type nests of vegetation during the wet season. This applies to most populations of the spectacled caiman as well (also see Medem 1983). In the Cusewijne population, nests are constructed on small islands bordering the creeks that drain the swamp. These nests are neither the hole-type nor the mound-type nest: the egg chamber is in a hole below ground level, covered by a small pile of dry leaves (Fig. 3). Probably, there is not enough material available for a big pile. In a real hole-type nest on the other hand, the eggs would be too close to the water level, increasing the chance of flooding.

Nests containing eggs were only found during the rainy season. In 1982/83 pods of newborn caimans were found between the end of August and the middle of November (Table 1), placing the time of nesting in the long rainy season. In later years newborn hatchlings were also found following the short rainy season. According to Crawshaw and Schaller (1980), nesting during the rainy season has several advantages: 1) as large parts of the area are flooded, females can select nest sites where the chances of egg predation are reduced; 2) hatchlings can avoid intraspecific competition for food during the first months of their lives; 3) in the rainy season humidity is high and temperatures vary little, which has a positive influence on hatching success. For the first factor there is evidence

Figure 3. Cross-section of a nest and nesting site.

Table 1. Age and calculated date of birth of 19 pods in the study area

Pod no.	Date first found	Mean size of hatchlings	Age (days)	Calculated date of hatching
1	10-10-1982	26.5 ± 1.1	15	25-09-1982
2	27-10-1982	25.7 ± 1.0	12	15-10-1982
3	20-11-1982	23.3 ± 0.2	4	16-11-1982
4	19-01-1983	29.6 ± 1.2	–	03-11-1982
5	22-01-1983	45.6	–	03-09-1982
6	28-03-1983	31.1	–	09-11-1982
	08-05-1983	42.9 ± 2.5	–	25-09-1982
7	20-09-1983	30.4 ± 1.4	30	20-08-1983
8	21-09-1983	25.6 ± 1.1	12	09-09-1983
9	25-09-1983	30.5 ± 0.7	30	25-08-1983
10	26-09-1983	30.6 ± 0.8	30	26-08-1983
11	26-09-1983	25.6 ± 0.8	12	14-09-1983
12	27-09-1983	28.4 ± 1.7	22	05-09-1983
	09-10-1983	29.9 ± 1.3	28	12-09-1983
13	09-10-1983	25.5 ± 0.7	12	27-09-1983
14	07-10-1983	32.1 ± 0.8	40	28-08-1983
1	21-10-1991	22.7	1	21-10-1991
		29.6 ± 0.86	35	16-09-1991
2	22-10-1991	32.6 ± 0.71	50	03-09-1991
3	20-11-1991	25.65 ± 0.37	16	05-11-1991
		31.83 ± 1,93	46	05-10-1991
4	22-11-1991	30.0	37	16-10-1991
5	23-11-1991	25.86 ± 0.79	16	07-11-1991

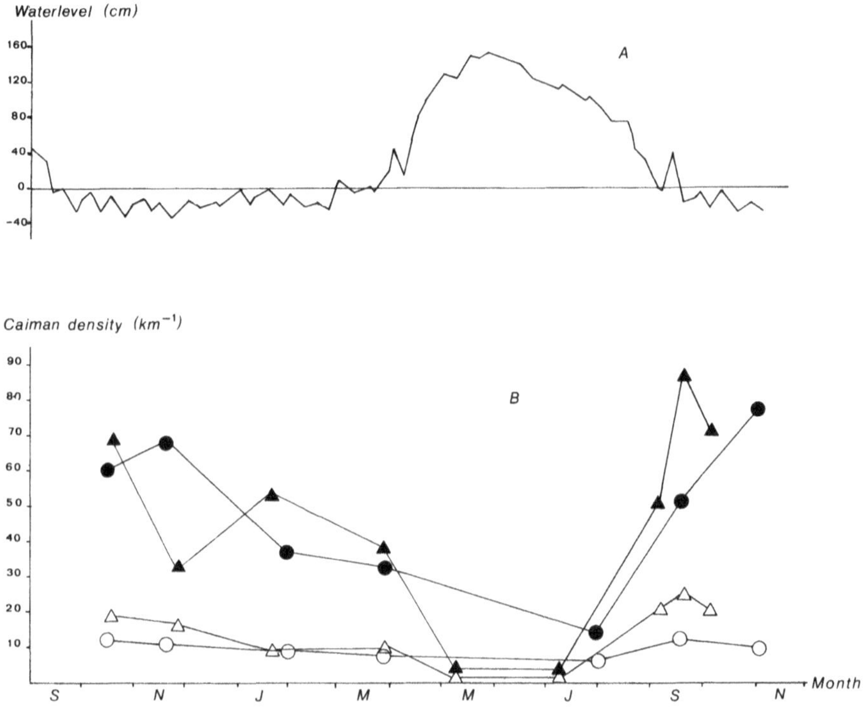

Figure 4. A. Water level with regard to Cusewijne shore level (for 1982/83, at the mouth of Zeekoe Creek). B. Number of caimans counted in 1982/83 in the Cusewijne River near Zeekoe Creek (open triangles); in 1982/83 in the Zeekoe Creek (closed triangles); in 1991/92 in the Cusewijne River near Kaaiman Creek (open circles); in 1991/92 in the Kaaiman Creek (closed circles).

that it is important in the Cusewijne population: the few data available on hatching rate implicate only limited predation on eggs compared to other populations of *C. crocodilus* (Ouboter and Nanhoe 1989).

4. Seasonal Changes in the Number of Caimans

During all studies on the natural history of freshwater aquatic animals, it is obvious to the observers that the density of specimens during the rainy season decreases drastically. This phenomenon is caused by the dispersal of the animals in the inundated swampy areas and tributaries surrounding or connected with the main river channel. The results of counts during the various seasons clearly illustrates the decrease in the number of spectacled caimans which can be spotted during the long rainy season (Fig. 4). Spectacled caimans, although highly aquatic, usually prefer to have some kind of a substratum under their feeth (or tail). With rising water levels they follow the retreating shore line, thereby entering the swamps and swamp forests. In the following long dry season, with the lowering of the waterlevel the numbers of caimans observed increase again. Ouboter and Nanhoe (1988) report on the movement of caimans entering Zeekoe Creek with rising water levels. Movement up the creek took place in such a short period (11–13 January), that aggregations of caimans were formed in front of narrow parts of the creek. Circumstantial evidence for movement of caimans into and out of the creek is provided by the counts (Fig. 4). At the end of the long dry season (November), the water level in Zeekoe Creek became so low that many caimans left the creek, thereby entering the main river. As soon as the water level rose again the number of caimans inside the creek increased, the number in the main river decreased.

Apart from the luxury of a substratum for body support, there are other reasons to enter the swamps. One reason is food. Caimans are sit-and-wait predators, which will have difficulties capturing prey while afloat. During the period of inundation, food is scarce in the main river due to the simultaneous dispersal of almost all other aquatic animals. A slight increase in the growth of first and second year caimans during the long rainy season (Ouboter and Nanhoe 1989), implicates that for this size class food is abundant in the swamps. Another reason for entering the swamps, is nesting (see above).

The number of caimans which can be spotted, is not affected by the water level only. During the long dry season caimans will have to remain inside the river and creek channels. Nonetheless, there is an obvious decrease in the numbers observed towards the end of this season (Fig. 4). We think that the decline is correlated to the availability of food, and activity. At the start of the dry season all aquatic organisms, that have reproduced during the long rainy season, will be concentrated in the river channels. Food is plentiful for piscivores, and many species prey on the abundance. As the dry season proceeds the number of fishes declines. Unfortunately, we have no quantitative data on this subject. Circumstantial evidence is provided by the food consumption of the caimans themselves: at the end of the long dry season most stomachs are empty (Ouboter in prep.).

Of course, the number of caimans which can be spotted is also determined by the actual numbers present. As stated before, spectacled caimans nest during the rainy seasons (especially the long rainy season) with the young hatching at the start of the dry seasons. This implies that caiman numbers will be highest during September/October, and decrease afterwards. Reproduction during the short rainy season could lead to a slight increase in March/April. This course coincides strongly with the path in Fig. 4. However, it can be shown that mortality only has a small share in the fluctuation of numbers observed. Table 2 shows the ratio between absolute numbers as determined by capture-recapture analysis according to Fisher and Ford (1947), and night counts. The enormous decline in numbers observed during the May/July rainy season is not supported by the capture-recapture figures.

In conclusion we can say that the number of caimans which can be spotted is determined by:

Table 2. Caiman numbers in the study area determined by capture-recapture analysis (Fisher-Ford) and by night counts. Water level with regard to Coesewijne shore level in cm

Month	Water level	Captures	Capture-recapture	Nights counts	Nights counts/ capture-recapture
October	- 15.05	134	–	230	–
November	- 19.75	77	216 ± 32	211	0.98
January	- 6.37	90	170 ± 25	133	0.78
March	+ 7.68	77	202 ± 30	139	0.69
May/July	+ 137.13	54	202 ± 30	22	0.11
September/ October	- 13.07	280	361 ± 54	422	1.17

1) Their actual number, which slightly decreases from the end of the long dry season on, mainly due to juvenile mortality.
2) The water level. With water levels rising, caimans disperse into the inundated swamps and swamp forests, and will not be visible from the river and creek channels.
3) Activity. Activity seems to be correlated with the availability of food. Food is abundant at the start of the long dry season, its availability declining towards the end of this season.

Summary

The number of spectacled caimans which can be observed fluctuates with the seasons. Density is highest at the start of the long dry season, due to reproduction in the preceding period (long rainy season). Subsequently, the numbers which can be spotted decrease, probably because availability of food, and hence activity become less. During the long rainy season the caimans disperse into the inundated swamps and swamp forests, and only few are visible from the river- and creek channels. Nesting occurs during the rainy seasons (mainly during the long rainy season). The young hatch at the start of the long dry season.

Acknowledgements

The 1982/83 study was supported financially by grants of the Beyerinck-Popping Fund, The Foundation for the Advancement of Herpetology, the Foundation for Scientific Research in Suriname and the Netherlands Antilles, the Netherlands Foundation for International Nature Protection (Van Tienhoven Stichting), the Society for Scientific Research in the Tropics (Treub-Maatschappij) and the World Wildlife Fund (Dutch Section). Caiman studies from 1987 on, were financed by the Netherlands Foundation for the

Advancement of Tropical Research (WOTRO) and the Netherlands Foundation for International Nature Protection (for project WAO 88/07).

The Foundation for Nature Preservation in Suriname (STINASU) and the Suriname Forest Service (LBB) gave logistic and material support.

We benefitted from discussions with many colleagues of whom we especially like to mention Marinus S. Hoogmoed and Jan Mol. Finally, we like to thank Djoe Franz Adoe who proved to be an indispensable assistant in the field.

References

Cott, H.B., 1961. Scientific results of an inquiry into the ecology and economic status of the Nile crocodile (*Crocodilus niloticus*) in Uganda and Northern Rhodesia. Trans. Zool. Soc. London 29: 211–356.

Crawshaw, P. and G. Schaller, 1980. Nesting of Paraguayan caiman (*Caiman yacare*) in Brazil. Papéis Avulsos Zool., S. Paulo 33: 283–292.

Fisher, R.A. and E.B. Ford, 1947. The spread of a gene in natural conditions in a colony of the moth *Panaxia dominuba* (L.). Heredity 1: 143–174.

Lowe-McConnel, R.H., 1987. Ecological studies in tropical fish communities. Cambridge Univ. Press, Cambridge, 382 pp.

Medem, F., 1983. Los Crocodylia de Sur America. Volumen II. Venezuela-Trinidad-Tobago-Guyana-Suriname-Guayana Francesa-Ecuador-Perú-Bolivia-Brasil-Paraguay-Argentina-Uruguay. Bogotá, Colciencias, 270 pp.

Ouboter, P.E. and M.S. Hoogmoed, in press. The distribution and abundance of crocodilians in Suriname. Zool. Meded., Leiden.

Ouboter, P.E. and L.M.R. Nanhoe, 1987. Notes on nesting and parental care in *Caiman crocodilus crocodilus* in northern Suriname and an analysis of crocodilian nesting habits. Amphibia – Reptilia 8: 331–348.

Ouboter, P.E. and L.M.R. Nanhoe, 1988. Habitat Selection and Migration of *Caiman crocodilus crocodilus* in a Swamp and Swamp-Forest Habitat in Northern Suriname. J. Herpetol. 22(3): 283–294.

Ouboter, P.E. and L.M.R. Nanhoe, 1989. Notes on the Dynamics of a Population of *Caiman crocodilus crocodilus* in Northern Suriname and its Implications for Management. Biol. Conserv. 48: 243–264.

PART III: The human impact

The population of Suriname amounts to a mere 400,000 inhabitants, i.e. less than $0.5/km^2$. The population is concentrated in and around Paramaribo, and most of the human activities encroaching upon nature are concentrated in the Coastal Plain; most of the Interior has less than 0.2 inhabitants/km^2. Consequently, many areas are in a virtually pristine condition. However, in areas of economic development the human impact is considerable. A preliminary overview of environmental problems is presented by Ouboter *et al.* (in press), but this is hardly exhaustive. The following listing of environmental issues concerning freshwater will be incomplete as well, but presents a beginning for further, more exhaustive surveys.

1) Changes in the hydrology
   a) upsetting the daily or seasonal hydrological cycle
   b) construction of reservoirs
2) Pollution
   a) pesticides
   b) heavy metals
   c) eutrophication
   d) other pollution
3) Changes in the biotic community
   a) decline of ecological key species
   b) introductions
4) Water-related health problems

ad 1a. In Northwestern Suriname a road was constructed on the border of freshwater and brackish-water swamps, interrupting the connection between these two habitats. Consequently, the inflow of freshwater in the coastal swamps was hindered, as a result of which the salinity of these swamps increased, and much of the vegetation died off (also see Teunissen, Chapter 5). Improvement of the road construction, increasing the number of culverts, is in progress.

In Northwestern Suriname rice-growing is the main economic activity. In order to expand the rice area and to supply sufficient freshwater, the Multi-

Paul E. Ouboter (ed.), Freshwater Ecosystems of Suriname, 223–226.

purpose Corantijn Project (MCP) was initiated. Upsetting the hydrology by the construction of dams lead to inundation of extensive swamp areas. Although some trees could adapt to this situation for some time, most died off after some months (also see Teunissen, Chapter 5).

Regulation of the water to a constant level is disastrous to the fish stock in all areas of the humid tropics. This is caused by the reproductive cycle of the fish in those areas, which is adapted to the cycle of wet and dry seasons. Most of the species breed during the rainy season in the inundated swampy areas (e.g. Lowe-McConnel 1987). If inundation holds off or becomes permanent, breeding ceases. This happened in the lower Suriname River after the construction of the Brokopondo dam, which lead to a decrease in fish stocks, including many commercially important species (Panday-Verheuvel 1982).

ad 1b. The man-made Brokopondo Lake was the first large reservoir to be created in South America. Its construction caused many ecological problems which were at the time unforeseen (e.g. anoxia due to the decay of organic material, aquatic macrophytes covering large areas of the lake, etc.). Chapter 13 focuses on limnological changes during the filling of the lake, and touches upon the environmental problems involved.

The construction of several other reservoirs is planned in Suriname. Plans for a reservoir in the Kabalebo area are in an advanced stage. Several studies were made to access the environmental impact of the project (e.g. Werkgroep Milieustudie West-Suriname 1976).

Smaller lakes were created by bauxite mining. After an area was mined out and the pumping out of water was stopped, the hole filled with groundwater. Until now, the area covered by those lakes is limited. The bauxite companies only recently started developing rehabilitation plans.

ad 2a. Only one study has focused on the environmental impact of pesticides. Vermeer *et al.* (1974) studied the effects of pesticides on fishes and birds in rice paddies of Northwestern Suriname. Between 1965 and 1971 thirteen different pesticides had been used. The application of sodium pentachlorophenate (NaPCP) against snails caused the death of snail kites and thousands of fishes. High levels of pentachlorophenol and endrin residues in the tissues of sick and dead snail kites, egrets, herons and jaçanas demonstrated the adverse effects of the extensive application of these chemicals. At present, the use of these pesticides is discontinued. The influence of the pesticides which are nowadays applied in the mechanized rice cultivation needs to be investigated.

ad 2b. In the traditional methods to recover gold, mercury is used to bind this metal. Gold winning in neighbouring countries has shown that serious pollution of the aquatic ecosystem occurs, and results in the poisoning of the consumers of fish (e.g. Martinelli *et al.* 1988). The effects of mercury on the environment and the health of the local people has not been investigated in Suriname.

Red-mud was recently analysed to investigate its potential to neutralize acid clay. During this analysis, several heavy metals were detected (lead, mercury, zinc, nickel, cadmium, titanium and molybdenum). It is not known if any of these are transported into the surrounding swamp ecosystems during the periods of overflow.

ad 2c. The effect eutrophication can have on aquatic ecosystems was clearly demonstrated at the time the Brokopondo Reservoir was filling, when the release of a relatively large quantity of nutrients led to some aquatic macrophytes becoming a pest. The same happens every year in many of the black-water creeks in a natural way (see Werkhoven & Peeters, Chapter 6). The rapid overgrowth of trenches and canals in the agricultural areas of the Coastal Plain is, among others, caused by eutrophication.

ad 2d. Other forms of pollution include increased turbidity and caustic soda. Increased turbidity is usually caused by mining, i.e. bauxite mining and gold mining. The impact of this form of pollution on the environment has never been studied in Suriname.

Caustic soda (NaOH) is used in the production of alumina, and considerable quantities are present in the disposed red-mud from where it leaks into the environment. Chapter 14 will describe *a swamp* (changed into a lagoon) *affected by caustic soda.*

ad 3a. Because of the extensive areas of pristine environment in Suriname, it is very unlikely that any species has become extinct up to now. Riverine species, however, are quite vulnerable to hunting or extensive habitat changes. In 1978/79 a minimum of 2,000 spectacled caimans was killed during commercial hunting operations in the Coesewijne River. As a result of publications by STINASU (Foundation for Nature Preservation in Suriname), the public opinion turned against this form of exploitation, and the killing stopped. After the killing, fishermen reported a reduction in fish numbers (a similar effect was reported from Brazil (Fittkau 1970)).

ad 3b. In 1955 the tilapia (*Oreochromis mossambicus*) was introduced in Suriname for aquacultural trials (Lijding 1959). This species entered the coastal ecosystems and is now dispersed over much of the Young Coastal Plain. In many habitats it seems to dominate, eliminating some of the local fish species (e.g. armoured catfish).

ad 4. Many diseases are water-related. Although some are not necessarily human induced, many are. For instance some forms of malaria occur in the country naturally, and the mosquitos breed in waterholes. Many of the naturally occurring water-related diseases, however, affect only a limited area and may have spread due to human actions. In this respect a good water supply and sewerage are of the utmost importance. In addition, all hydrologic changes, notwithstanding their importance for the economic development of the country, need to be evaluated well with regard to the possible dispersal of vectors. Chapter 15 *Water and health* gives an overview of water-related diseases and describes the history of water supply and sewerage.

The human impact on nature and the human environment need not only be negative. *Conservation* is man's final attempt to preserve (selected) parts of the naturally occurring biodiversity. In the past it was usually concentrated on the survival of key species and their habitats; invertebrates and lower vertebrates, notwithstanding their role in the foodchain, were of minor importance. Consequently, freshwater areas were usually not selected to be protected. Also in Suriname, the conservation of freshwater ecosystems is based on a number of protected, mainly terrestrial areas, incidentally including some rivers and creeks. Enhancement of conservation measures, tailored to aquatic ecosystems, seems necessary (Chapter 16).

Suriname is a country with plenty of freshwater. As its population is small, it is in the unique position to conserve this natural resource, without too much efforts, for the well-being of both man and nature.

References

Fittkau, E.J., 1970. Role of Caimans in the Nutrient Regime of Mouth-lakes of Amazon Affluents (An Hypothesis). Biotropica 2(2): 138–142.

Lowe-McConnell, 1987. Ecological studies in tropical fish communities. Cambridge Univ. Press, Cambridge, 382 pp.

Lijding, H.W., 1959. Proeven met Tilapia in Suriname. Surinam. Landbouw 6: 183–194.

Martinelli, L.A., J.R. Ferreira, B.R. Forsberg & R.L. Victoria, 1988. Mercury contamination in the Amazon: A gold rush consequence. Ambio 17(4): 252–254.

Ouboter, P.E., P.C. Dixit, M.C.M. Werkhoven, J.H.A. Mol & M. Khodabaks, in press. Some aspects of environmental research in Suriname with special emphasis on biodiversity. In: Environmentally Sound Socio-Economic Development in the Humid Tropics. UNESCO, Paris.

Panday-Verheuvel, M.P., 1982. Factors influencing water quality and life in the Suriname River. Unpubl. report, Hydraulic Research Division, Paramaribo.

Vermeer, K., R.W. Risebrough, A.L. Spaans & L.M. Reynolds, 1974. Pesticide effects on fishes and birds in rice fields of Surinam, South America. Environ. Pollut. 7: 218–236.

Werkgroep Milieustudie West-Suriname, 1976. Ecological consequences of the first two phases of the Kabaleboproject. Paramaribo, 76 pp.

13. The man-made Lake Brokopondo

P. LEENTVAAR

1. Introduction

Lake Brokopondo (officially named Prof. Ir. W.J. van Blommensteinmeer) has been constructed for generating electricity, which is used for smelting aluminium from bauxite. Near Afobaka a dam was built across the Suriname river, and after closing the gate on February 1, 1964 the lake began to fill. An area of approximately 1,560 km² of tropical forest was inundated (Fig. 1). The implications for water quality, fish fauna and vegetation were studied during the filling up of the lake. In this chapter some of the results of the hydrobiological observations are given. A short summing up of the events will be given here. In the first years of water stagnation an enormous development of floating vegetation (*Eichhornia crassipes*) occurred, which has been controlled by spraying the herbicide 2-4-D. Characteristic plankton species from the flowing Suriname River did not survive after stagnation. Also bottom dwelling macrofauna and fish died as a result of oxygen depletion by the rotting forest. Navigation on the lake became dangerous as subsurface tree trunks or floating logs became a risk. Wind action increased causing high waves. Below the dam in the lower course of the Suriname River the salinity increased as the sea water in the tidal region reached higher up. At the port silt sedimentation increased. As to the use of water from the lake, it should be mentioned that the turbines were affected by the aggressive, acidic water. This also caused corrosion of the shells of snails and bivalves in the river and lake. The snail *Biomphalaria glabrata*, which is a host for bilharzia, is absent in this lime poor environment. Mosquito larvae, however, increased in numbers after construction of the lake, favoured by the expanding swampy areas. The lake as an acidic oligotrophic environment shows an interesting development of characteristic species dominated by Desmids after an initial biological self-purification by green flagellates and green algae. Phytoplankton data are not presented in this chapter; the reader is referred to Leentvaar (1975).

Paul E. Ouboter (ed.), Freshwater Ecosystems of Suriname, 227–237.

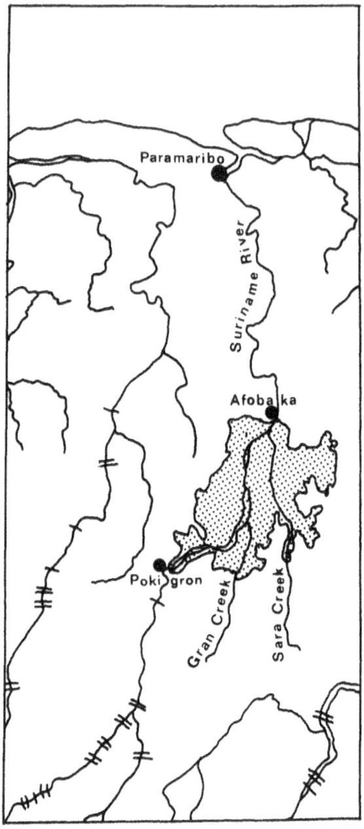

Figure 1. The area of the Suriname River and Brokopondo Lake.

2. Hydrobiological Situation Before the Closing of the Dam

2.1. The Suriname River

From November 1963 till February 1, 1964, the physico-chemical conditions of the water, the composition of the plankton and the bottom fauna in the undisturbed Suriname River could be observed at Afobaka. Due to the closing of the dam on February 1 it was not possible to obtain a picture of a complete year cycle of the flowing river at this station. This, however, could be done upriver at Pokigron; from the observations it can be concluded that no essential differences occur between the two sampling stations, so that a year cycle at Pokigron can be compared with a year cycle at Afobaka. Yet, it should be kept in mind that from Pokigron to Afobaka, a distance of some 90 river kilometers, a gradual, though very slight increase of pH, temperature, amount of electrolytes and of plankton occurred, and also that the river flow at Afobaka was larger than at Pokigron. The river-bed of this stretch was more or less uniform. The bottom consisted of hard rocks, with hardly any silt; at places where the current was negligible sandbanks were located between the rapids. The rain forest directly bordered the river; other types of shore vegetation were of minor importance.

Observations, also from other rivers in Suriname, show that the river-water generally is nearly saturated with oxygen from surface to bottom. This favourable situation is brought about by the good aeration in rapids and falls; at the same time very little reducing matter is present. There is no pollution by human influences. The purity of the water is also shown by the very low amount of electrolytes, which becomes clear in the conductivity. The river is oligotrophic. As in other river-systems in the Guianas and the Amazon region the scarcity of electrolytes is due to the nature of the soil, which consists of a very old thoroughly weathered layer on top of poor alkaline rocks of the crystalline basement (see Noordam, Chapter 2). Therefore, only very few electrolytes are washed into the river from the surrounding forest area. The bottom is also poor in lime so that the water contains only a few milligrammes of calcium per liter. The water is soft and the pH is approximately 6.5. During longer periods of rain it dropped after a short initial rise. A lowering of the pH might be caused by CO_2 produced by decaying material which was washed into the river. The initial rise could be caused by silt stirred up from the bottom of the river when the rains began. This short rise in pH must therefore be considered to be the effect of turbulence on the river-bed, by which the mineral content increased. Humic acids account for a more or less brownish colour of the water. Furthermore, chemical analysis showed that the brown colour of the Suriname River is also partly caused by a high iron amount. Iron and silicates are the main components of the water, in colloidal form; together with some organic matter they cause brown turbidity of the water, intensified in the rainy season by transported material from the forest floor. Measurements with the Secchi-disc show that the transparency is not very high, viz. 1 to 2 m. Nitrogen and phosphorus compounds, as well as sulphate, could not be analyzed from the water; only sometimes they were found in very small amounts at the changing of dry and rainy periods. In the Suriname River the temperature varied between 27 °C and 31 °C.

2.2. Other Creeks

Apart from the sampling stations Afobaka and Pokigron in the stretch of the Suriname River, attention was paid to the characteristics of the creeks that flowed into the Suriname River in the inundation area. The most important of these is the Sara Creek. The amount of electrolytes and iron in the water was higher than in the Suriname River, and the water was of a deeper brown colour. Due to a less broad creek the water was more shaded, and had a lower temperature than the open water of the Suriname River. The oxygen content was also lower than that of the main stream. The plankton of the Sara Creek was scarce in number of species, and also differed in composition from that of the Suriname River. The Gran Creek showed much similarity to the Sara Creek.

The very small creeks in the inundated area, completely surrounded by trees and protected from the sun, contained almost no plankton; where present, the plankton was principally composed of bottom dwelling species.

3. Hydrobiological Situation After the Closing of the Dam (Period February 1–September 1964)

3.1. Rise of the Water Level

In Fig. 2 the rise of the water level is shown. Immediately after closing the dam the water level rose rapidly, as the existing river-bed filled with water. After that, the water rose less quickly, because the banks were overflowed and a larger area was inundated. Later, rapid rises were caused by rainy periods at the end of March and during the main rainy season from the end of May until September.

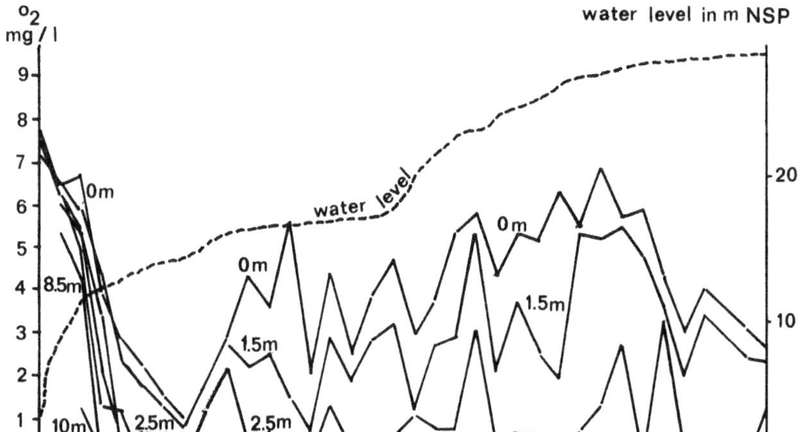

Figure 2. Dissolved oxygen content at different depths at sampling point Afobaka, January 31–September 30, 1964. The dam was closed on February 1, 1964. Periods of rainfall were the short rainy season in March and the main rainy season in May–August.

This rapid rise determined largely the physico-chemical and biological developments in the forming lake. In a short time a large area of the forest floor was covered with water, and with it also the layer of decaying leaves and other organic matter of the tropical forest. From that time onward the oxygen source for decomposition processes had to be found in the water instead of in the air; the originally well-aerated river-water soon lost most or even all of its oxygen. The biological consequences of the inundation were particularly catastrophic in the initial phase as the total volume of water was still small compared with the inundated area, and complete mixing took place by the force of the inflowing river-water.

3.2. Physico-chemical changes

3.2.1. Oxygen

The inundation resulted in a rapid decrease of the oxygen content of the water at Afobaka (Fig. 2). On January 31 the oxygen was about equally distributed in the flowing river, from the surface down to the bottom. Although the stream had ceased to flow in a specific direction, and the water was already 15 m deep on March 18, there was still enough turbulence until this date to cause a decrease in oxygen content of the top-layers; first at almost the same rate as that of the water in contact with the forest soil. Eventually the oxygen content of the deeper layers dropped to zero, but the superficial layers still kept some oxygen. In the hypolimnion the organic matter could not be completely mineralized due to a lack of oxygen; bacterial anaerobic decomposition took place there, which resulted in evil smelling compounds, such as hydrogen sulphide.From March 18 onwards the oxygen content in the upper layers increased and the situation there, although strongly fluctuating, remained rather favourable. The improvement of the oxygen content cannot be explained by a more abundant development of the plankton, but has probably been caused by a change in the vertical distribution of the water temperature.

The short rainy season at the end of March (increased rise of the water level) provided some deeper layers with oxygen by mixing (March 25). In the long rainy season, however, the lake had expanded so much, that the greatly increased inflow of river-water exerted no marked influence on the oxygen content, apart from a small supply of oxygen in the 4.5 m layer (June 3 and 10).

3.2.2. Temperature

After the closing of the dam the temperature was mostly uniform from top to bottom for several weeks (Fig. 3). The water was apparently well-mixed, and gradually the currents and turbulences subsided. Isotherms show that from March onwards the top layer of 0–3 m became warmer, while the deeper water gradually became colder. This development was interrupted during the rainy season, starting on March 11, at a moment when the volume of the lake was still fairly small. The temperature was markedly affected from top to bottom. Later, on May 27 and in August, the rains had only a mild effect on the top layer of 3 m, but deeper water showed a stronger response. Undercurrents may have interfered. Diurnal fluctuations became notable in the top layers, but were less distinct in deeper layers as heat absorption was restricted almost exclusively to the top layer. A thermocline with corresponding epi-, meta- and hypolimnion was not found at all, or only temporarily during the day. Temperature inversions were recorded, as on May 6, but no permanent thermocline developed in deeper water. The wind action as a downward acting force was subordinate to differences in density, which resulted in density stratification restricted to the top layers.

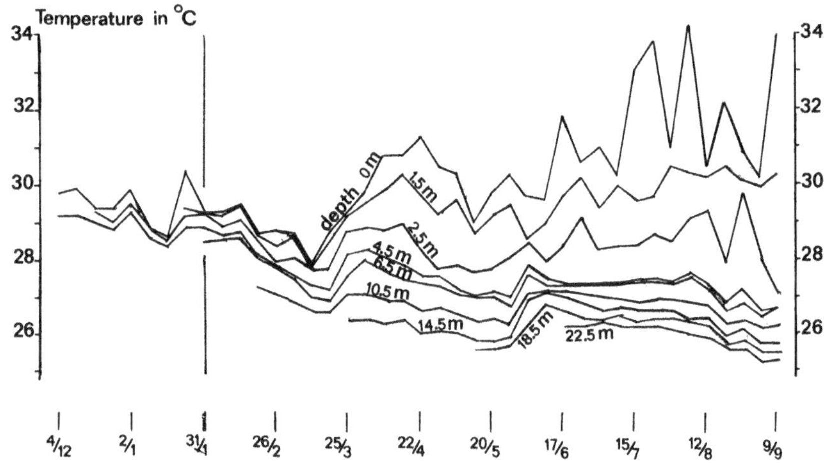

Figure 3. Temperature at different depths at sampling point Afobaka, December 4, 1963–September 9, 1964. The dam was closed on February 1, 1964.

The appearance of the 27° and 26° isotherms in higher water-levels in July and August clearly shows that the deep water gradually cooled off. This means also that the water was not stabilized thermically and that the heat exchange had not yet reached an equilibrium. The lowest temperature to be reached near the bottom will probably be about 25 °C, which is also the lowest temperature recorded in tributaries not exposed to sunlight.

3.2.3. Conductivity

The conductivity at the surface showed a tendency to rise until the long rainy season with two marked drops at the beginning of the two rainy seasons (March 25, June 3). From June onwards the value remained more or less constant (Fig. 4). As time passed the conductivity increased with the depth. At the bottom, values up to 70 μS (micro-Siemens, micro-mho, 10^{-6} Ohm) were found. However, the decrease as a result of the beginning rainy season was greater with increasing depth.

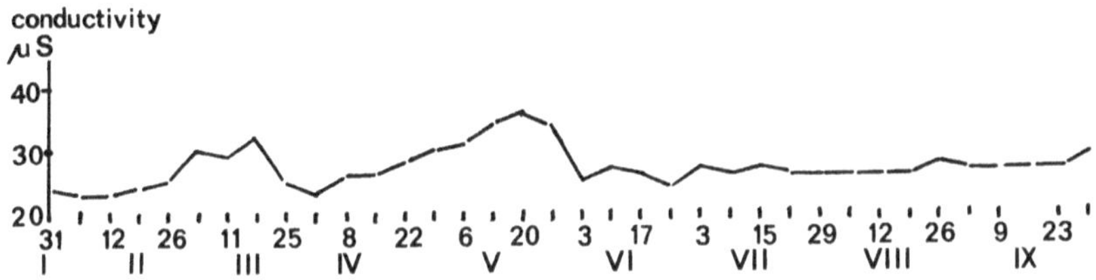

Figure 4. Conductivity at the surface at sampling point Afobaka, January 31–September 30, 1964. The dam was closed on February 1, 1964.

3.2.4. pH

After the closing of the dam the pH dropped to about 5.5, and after February 19, no pH higher than 6.0 was found at any depth. In a vertical direction the pH values did not vary significantly. During the rainy periods the pH fluctuations were of the same order in all water-levels, becoming constant again afterwards. This occurred after an initial decrease in mineral content. The sudden pH increase seemed therefore to be connected with a sudden change in the water at the beginning of the rainy season. As the water was quite deep at that time a bottom effect can be ruled out; so the addition of suspended organic matter from the flooded forest appeared to be the most likely cause.

It is remarkable that in the 0–3 m layer the temperature fluctuations (and also oxygen fluctuations) were very large, but fluctuations in conductivity were negligible. This means, that the chemical situation in the surface layers was more stable than temperature and oxygen. Some fluctuations in conductivity did occur at the surface, but they soon moved downwards. This may be caused by chemical processes in the top layers.

4. Bottom Flora and Fauna

4.1. The Suriname River Before the Closing of the Dam

The bottom in the upper reaches of the Suriname River was composed of rocks and sand. Soft mud was absent, apart from a few places near the outflow of rivulets, and in places where the water was more or less stagnant. In such places many tree leaves could be found also.

The most common type of aquatic macrophytes found were Podostemaceae, occurring in the rapids, *Mourera fluviatilis* being the most common species. In one case a few specimens of *Elodea* were found in a stagnant pool near Djoemoe during the dry season. The rivers were not suitable as an environment for floating and submerged plants due to the fluctuations in water-level, and the velocity of the current. For this reason *Eichhornia crassipes* was only found in sheltered places.

The bottom fauna of the river was also very poor. Bottom samples taken with a grab contained only sand. During the dry season, however, several small piles of excrements produced by a species of Oligochaeta could be found along the edges of the sandflats. At this time frogspawn and tadpoles could also be found, deposited by tree frogs in burrows containing water. In the stagnant parts of the river with a muddy bottom and decaying leaves only midge larvae were found. Most organisms seemed to concentrate in the flowing parts of the river especially in the rapids and the falls. Several organisms adapted to strong currents were found in between the Podostemaceae. Simulidae and Ephemeroptera were common; Trichoptera, Odonata, Sialoidea *(Corydalis)* and Lepidoptera *(Cataclysta)* were found as well. Freshwater shrimps *(Macrobrachium)* and freshwater crabs *(Potamocarcinus latifrons)* lived in fissures and quiet parts of the rapids. Small Coleoptera were found mining the leaves of Podostemaceae. Encrusting freshwater sponges were also found. On the rocks snails of the genera *Pomacea* and *Doryssa* were present; their tops were invariably eroded by the aggressive water. These snails were found together with small fish, tadpoles, waterbugs and waterstriders in small pools on the rocks during the dry season. Apart from the few organisms living in stagnant pools, all specimens belonged to groups normally living in running water.

The presence of a veliger-type of larvae and zoecia in plankton samples indicated that Lamellibranchiata and Bryozoa must be present in the river. Adult colonies of Bryozoa were not found, however. Only a few shells of *Diplodon voltzi, Diplodon granosus*, and *Castalia ambigua* were found occasionally. The veliger larvae in the plankton probably belonged to *Diplodon voltzi*, the most common lamellibranch in the Suriname River. Shortly after the dam at Afobaka was closed a large number of *Diplodon voltzi* could be collected from the dry bottom of the lower course. The mussels were especially present in the soft muddy parts near the outflow of small tributaries. Freshwater shrimps were concentrated in the shallow parts of the dry river-bed filled with stagnant water; their presence was betrayed by numerous little pits in the sandy bottom. Worms could be collected easily in places where excrements were found, piled on the surface.

The observations in the dry river-bed gave a certain indication of the food chain of fish in the river. The scarcity of food organisms was obvious. Probably only the freshwater shrimps were of some importance as a source of food as they were most numerous and formed an easy prey. The majority of the fish were predators. Insects blown into the water from the banks may have formed another source of food, though I never observed fish feeding on floating insects.

4.2. The Lake After the Closing of the Dam

After the closing of the dam the water became stagnant and all bottom organisms must have died rapidly through lack of oxygen. This happened for instance with the catfish, *Hypostomus plecostomus*, which was fully dependent on the rapids. Other bottom-dwelling organisms which could not escape to more favourable places also died. Some sponges, freshwater shrimps, ephemerids and dragonflies could be found on floating substances or other places where oxygen was present. Several organisms, common in stagnant water, also invaded this substrate. This is apparent from the fact that in the Sara Creek region extensive fields of *Lemna valdiviana, Spirodela biperforata* and *Ceratopteris pteridoides* developed, which must have originated in stagnant water. In the flooded forest, dense algal mats (*Spirogyra, Mougeotia*) developed, together with many organisms from stagnant water like the snails *Drepanotrema anatinum, Aplexa marmorata, Acroloxus sp., Gundlachia sp., Taphius kühnianus;* the crustaceans *Euryalona occidentalis, Chydorus sp., Cyclestheria hislopi*; ephemeropterids, Odonata and waterbugs. In the Sara Creek, and other shallow rivulets with mainly stagnant water *Ranatra linearis* and other Rhynchota could be found (*Belostoma, Microvelia*). The snails mentioned were only present at certain localities from which they dispersed with the floating material. During the rainy season this bio-community did not survive, many dead *Pomacea glauca* were found floating between the duckweed.

The dead leaves of the swamped trees were soon covered with green masses of filamentous algae (*Spirogyra, Mougeotia*), which eventually died when the water level rose. On the dead leaves of the trees many muddy tubes of Oligochaeta were visible (*Dero, Aulophorus, Aeolosoma* and *Pristina)*).

Where filamentous algae developed, bubbles of gas kept the mats of algae floating. In places where dense mats of Lemnaceae developed the algae died from lack of oxygen and light. In the former Sara Creek region where oxygen was absent from top to bottom, the floating algae formed oxygen-producing islands favourable to many organisms. Crustaceans and young fish were seen just around these patches of algae.

Among the roots of water-hyacinths a rich community of organisms could be found. It was composed mainly of crustaceans, filamentous algae, desmids, midge larvae, Odonata, ephemerids, shrimps, waterbugs, and small fish. Larger fish were also found, such as *Gymnotus*, hiding among the roots from where they could easily reach the surface.

4.3. Comparison of the Fauna and Flora Before and After the Closing of the Dam (November 1963 «» September 1964)

Suriname River (running water)	**Brokopondo Lake** (stagnant water)
Plankton mainly composed of phytoplankton	Plankton of another type: other phytoplankters dominant; Cladocera and Copepoda in great numbers in oxygenated top layers; new species from stagnant waters *(e.g. Cyclestheria, Euryalona; Lacinularia, Octotrocha, Sinantherina)*.

Suriname River (running water)	**Brokopondo Lake** (stagnant water)
Blue-green algae on firm substrates.	Blue-green algae abundant on roots of *Eichhornia* and on other floating substrates.
Filamentous algae scarce.	Filamentous algae locally forming mats.
Duckweeds not recorded.	Duckweeds *(Lemna, Spirodela)* locally covering large areas.
Eichhornia and *Ceratopteris* in only a few specimens.	*Eichhornia* and *Ceratopteris* covering large areas.
Podostemaceae common in rapids.	Podostemaceae absent.
Sponges on rocks.	Sponges on floating substrates only.
Turbellarians not recorded.	Turbellarians numerous in the plankton *(Catenula, Mesostoma)*.
Earthworms common; no Naididae recorded.	Earthworms not recorded; *Dero* and *Aulophorus* on decaying substrates.
River molluscs *(Pomacea, Doryssa; Castalia, Diplodon)*.	River molluscs absent; species from stagnant water locally and temporarily *(Acroloxus, Aplexa, Drepanotrema, Gundlachia, Taphius)*.
Crabs and shrimps *(Macrobrachium)*, and other bottom-dwelling organisms.	Crabs and shrimps, and other bottom-dwelling organisms only near the oxygenated surface.
Ephemeroptera common on firm substrates.	Ephemeroptera not common; only a few species *(Asthenoplus, Callibaetes)* recorded.
Veliidae, Gerridae, and Belostomatidae scarce.	Veliidae *(Microvelia)* in great numbers; Gerridae not observed; Belostomatidae on floating material only.
Trichoptera common on rocks in the rapids.	Trichoptera not recorded.
Megaloptera, Lepidoptera *(Cataclysta)*, and Simulidae.	Megaloptera, Lepidoptera, and Simulidae not recorded.
Fishes in many species and specimens.	Fishes in only a few species and specimens; surface dwellers *(Leporinus)* in oxygenated top layers; Gymnotidae among floating plants.

5. Conclusions

The inundation of a tropical rainforest is a hazardous experiment, as no one can predict the long term effects on nature and the human environment. The water in the lake is of bad quality and the lower course of the river is influenced by the discharges of the turbines. Fluctuations in quality and quantity occur, related to turbine operations. In order to avoid stagnation of water circulation in the lake, and related oxygen depletion, obstacles like tree-trunks should have been removed before filling up the lake. As it is impossible to clear a projected lake bottom as large as the Brokopondo Lake, many functions of the lake are lost. Navigation is hampered by the submerged tree trunks, fishery is restricted to locations in the former river-bed or to locations cleared from trees. Control of floating vegetation is of permanent concern. The substitution of the large area of rainforest by a smooth water surface caused local climate changes with more wind action and rainshowers. The hydropower generated per inundated area of Brokopondo is low compared to the forest area lost. Probably from an environmental point of view it would have been more beneficial if no lake construction had taken place, but as an alternative, exploitation of the forest had been performed in relation to the use of fire wood for the turbines.

Summary

The filling up of Lake Brokopondo after the closing of the dam on February 1, 1964, led to considerable changes in several physico-chemical parameters and in the biotic community. Due to the decomposition of the inundated forest the oxygen content dropped, in the deeper layers to zero, in the upper layers an increase in oxygen content was observed again after some months. Changes in temperature, conductivity and pH were recorded, but are of less importance. Many plant and animal species disappeared, which were sometimes replaced by species from stagnant waters. Some floating macrophytes covered large areas of the lake. From an environmental point of view, the creation of Lake Brokopondo did have many negative impacts on the freshwater ecosystems involved.

Reference

Leentvaar, P., 1975. Hydrobiological observations in Surinam, with special reference to the man-made Brokopondo Lake. Stud. Fauna Suriname Guyanas 15(56): 1–173.

14. Changes in a polluted swamp

PAUL E. OUBOTER and BART P.E. DE DIJN

1. Introduction

In the coastal area of Suriname, bauxite exploitation has become the major industrial activity ever since the Second World War. At that time, the war machinery of the US demanded enormous amounts of aluminium. A large bauxite processing plant was later built at Paranam, some 25 km inland from Paramaribo. The bauxite refinement process involves the use of caustic soda. The voluminous residue resulting from this processing is a reddish alkaline slurry, better known as "red mud". As elsewhere in the world, the red mud is dumped within the confines of open man-made reservoirs, red-mud lakes. Not far from the Paranam plant, there now exists a large red-mud-lake complex at the old mining site of Topibo. The swamps immediately surrounding this red-mud-lake complex are heavily polluted by the alkaline water overflowing the dykes during periods of high rainfall. This type of alkaline pollution goes largely unnoticed because its effects show most clearly on terrains that are only accessible to personnel of the bauxite companies. In the course of 1990 we noticed that changes must have occurred in a swamp receiving polluted water indirectly from the red-mud-lake complex. The most apparent change that seems to have taken place is a change from a lushly vegetated natural swamp to an open lagoon, bordered by vegetation that is clearly not doing all too well. Because of the physical aspect of this swamp, we baptised it "Topibo Lake". Limited investigations have been carried out since the beginning of 1990 into the aquatic ecology of Topibo Lake. In this chapter we will present data resulting from these investigations, and we will try to describe the effects of pollution by alkaline red-mud-lake overflow on these swamps.

2. Topibo Lake and its Surroundings

Topibo Lake is situated in the coastal lowlands of Suriname, well within the large complex of freshwater wetlands (Fig. 1). Most grass swamps are situated near rivers or creeks, but are separated from them by swamp forests that

Paul E. Ouboter (ed.), Freshwater Ecosystems of Suriname, 239–260.

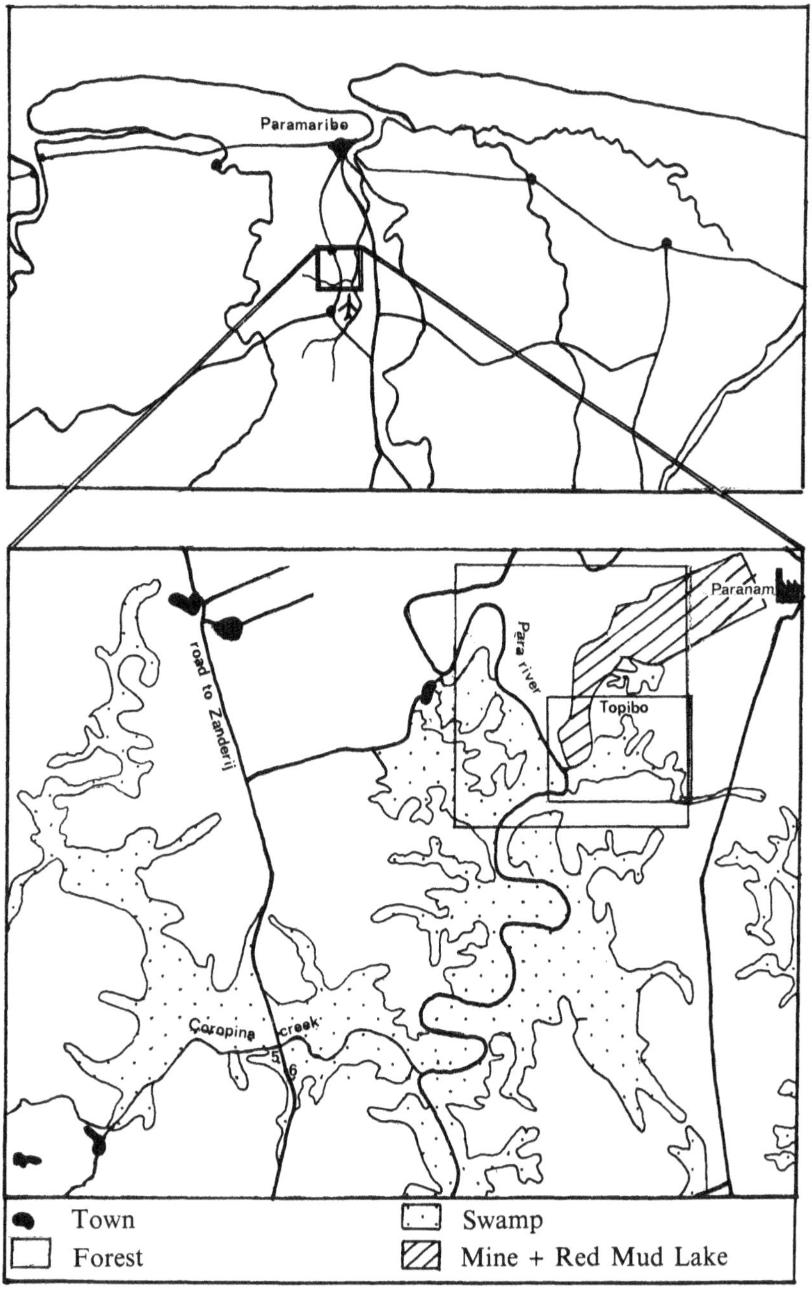

Figure 1. The Topibo Lake and Coropina Swamp area, as situated in the Coastal Plain of Suriname. Numbers 5 and 6 in the bottom map indicate the invertebrate sampling sites at Coropina Swamp (redrawn from Suriname Planatlas and aerial photograph by Centraal Bureau luchtkartering, Paramaribo).

dominate the banks of these streams. Other forest types in this area are marsh forests, covering large stretches of the low-lying terrain, and high dryland forests occurring in the form of vegetation islands on elevated terrain. The river nearest to Topibo Lake is the black-water Para River which originates in the Savanna Belt, south of Zanderij.

The photograph (Fig. 2) showing the Topibo area and its surroundings gives a good idea of the landscape, and the extent of alkaline pollution. East of the meandering Para River is the old mining site of Topibo, now used as a dump site. Extending to the northeast from the old mine is a red-mud-lake complex containing the alkaline slurry from the Paranam processing plant. To the north and the south of the red-mud-lake complex are heavily polluted swamps. The alkaline pollution is clearly noticeable, because of the absence of vegetation and

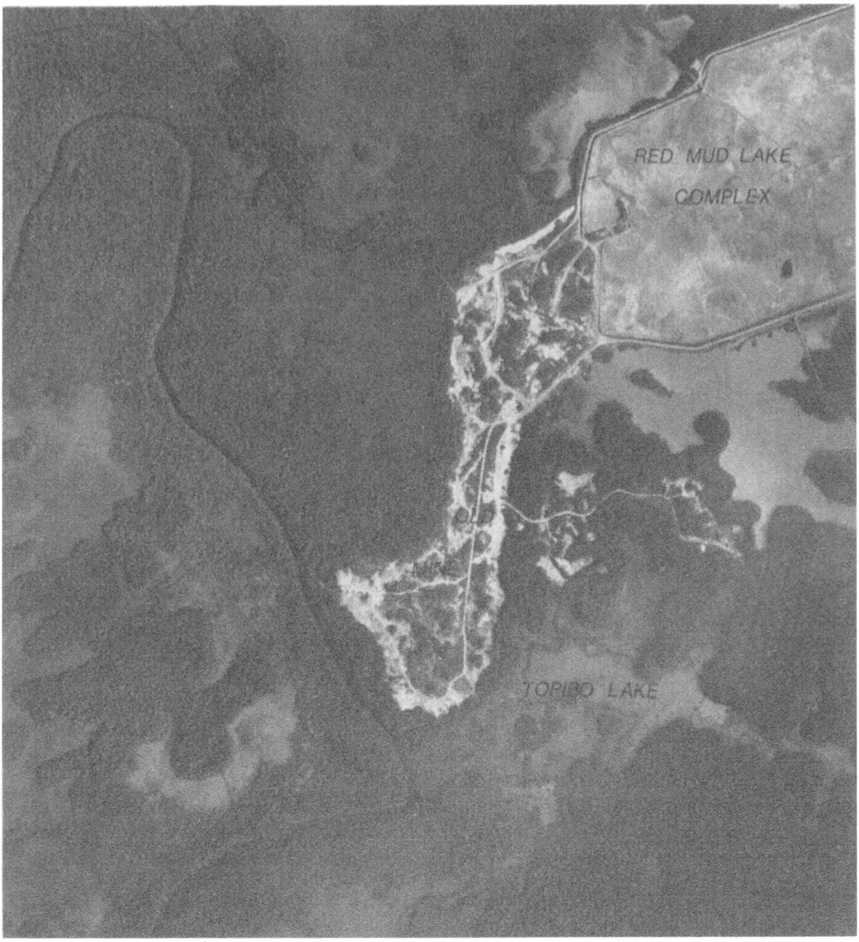

Figure 2. Aerial photograph of the Topibo area (photograph by Centraal Bureau Luchtkartering, Paramaribo).

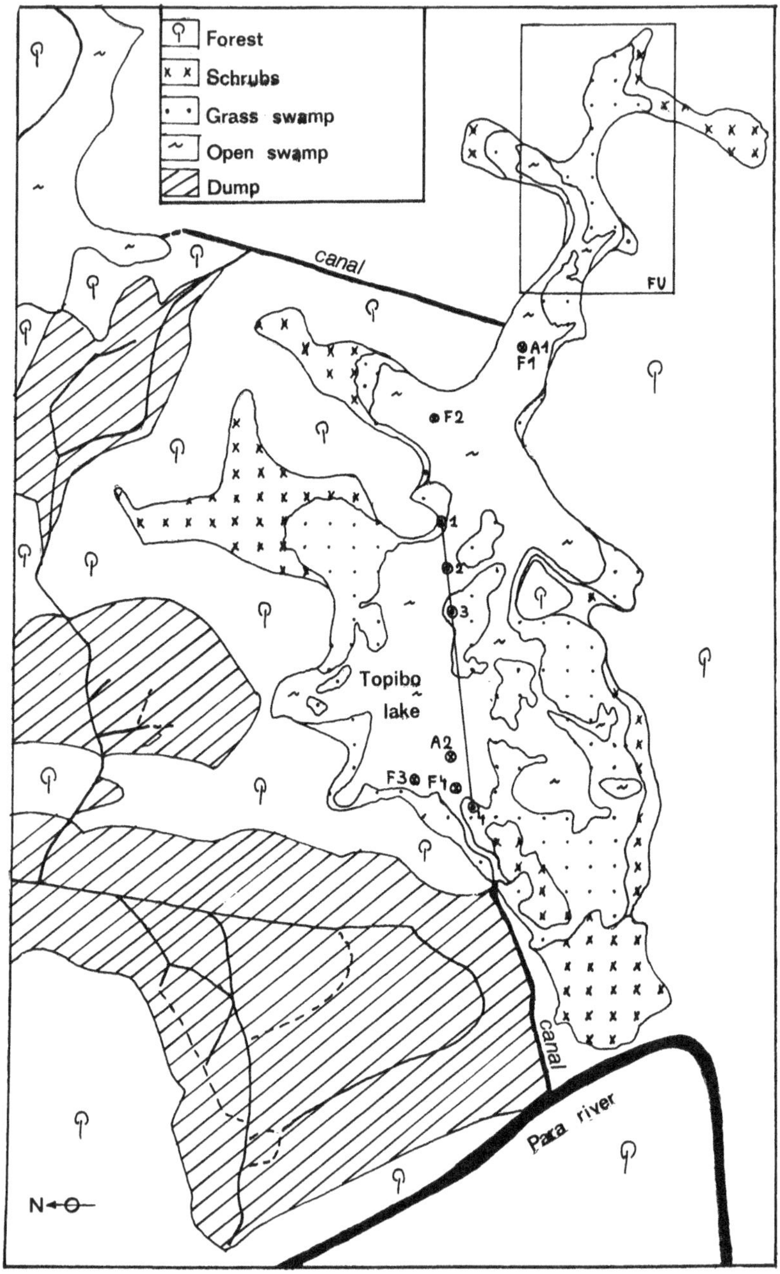

Figure 3. Topibo Lake area with indication of the sampling sites for water (A 1–2), and fishes (FU, F1–4), and of the invertebrate sampling sites along an east-west line transect (1–4) (drawn from aerial photograph by Centraal Bureau Luchtkartering, Paramaribo).

the silky whitish colour of the polluted water. Also noticeable is the contrast with natural swamps west of Topibo, where swamps have a continuous vegetation, and do not have the aspect of lakes. Topibo Lake is situated east of the southern tip of the old mining area, it forms the central part of the branched Topibo swamp area extending from the Para River to the east (Fig. 3). The eastern part of Topibo Lake is connected by means of a canal to a heavily polluted swamp bordering the red-mud-lake complex. In the west it has a direct connection to the Para River via another canal. As a consequence of these connections, alkaline water flows into Topibo Lake in the northeast, and is drained off to the west into the Para River. The part of the swamp through which the polluted water flows is almost entirely devoid of vegetation. The outer eastern part of the swamp and parts of the swamp further away from the alkaline current, basically still look like natural swamps. Complicating the situation, however, is the Para River current which is under heavy tidal influence: at low tide polluted water is indeed discharged into the Para River via the canal connecting Topibo Lake to the Para River, but at high tide the dark-coloured water from the Para River is pushed via the same canal into the western part of the lake. The water level in Topibo Lake is mainly determined by rainfall patterns, but also by the tides. During the long dry season the largest part of the lake falls dry, even at high tide. During the rainy season the lake always has water, the central part being 50 to 100 cm deep. At other times, during periods of intermediate rainfall intensity, more or less extensive parts of the lake fall dry, especially during low tide.

Comparisons are made with the Para River, a swamp along the Para River, and especially with a swamp along the Coropina Creek (see Fig. 1). The Coropina Creek is a black-water tributary of the Para River. Both swamps are good natural counterparts to compare Topibo Lake with.

3. Water Quality and Primary Production

In Table 1 some physical and chemical parameters are presented for Topibo Lake, Coropina Swamp, and the Para River. For many parameters the figures are almost similar. Deviations only appear for pH, alkalinity, humic acids and conductivity. The increase of the pH from less than 5 to more than 9 is a very obvious effect of the inflow of NaOH. The increase in alkalinity and conductivity can also be explained by the inflow of NaOH. Humic acids are present in amounts of more than 3 mg/L in black-water rivers which drain podzolic soils. In swamps they directly originate from the decomposing vegetation. The lack of (semi-)aquatic vegetation in Topibo Lake explains the decrease in humic acid content. The same happens in swamps that burn frequently.

Primary production was measured using two methods. The method with a light and a dark bottle gives an estimate of primary production by phytoplankton only. Diel oxygen change includes aquatic macrophytes as well.

Table 1. Chemical and physical water parameters.

Locality	Month	Temp. (C)	pH	Alkal.	Hardn. (mg/L $CaCO^3$)	NH^4 (mg/L)	NO^2 (mg/L)	NO^3 (mg/L)	ortho PO^4 (mg/L)	Cl (mg/L)	Humic acids (mg/L)	Conduc-tivity	Turbi-dity	Colour	Secchi depth (m)
Topibo Lake	May	28.0	9.66	340	7.0	0.39	0	0	0.1	10.8	1.8	800	10	80	0.4
	July	35.7	5.30	–	–	–	0	0	0.3	9.0	–	28	4	300	–
	Sept	–	9.94	–	–	–	–	–	–	–	–	–	–	–	–
	Oct	37.2	9.40	–	–	–	–	–	–	–	–	–	–	–	–
Coropina Swamp	July	28.0	4.72	–	2.7	1.56	0	0	0.5	10.0	3.5	25	4	175	–
Para River	July	25.5	4.49	ca.6	–	1.43	0	0	0.4	9.0	3.6	24	4	350	0.8
	Sept	27.0	4.23	–	–	1.60	–	–	–	5.6	–	–	–	–	–

For both methods primary production in Topibo Lake is rather high compared to the Para River (Table 2). Natural swamps give fluctuating results for primary production. Coropina swamp yielded a lower figure for diel oxygen change, but a swamp along the Para River gave results comparable to those for Topibo Lake. There is, however, one remarkable difference: oxygen content is much higher in Topibo Lake. Already at dawn the oxygen content is at least as high as for the other aquatic ecosystems at dusk. At dusk oxygen can be supersaturated, and bubbles of oxygen escape to the surface. The relatively high primary production can be explained by the ideal circumstances of shallow water, high light incidence, and the development of algal mats on the bottom. On the other hand, primary production is restricted by a lack of nutrients; those available are probably mostly fixed by the algal mats. We have no explanation for the much higher oxygen content.

Table 2. Dissolved oxygen contents and primary production measured by diel oxygen change, and the light-dark bottle method

Locality	Month	DO sunrise (mg/L)	DO sunset (mg/L)	Diel 0_2 change surface	L-D bottle method surface	L-D bottle method 0.5 m depth
Topibo Lake	May	-	-	-	0.30	0.20
	June	6.37	9.22	2.85	-	-
		2.62	5.65	3.03	0.57	-
Coropina Swamp	July	0.37	1.05	0.68	-	-
Para Swamp	June	0.47	2.87	2.40	1.40	-
Para River	July	0.60	0.67	0.07	-0.04	0.3

4. Lake Bottom and Vegetation

The vegetation was not studied intensively, but four vegetation samples were taken on a line-transect for invertebrates. The transect started at the vegetated central eastern side of the lake, and ended at the vegetated western side of the lake, thus crossing the open lake and a shallow mudbank with sparse vegetation (Fig. 3). When present, vegetation at these sites consisted exclusively of semi-aquatic macrophytes, truly aquatic macrophytes were not seen at all. General physical and vegetational characteristics at the sampling sites are as follows (samples were taken March 12, 1993, during the short rainy season):

- site 1: shallow lake edge, with soft muddy bottom at 10–30 cm below high-tide water level, with grass-dominated vegetation, consisting of two species of Cyperaceae (one *Rhynchospora* sp.), two species of Gramineae and *Montrichardia arborescens*, and with a vegetation coverage between 20 and 50%.

- site 2: shallow centre of lake, with soft muddy bottom at 30 to 50 cm below high-tide water level and without macrophytic plants.
- site 3: shallow mudbank, with very soft muddy bottom at 20–40 cm below high-tide water level, with only *Montrichardia arborescens*, and with a vegetation coverage between 10 and 20%.
- site 4: shallow lake edge, with a moderately soft, loamy bottom at 20–40 cm below high-tide water level, covered in part by root mats, with grass dominated vegetation, consisting of two species of Cyperaceae, one species of Gramineae and *Borreria* cf. *capitata* (Rubiaceae), and with a vegetation coverage between 30 and 70%.

To draw a comparison, we took some vegetation samples at two other invertebrate sampling sites in the grass swamp bordering Coropina Creek (see Fig. 1). Other swamps nearer to Topibo Lake were not accessible to us at the time of sampling. The sites at Coropina Swamp can be characterized as follows (samples were taken March 14, 1993):

- site 5: moderately deep swamp, with bottom at 80–100 cm below high-tide water level, consisting of a firm loamy soil covered with a continuous root mat and dead vegetation remains, with grass dominated vegetation consisting of two species of Gramineae, one species of Cyperaceae, and a species of Convolvulaceae, and with a vegetation coverage between 50 and 80%.
- site 6: moderately deep swamp, with bottom at 90–120 cm below high-tide water level, consisting of a firm, loamy soil covered with a continuous root mat and dead vegetation remains, with grass dominated vegetation consisting of one species of Gramineae and *Eleocharis interstincta* (Cyperaceae), and with a vegetation coverage between 50 and 80%.

Even from these limited data, we can conclude that vegetational changes have occurred at Topibo swamp. The fact that natural swamps in the area always have at least some vegetation cover, in contrast with an absence of vegetation in most parts of Topibo Lake, lead to the conclusion that the vegetation that formerly must have existed at Topibo Lake has died off under the influence of the alkaline water from the red-mud lakes. This deduction is confirmed by the observation that the vegetation at the border of Topibo Lake is distinctly below average in size, and that at times the vegetation bordering the lake seems to have died off partly. At the same time the largely or totally devegetated swamp soil of the lake seems to have become soft and muddy, in contrast to the firmer, root-mass-covered loamy soil which is typical of nearby natural swamps. Characteristics of the lake bottom and vegetation at the edge of the lake seem to indicate the existence of a tempering influence on the effects of alkaline pollution. This influence probably comes from the dilution of the alkaline water by inflowing water from less polluted parts of the swamp, and probably also from the Para River in the western part; the photograph (Fig. 2) also suggests the presence of a less polluted "border" around the polluted water in Topibo Lake.

5. Fauna

5.1. Invertebrates

5.1.1. Methods Employed

Sampling of aquatic invertebrates was done at the six sampling sites already mentioned for vegetation sampling. Samples were taken with a fine-meshed insect net which was pulled through the water from bottom to surface level for about a minute. The aim was to collect a representative sample of the aquatic fauna, containing invertebrates that are "pelagic", benthic and surface-dwelling. The animals present in the samples were sorted out and identified at two different levels: at a high level, where all animals were assigned to a relatively high-level taxonomic entity (from phylum to family, see Appendix 1), and at a species level, where aquatic insects were identified to the lowest taxonomic level possible (with the available identification keys), and then arbitrarily split up into probable species (Appendix 2). The identifications were based on the keys contained in Ward and Whipple (1959), diagnostic descriptions in zoological and entomological systematics handbooks, and more detailed keys for water bugs (Heteroptera) of the region (Cobben 1960; Drake and Cobben 1960; Nieser 1970, 1975). Caution is necessesary with respect to misidentifications below the family level.

Concerning the calculation of richness values a word of caution is important here regarding the number of insect species distinguished: because of the presence of many larval insect stages, it is often very difficult to assign specimens to clearly distinguishable species, or to assign larvae to (supposedly) conspecific adult developmental stages. Thus it will be clear why the species enumerated in Appendix 2 are termed "probable" species, and why one such probable species is often represented only by specimens belonging to the same developmental stage. As a consequence pairs of probable species in different developmental stages may in fact belong to single biological species, and probable species may in fact be complexes of species that are difficult to discriminate on the basis of the morphology of the specimens. When calculating richness values the former will lead to overestimation of richness (error 1), because of the double-counting of larvae and adults or pupae that in fact belong to the same biological species; the latter will lead to underestimation of richness (error 2). It seems preferable to avoid overestimation of richness; to achieve this a "correcting" procedure has been introduced to eliminate error 1. Correcting implies disregarding probable species (in a particular developmental stage) that are potentially conspecific with other probable species (in another developmental stage). To estimate the richness of sets of samples, a total richness value has been determined; three sets of samples are considered here: the set of all samples from Topibo Lake (4 samples), from Coropina Swamp (2 samples), and from both Topibo Lake and Coropina Swamp (all 6 samples). The total richness value is calculated as the sum of richness scores per taxon. For a set of samples from the high-level data set, the richness score is "1" for a taxon with representatives in at least one of the samples, and is "0" for one with no representatives in any of the samples. For a set of samples from the species-level data set, the richness score per taxon equals the number of species present in the sample with the highest number of species (for that taxon); we call the latter type of richness scores "maximal". These maximal scores and total values for richness are also used to calculate maximal (per taxon), and total richness ratios of Topibo Lake over Coropina Swamp.

5.1.2. Faunal Similarity

Before discussing richness estimates, we will present data on faunal similarity between Topibo Lake and Coropina Swamp. We will only consider presence or absence of supra-specific taxa, even when using the so-called species-level data set, because even for that data set, we only discriminated between probable species within each sample, without comparing those from one sample with those of another. We apply Jaccard's Similarity coefficient (Sj, as in Krebs (1989)) to these data, which yields Sj = 0.46 for the high-level data set (35 aquatic invertebrate taxa, taken from Appendix 1), and Sj = 0.26 for the species-level data set (34 supra-specific aquatic insect taxa, taken from Appendix 2). The value of 0.26 can be considered to indicate a clear dissimilarity, whilst that of 0.46 can be interpreted as indicating weak similarity. This leads us to the conclusion that in general the aquatic invertebrate faunas of both swamps are similar. The insect faunas, however, are clearly different.

5.1.3. Richness

At a high taxonomic level there is a clear contrast in richness between Topibo Lake and Coropina Swamp (Table 3): Topibo Lake scores consistently lower than Coropina Swamp. Scores for individual samples from Topibo Lake are clearly lower than those for individual Coropina Swamp samples. The total richness ratio (Topibo Lake over Coropina Swamp) of 0.74 also indicates that high-level taxon richness in Coropina Swamp is higher than in Topibo Lake. Based on the high-level data set, affinities with respect to alkaline pollution can be determined in a tentative way: taxa only present in samples from Coropina Swamp are considered pollution intolerant, taxa present only in samples from Topibo Lake are considered pollution tolerant, with a preference for pollution, and taxa present in both Coropina Swamp and Topibo Lake are considered pollution tolerant, without preference for pollution. Underlined taxa in the table are those for which the affinities seem to be particularly clear. Thus, we may draw the conclusion that there is a distinct tendency on a high taxonomic level for alkaline pollution preference with Stratiomyidae, Hebridae and Mesoveliidae, whilst Chironomidae, Ceratopogonidae, Hydrophilidae and Libellulidae tend to be indifferent to it, and Porifera, Nematoda, Oligochaeta, Crustacea, Sminthuridae, Ephemeroptera, Helotrephidae, Hydroptilidae and Sciomyzidae tend to be intolerant to alkaline pollution. The complete absence of such taxa as Nematoda, Oligochaeta, and Crustaceae from Topibo Lake was not at all expected. These taxa are normally a major component of the fauna of swamps in Suriname, and their members are major primary consumers of micro-organisms in the water (see also Mol, Chapter 7). It seems that alkaline pollution has eliminated these taxa completely from the swamp.

Table 3. Richness in macroscopic aquatic invertebrates of samples from Topibo Lake and Coropina Swamp.
Columns 'High-Level' and 'Species-level' respectively indicate richness in high-level taxa of aquatic invertebrates, and richness in probable species of aquatic insects; 'tot. (Topi+Coro)' indicates the total richness of the set of all six samples, 'tot.Topi/tot.Coro' indicates the total richness ratio of Topibo Lake over Coropina Swamp. Further explanation in text, see also Appendix 1 and 2

Richness		High-level	Species-level
Topibo Lake	sample 1	13	20
	sample 2	4	6
	sample 3	7	12
	sample 4	15	21
	total	20	33
Coropina Swamp	sample 5	25	32
	sample 6	25	33
	total	27	42
tot. (Topi+Coro)		33	63
tot. Topi/tot. Coro		0.74	0.79

We will now try to refine this picture by means of the species-level data set of aquatic insects present in the samples. Table 3 also lists the richness of aquatic insect species; the tendency already present at a high taxonomic level persists: relatively poorer samples in Topibo Lake, as compared to Coropina Swamp, Topibo also being poorer in general. One might have expected to find a sharper difference in species richness than in high-level taxon richness, but this is not true: the total richness ratio is 0.79 for insect species richness, which is even slightly higher than the one for high-level invertebrate taxa. Table 4 lists the maximal species richness scores for taxa or partial taxa of aquatic insects (partial taxa, indicated by (part.), being taxa from which those subtaxa are excluded that are treated separately); taxa with only one representative in just one sample are excluded from analysis[1]. The richness ratios for most taxa in Table 4 are not well interpretable, because of the limited number of species per taxon present in the samples. For three groups with a larger number of species, Entomobryomorpha, Chironomidae, and Dytiscidae (part.), Coropina Swamp seems to be richer in species than Topibo Lake; for two also speciose groups, Ceratopogoninae and Hydrophilidae (part.), it seems that species richness is higher in Topibo Lake. In

[1] Partial taxon Noteridae (part.) does not include *Pronoterus*, which is also not included for analysis as a separate taxon, because only one *Pronoterus*-specimen has been found.

Table 4. Maximum richness in probable species per aquatic insect taxon. Explanation in text, see also Appendix 2

maximum richness (species-level)	Topibo Lake	Coropina Swamp	Topi/ Coro
Sminthuridae	0	1	0.00
Entomobryomorpha	1	4	0.25
Poduromorpha	1	0	∞
Baetidae	0	1	0.00
Caenidae	0	2	0.00
Agrionidae	0	1	0.00
Libellulidae	1	2	0.50
Microvelia	1	2	0.50
Mesovelia	1	0	∞
Hebrus	1	0	∞
Helotrephidae	0	1	0.00
Oxyethira	0	1	0.00
Culicinae	1	1	1.00
Chironomidae	1	10	0.10
Ceratopogoninae	5	3	1.67
Eulalia	2	0	∞
Sciomyzidae	0	1	0.00
Brachydeutera/Paralimna	2	0	∞
Ephydra/Cetacera	1	0	∞
Notiphila	2	0	∞
Dytiscidae (part.)	0	4	0.00
Bidessonotus	1	0	∞
Noteridae (part.)	0	1	0.00
Hydrophilidae (part.)	4	1	4.00
Berosus	1	0	∞
Helodidae	2	2	1.00
Curculionidae	2	1	2.00

principle, it should be possible to interpret such richness-ratios as richness-based measures of pollution affinities (intolerance to or tolerance for, and preference or lack of preference for pollution) for specific taxa. Taxa could be considered pollution intolerant when their species-richness ratios are considerably lower than one, while they could be considered pollution tolerant and at the same time preferring pollution when their species-richness ratios are considerably higher than one; taxa that are tolerant to pollution, but without preference for it would then have species-richness ratios that are near one. The result of such an interpretation exercise applied to our data set is given in Table 5; except for the more speciose taxa already mentioned, it might prove to be of little but indicative value. When comparing these species-level affinities with the high-level ones, it can be seen that some refinement has indeed occurred: most taxa are confirmed in their affinities, whilst others have switched or split up with respect to their affinities. Dytiscidae has split up into the alkaline pollution tolerant and preferring *Bidessonotus*, and the alkaline pollution intolerant

Table 5. Affinities with respect to pollution of invertebrate taxa sampled at Topibo Lake and Coropina Swamp. Explanations in text, see also Appendices 1 and 2

a) of aquatic invertebrate taxa from the *high-level* data-set

tolerant preference	tolerant no preference	intolerant
Mesoveliidae	Libellulidae	Porifera
Hebridae	Chironomidae	Nematoda
Stratiomyidae	Ceratopogonidae	Oligochaeta
Poduromorpha	Hydropthilidae	Crustacea
Tipulidae	Mollusca	Sminthuridae
Ephydridae	Pseudoscorpiones	Ephemeroptera
	Entomobryomorpha	Helotrephidae
	Veliidae	Hydroptilidae
	Naucoridae	Sciomyzidae
	Culicidae	Agrionidae
	Dytiscidae	Haliplidae
	Noteridae	Platyhelminthes
	Helodidae	
	Curculionidae	

b) of aquatic insect taxa from the *species-level* data-set

tolerant preference	tolerant no preference	intolerant
Mesovelia	Culicinae	Sminthuridae
Hebrus	Libellulidae	Baetidae
Eulalia	*Microvelia*	Caenidae
Brachydeutera/Paralimna	Ceratopogoninae	Agrionidae
Ephydra/Cetacera	Helodidae	Helotrephidae
Notiphila	Curculionidae	*Oxyethira*
Bidessonotus		Sciomyzidae
Berosus		Dytiscidae (part)
Poduromorpha		Noteridae (part)
Hydrophilidae		Entomobryomorhpha
		Chironomidae

remaining Dytiscidae (part.). Because of relatively high species richness at Coropina Swamp, Chironomidae and Entomobryomorpha have switched from tolerant (without preference) to intolerant; the same switch can be seen with the Noteridae, but this is a result of the elimination of *Pronoterus* from the analysis. Other minor changes in affinities are apparent for other taxa, such as the split up of Hydrophylidae into the alkaline pollution preferent *Berosus,* and the remaining Hydrophilidae (part.),showing less preference. Probably of more value for this type of interpretation is the total species richness ratio from Table 3: the value of this ratio is not much lower than one, which leads to the conclusion that, from the point of view of richness, aquatic insects as a group tend to be tolerant to alkaline pollution, without having a preference for it.

We will compare these results with the very limited information available in the literature on the effects of caustic pollution on aquatic invertebrates. The lack of Crustacea in Topibo Lake seems to agree with observations by Anonymus, cited by Hobbs and Hall (1974), who reports the complete absence of crayfish in a tributary polluted by caustic slurry (river water pH = 11.4), whilst other tributaries have an abundant crayfish fauna. For aquatic insects a comparison can be made with an "extreme" tolerance list of Roback (1974) for species found at a pH greater than 8.5. Observations of Roback concerning Odonata, that only one agrionid and no libellulids were found at high pH, do not seem to conform with the presence of two species of Libellulidae (and no Agrionidae) in Topibo Lake. Roback found Ephemeroptera in caustic environments, whilst we found none in Topibo Lake. For Hemiptera and Coleoptera, which are reported to be less affected by pollution in general (especially those with semi-aquatic adults) there is better correspondence: like we do, Roback lists tolerant Hemiptera (e.g. Veliidae), and many tolerant Hydrophylidae, such as *Berosus*, *Tropisternus* and *Enochrus*. An absence under caustic circumstances confirmed by Roback, is that of case-making Trichoptera, such as the Hydroptilidae. For the Diptera, Roback exclusively lists a large number of tolerant Chironomidae, in contrast to the rather limited number of Chironomidae we found in Topibo Lake, where also representatives of other dipteran families seem to be tolerant to alkaline pollution.

5.2. Vertebrates

The tetrapod vertebrates present in the area, are listed in Table 6. Aquatic mammals are absent or very rare in the area of the Para River. The manatee (*Trichechus manatus*), and the giant otter (*Pteronura brasiliensis*) are numerous in most black-water rivers; these species were never seen by us in the Para River. The very shy swamp otter (*Lutra enudris*) will probably be present in certain localities. A comparison of aquatic mammals for polluted and unpolluted areas is hardly possible. The fish-eating bat (*Noctilio leporinus*) is not really aquatic, but depends completely on the aquatic habitat for its food. It is seen on the Para River, as well as on Topibo Lake. It is absent from unpolluted swamps since it needs open water to catch fishes.

Birds sighted along the Para River, and in unpolluted swamps are compared with sightings in Topibo Lake in Table 6. Many species seem to be restricted to a riverine habitat, and are absent from both polluted and unpolluted swamps. Not a single species is absent from Topibo Lake only. On the contrary, several species are exclusively found at Topibo Lake (semipalmated sandpiper, common stilt, black skimmer). These are species usually found near the coast. It seems that the lagoon-like, alkaline habitat of Topibo Lake has much in common with the marine mudcoast habitat.

Watersnakes lead a very secretive life. Consequently, it is almost impossible to be sure that they are absent from an area. From the three species of caimans

Table 6. Aquatic Tetrapod vertebrates known from the Para River, Topibo Lake and unpolluted swamps in the area

Species	Para River	Unpolluted swamp	Topibo Lake
Mammals	1	0	1
Fish-eating bat (Noctilio leporinus)	+	–	+
Birds	11	6	9
Anhinga (Anhinga anhinga)	+	–	–
White-necked heron (Adea coccoi)	+	–	–
Great egret (Casmerodius albus)	–	+	+
Snowy egret (Egretta thula)	+	+	+
Striated heron (Butorides striatus)	+	+	+
Boat-billed heron (Cochlearius cochlearius)	+	–	–
Rufescent tiger-heron (Tigrisoma lineatum)	+	+	+
Green ibis (Mesembrinibis cayennensis)	+	–	–
Muscovy duck (Cairina moschata)	+	–	–
Fishing hawk (Busarellus nigricollis)	+	+	+
Sungrebe (Heliornis fulica)	+	–	–
Wattled jacana (Jacana jacana)	+	+	+
Semipalmated sandpiper (Calidris pusilla)	–	–	+
Common stilt (Himantopus himantopus)	–	–	+
Black skimmer (Rynchops nigra)	–	–	+
Reptiles	5	(2)	(2)
Anaconda (Eunectes murinus)	+	(–)	(–)
Water snake (Helicops angulatus)	+	+	(–)

Table 6 (continued)

Species	Para River	Unpolluted swamp	Topibo Lake
Reptiles (continued)			
Spectacled caiman (Caiman crocodilus)	+	+	+
Cuvier's smooth-fronted caiman (Paleosuchus palpebrosus)	+	(–)	+
Schneider's smooth-fronted caiman (Paleosuchus trigonatus)	+	–	–

inhabiting the Para River, two also occur in Topibo Lake. Schneider's smooth-fronted caiman is very rare in the downstream and upstream parts of the river. Thus, its occurence in Topibo Lake would be very unlikely.

Amphibians were not found at Topibo Lake. However, we heard frogs calling from the trees near the lake edge.

Fishes caught at Topibo Lake are listed in Table 7. Only 14 species were found. From the Para River system 72 species are known (see Ouboter and Mol, chapter 8), so only 19% is present in the polluted Topibo Lake. Unfortunately, we were not able to sample an unpolluted swamp for comparison. Not all fishes present in the river will enter swamps. Topibo Lake is devoid of vegetation, so fishes are not obstructed by physical structures to enter it. This would make it likely that in the absence of pollution, it is entered by more species than a natural swamp. Of the 72 species known from the river system we excluded 10 which are mostly confined to the main streams. With this correction, still only 23% of the fishes potentially able to enter Topibo Lake really did so.

The fish fauna of Topibo Lake belongs to several families. Important families not represented are the Lebiasinidae and Rhamphichthyidae. Typical swamp species also absent from the Topibo Lake samples are the Callichthyids *Hoplosternum littorale* and *H. thoracatum*, and the Auchenipterid *Parauchenipterus galeatus*.

Some of the fish species present in Topibo Lake occur there in high numbers (see also Table 7). This affirms the rule that decreased diversity coincides with a high number of specimens per species.

From the observations mentioned above it seems that the pollution by caustic soda in Topibo Lake forms a barrier for several fish species to enter the area. For fish, whose skin is more permeable than that of reptiles, birds and mammals, and who need the aquatic medium to breath, the pollution with NaOH seems to have more influence than for the higher vertebrates.

Table 7. Fishes collected at Topibo Lake.

Taxa	Number of specimens
Characoidei	
Anostomidae	
Leporinus friderici	1
Characidae	
Astyanax bimaculatus	29
Charax gibbosus	3
Hemigrammus boesemani	6
Curimatidae	
Curimata cyprinoides	41
Curimatella alburna	4
Erythrinidae	
Hoplerythrinus unitaeniatus	1
Hoplias malabaricus	26
Gymnotidae	
Gymnotus carapo	1
Siluriformes	
Pimelodidae	
Pimelodus spec.	1
Perciformes	
Cichlidae	
Aequidens tetramerus	34
Crenicichla saxatilis	*1*
Nandidae	
Polycentrus schomburgki	1
Atheriniformes	
Poeciliidae	
Poecilia reticulata	2

Summary

Several physical, chemical and biological aspects are discussed of a swamp indirectly polluted by alkaline industrial waste from the bauxite processing industry in Suriname. The polluted swamp seems to have changed from a continuously vegetated swamp with a firm soil into a largely open, vegetationless swamp with a soft muddy soil. Due to the inflow of caustic soda, pH, alkalinity and conductivity have increased considerably. Primary production is relatively high. Changes in the aquatic invertebrate fauna, especially the insect fauna, seem to indicate that the alkaline swamp has become a habitat different from the natural swamp; the total richness in invertebrate taxa and insect species is not much lower in the alkaline habitat, but faunal similarity between the two swamps is moderate to low (for insects), whilst different insect taxa reach their highest richness in one of both habitats. Higher vertebrates seem to be hardly affected by the alkaline pollution; the changed habitat even attracts some coastal bird species. Fish diversity is seriously decreased, although fish biomass seems to be high.

Appendix 1. Presence or absence of macroscopic high-level invertebrate taxa in the samples from Topibo Lake and Coropina Swamp.
Column "AQ/NAQ" indicates whether the specimens found in the samples have aquatic life stages (AQ = having an aquatic life stage, NAQ = not having an aquatic life stage); the "FIELDSAMPLES"-columns indicate absence (0) or presence (*) of taxa in the six samples seperately; ? indicates doubtful indentification or life stage characterisation.

MACROSCOPIC ANIMAL TAXA present in at least one sample	AQ/NAQ	FIELDSAMPLES Topibo				Coropina	
		1	2	3	4	5	6
Phylum Porifera	AQ	0	0	0	0	*	*
Phylum Platyhelminthes	AQ	0	0	0	0	0	*
Phylum Nematoda	AQ?	0	0	0	0	*	*
Phylum Mollusca	AQ?	0	0	0	*	*	*
Phylum Annelida							
Classis Oligochaeta	AQ?	0	0	0	0	*	*
Phylum Arthropoda							
Subphylum Chelicerata							
Classis Arachnida							
Ordo Pseudoscorpiones	AQ?	0	0	0	*	*	*
Ordo Araneae	AQ+NAQ	*	*	*	*	*	*
Ordo Acarina	AQ+NAQ?	*	*	*	*	*	*
Subphylum Crustacea	AQ	0	0	0	0	*	*
Subphylum Uniramia							
Classis Insecta							
Subclassis Apterygota							
Ordo Collembola							
Subordo Sympleona							
Familia Sminthuridae	AQ?	0	0	0	0	*	*
Subordo Arthropleona							
Entomobryomorpha	AQ?	0	0	0	*	*	*
Poduromorpha	AQ?	0	0	0	*	0	0
Subclassis Pterygota							
Ordo Ephemeroptera							
Familia Baetidae	AQ	0	0	0	0	*	*
Familia Caenidae	AQ	0	0	0	0	*	*
Ordo Odonata							
Subordo Zygoptera							
Familia Agrionidae	AQ	0	0	0	0	*	0
Subordo Anisoptera							
Familia Libellulidae	AQ	*	0	*	*	*	*
Ordo Orthoptera	NAQ	*	0	0	*	*	*
Ordo Isoptera	NAQ	*	0	0	0	0	0
Ordo Thysanoptera	NAQ	*	*	0	0	*	0

Appendix 1. (continued)

MACROSCOPIC ANIMAL TAXA present in at least one sample	AQ/NAQ	FIELDSAMPLES Topibo 1	2	3	4	Coropina 5	6
Ordo Hemiptera							
Subordo Homoptera	NAQ	*	0	0	*	*	*
Subordo Heteroptera	AQ?+NAQ	*	0	0	*	*	0
Gymnocerata							
Familia Veliidae	AQ	*	0	0	*	0	*
Familia Mesoveliidae	AQ	*	0	0	*	0	0
Familia Hebridae	AQ	*	0	*	0	0	0
Cryptocerata							
Familia Naucoridae	AQ	*	0	0	0	*	0
Familia Helotrephidae	AQ	0	0	0	0	*	*
Ordo Trichoptera							
Familia Hydroptilidae	AQ	0	0	0	0	*	*
Ordo Diptera	AQ?+NAQ	*	*	*	*	*	*
Subordo Nematocera							
Familia Tipulidae	AQ?	*	0	0	0	0	0
Familia Culicidae	AQ	*	0	0	*	*	0
Familia Chironomidae	AQ	*	*	*	0	*	*
Familia Ceratopogonidae	AQ	*	*	*	*	*	*
Subordo Brachycera							
Familia Stratiomyidae	AQ	*	*	*	*	0	0
Familia Sciomyzidae	AQ	0	0	0	0	*	*
Familia Ephydridae	AQ	0	0	*	*	0	0
Ordo Coleoptera	AQ?+NAQ	*	*	*	*	*	*
Subordo Adephaga							
Familia Dytiscidae	AQ	*	0	0	*	*	*
Familia Noteridae	AQ	0	0	0	*?	*	*
Familia Haliplidae	AQ	0	0	0	0	0	*
Subordo Polyphaga							
Familia Hydrophilidae	AQ	*	*	*	*	*	*
Familia Helodidae?	AQ	*	0	0	0	*	*
Familia Curculionidae	AQ	0	0	0	*	*	*
Ordo Hymenoptera	NAQ	*	*	*	*	*	*

Appendix 2. Probable species of aquatic insects present in samples from Topibo Lake and Coropina Swamp.
Column “DEVIL. STAGE” indicating per taxon the development stage(s) of the specimens in the samples (L = larva, P = pupa and A = adult); the “FIELDSAMPLES”-columns indicate per sample the presence of probable species by a “+” when represented by at least two specimens, or by an “o” when only one specimen is present; “?” indicates doubtful identification or development stage characterisation.

AQUATIC INSECT TAXA present in at least one sample	DEVEL. STAGE	FIELDSAMPLES Topibo 1	2	3	4	Coropina 5	6
Ordo Collembola							
Subordo Sympleona							
Familia Sminthuridae	?					+	+
Subordo Arthropleona							
Entomobryomorpha	?				+	+	+
						o	+
							o
							o
Poduromorpha	?				+		
Ordo Ephemeroptera							
Familia Baetidae	L					o	o
Familia Caenidae	L					+	+
						o	
Ordo Odonata							
Subordo Zygoptera							
Familia Agrionidae	L					+	
Subordo Anisoptera							
Familia Libellulidae	L	+		+	+	+	+
						o	+
Ordo Hemiptera							
Subordo Heteroptera							
Gymnocerata							
Familia Veliidae							
Microvelia	L+A	+			+		o
							o
Familia Mesoveliidae							
Mesovelia	L+A	+			+		
Familia Hebridae							
Hebrus	A	+		o			
Cryptocerata							
Familia Naucoridae							
Ambrysus?	L	o					
Cataractocoris?	L					o	
Familia Helotrephidae	L+A?					+	+
Ordo Trichoptera							
Familia Hydroptilidae							
Oxyethira?	L					+	+

Appendix 2 (continued)

AQUATIC INSECT TAXA present in at least one sample	DEVEL. STAGE	FIELDSAMPLES Topibo 1	2	3	4	Coropina 5	6
Ordo Diptera							
Subordo Nematocera							
Familia Tipulidae							
Subfamilia Limoniinae	L	o					
Familia Culicidae							
Subfamilia Culicinae	P	+			+		
Culex	L					+	
Familia Chironomidae	L	+	+	+		+	+
						+	+
						+	o
						o	
						o	
	P	+	+	+		+	+
						+	+
						+	o
						o	o
Subfamilia Tanypodinae	L					+	+
						+	+
						+	+
						+	o
							o
Familia Ceratopogonidae							
Subfamilia Ceratopogoninae	L	+	+	+	+	o	
						o	
	P	+	+	+	+		
		o			o		
Culicoides	L	+	+	+	+	+	o
	P	+	+	+	+		
					+		
					o		
Subfamilia Forcipomyiinae	L					o	
Subordo Brachycera							
Familia Stratiomyidae							
Eulalia?	L+P	+	+	+	+		
			o		o		
Familia Sciomyzidae	L					o	o
Familia Ephydridae							
Brachydeutera/Paralimna	L+P			+	o		
				o			
Ephydra/Cetacera?	L+P			o	+		
Notiphila?	P			o			
				o			

Appendix 2 (continued)

AQUATIC INSECT TAXA present in at least one sample	DEVEL. STAGE	FIELDSAMPLES Topibo 1	2	3	4	Coropina 5	6
Ordo Coleoptera							
Subordo Adephaga							
Familia Dytiscidae	L					+	+
							+
							o
Subfamilia Hydroporinae							
Bidessonotus?	L	+			+		
Bidessonotus	A				+		
Hydroporus?	L						+
Familia Noteridae	A					o	o
Pronoterus?	A				o		
Familia Haliplidae	L						o
Subordo Polyphaga							
Familia Hydrophilidae	L					o	+
	A	+					
		+					
Berosus	L	+	+	+	+		
Berosus	A	+		+	+		
Enochrus	A	+					
Tropisternus	A	+			+		
Familia Helodidae?	L	+				+	+
							o
	A	+					
		+					
Familia Curculionidae	L					+	
Brachybambus?	A				o		
Steremnius?					o		+

References

Cobben, R.H., 1960. The Heteroptera of the Netherlands Antilles – I, Foreword, Gerridae, Veliidae, Mesoveliidae (water striders). Stud. Fauna Curacao Carrib. Islands 11:1–34.

Drake, C.J. and R.H. Cobben, 1960. The Heteroptera of the Netherlands Antilles – II, Hebridae. Stud. Fauna Curacao Carrib. Islands 11:35–43.

Hobbs, H.H., Jr. and Hall, E.T., Jr., 1974. Crayfishes (Decapoda: Astacidae). In: Hart, C.W., Jr. and S.L.H. Fuller (eds.), Pollution Ecology of Freshwater Invertebrates. Academic Press, London:195–214.

Krebs, C.J., 1989. Ecological Methodology. Harper & Row, New York, 654pp.

Nieser, N., 1975. The water bugs (Heteroptera: Nepomorpha) of the Guyana region. Stud. Fauna Suriname Guyanas 16:1–308.

Ward, H.B. and G.C. Whipple, 1959. Freshwater biology. Wiley, New York, 1248pp.

Roback, S.S., 1974. Insects (Artropoda: Insecta). In: Hart, C.W., Jr. and S.L.H. Fuller (eds.), Pollution Ecology of Freshwater Invertebrates. Academic Press, London: 313–376.

15. Water and health

BALTUS F.J. OOSTBURG

1. Introduction

Water is a basic and essential component of life. It is essential to our health and economic prosperity. Water is needed for the preparation of food, for drinking, bathing, washing, in agriculture and fisheries to provide us with food, in industry, as drinking water for animals (livestock), for recreation, for hydro-electric energy, for fire-extinguishing, for the removal of human waste, as a solvent of all kind of products etc. Managing water resources and safeguarding them from pollution will ensure a healthy and sustainable aquatic ecosystem. Without water, none of the biological, chemical and physical processes which make up what we know as life, could take place. Water is responsible for the transformation and transportation of physical and chemical substances to provide nutrients to biota to ensure their survival and diversity. The availability of water in the earth system is a precondition to human existence.

Man pollutes the water with his own wastes and causes extensive and serious human health problems. Polluted water affects the health of some 1.2 billion people and contributes to the death of some 15 million children under 5 every year in developing countries. Community water supply and sanitation are traditionally a cornerstone of community health. The World Health Assembly of 1977 launched the activity "Health for all by the year 2000". On 12 September 1978 the International Conference on Primary Health Care in Alma Ata (Soviet Union) adopted the Alma Ata Declaration, stating that primary health care is the key to "Health for all by the year 2000". As far as water supply is concerned, the objective was: potable water in the house or within 15 minutes walking distance. The WHO ruled that this means that in urban areas potable water must be available within a distance not beyond 200 meters from the house and in rural areas it means that the housewife must not spend a disproportionate part of the day getting water for domestic use. The traditional health indicators (infant mortality, life expectancy, death rate etc.) will improve if water supply and sanitation are improved.

Paul E. Ouboter (ed.), Freshwater Ecosystems of Suriname, 261–273.

2. Water Supply and Sewerage in Suriname

2.1. History

The written history of Suriname goes back to the 17th century. In those days, there was no water supply system. In the yard around every house there was a well, dug in the hard shell-sand soil, which provided brackish water, intended for the negro-slaves, livestock and as washing water. The water was collected with a bucket, attached to a rope. The Europeans, who were the slave-owners, used rainwater, collected in cisterns through eaves-gutters along the houses. Sometimes rainwater was stored in large brickwork underground rainwater tanks. In water jugs made from clay, rainwater was kept cool in the houses, to be used for drinking in the hot climate. In the rural areas rainwater or water from the rivers was stored in barrels in front of the houses. In the dry season the availability of water was a real problem and at the beginning of the twentieth century, water was brought by train from Republiek, a village some 40 km South of Paramaribo. Upon arrival in Paramaribo, the water distribution took place at different locations.

2.2. Water Supply

On 27 April 1933 Suriname was able to provide its population with potable water. A general government regulation, adopted on 22 February 1938, forbids the collection and storage of well or rainwater in Paramaribo and in this way the people were compelled to have their houses connected to the distribution system of the Suriname Water Company (SWM). In 1948 the government of Suriname became sole share-holder of this limited liability company. In order to provide more rural areas with potable water, the government created a water supply section in the Ministry of Public Works in 1962. In a later stage this section was transferred to the Ministry of Natural Resources. This section is responsible for the development, operation and maintenance of water supply in rural areas.

According to the policy outline of the government of Suriname for 1983–1990, the government has fully adopted the "Mar del Plata Action Plan" for the "International Drinking Water and Sanitation Decade" (1980–1990). At present, there are 56 water supply systems operating in Suriname.

With the exception of the Suralco water supply system of Mungo, use is made of groundwater in all the water supply systems in Suriname. Surface water is usually more seriously affected by human activities than groundwater sources, which are more protected from pollution. The use of groundwater sources has a relatively low capital cost and, in most cases, little treatment is needed to attain the standards recommended in the drinking water quality guidelines of the WHO.

2.2.1. The Suriname Water Company (SWM)

Most of the urban areas are covered by the SWM, which is responsible for the water supply from Republiek to Paramaribo up to a distance of 500 meters left and right of the Kennedy Highway and the Indira Gandhi Road. On 7 October 1958 the SWM started a water supply system in the Nickerie District near the western border, and on 9 August 1960 the water supply of Albina, near the eastern border, was put into operation by the SWM. The population served by the SWM is approximately 65% of the total population of the country. It appeared that the production station at Republiek was not sufficient to serve the whole population. The SWM was forced to expand and hence the following production stations were constructed: 1958 William Kraanplein, 1972 Leysweg, 1981 Livorno, and 1982 Lelydorp. Furthermore, at Blauwgrond there is a distribution station with a reservoir capacity for 1400 m^3 water. Drillings in the northern part of Paramaribo and in the District of Commewijne showed that the groundwater in these areas has a high chloride contents and is not suitable for domestic use. The water of Republiek has the best quality and this water is mixed with water from the Paramaribo wells to provide the population of Paramaribo with potable water. Every day the SWM pumps into the distribution system some 2200 m^3/h of water, which is not sufficient (it should be at least 2800 m^3/h). Some 250,000 inhabitants in Paramaribo and surroundings have access to potable water. The average daily use of water is about 150 liters per person.

2.2.2. The Water Supply Service of the Ministry of Natural Resources

This service is in charge of some 40 systems, covering mostly rural areas. The task of this government water supply service can be summarized as follows:

a. development, operation and maintenance of government water supply systems;
b. research and planning for watersupplies in general;
c. provide water to the population, through private house connections or public taps or water containers.

Taking into account all the water supply systems in Suriname, we may conclude the following:

a. Almost 90% of the urban population is served by house connections and the remaining 10% has easy access to safe, potable water.
b. In rural areas approximately 60% of the population is served by house connections, whereas about 20% has easy access to safe, potable water, leaving only 20% unserved.

2.3. Sewerage

The concern of the government of Suriname is not only to provide safe, potable water to the population, but also to remove waste water, largely in the form of domestic sewerage and industrial waste. The disposal of this water must take place by sanitary means in order to improve the public health. The removal of rainwater from polluted areas also causes special problems in Suriname, especially in the more densely populated coastal areas, where the flat terrain makes natural drainage difficult. Thus in the rainy season parts of the country, especially Paramaribo, are flooded and this can cause some diseases (e.g. bilharzia).

2.3.1. History of Excreta Disposal in Suriname

In the 17th century the excreta disposal in Paramaribo took place in privies in the backyards. When these privies were full, the excreta were removed by "faecal trucks", with a suction apparatus. The contents of these trucks were then dumped in the Suriname River at a particular spot, indicated by the government. In the homes the faeces were collected in chamber pots, mostly at night, and in the morning these were emptied in the privy in the backyard. There was an open sewerage system and the sewerage drained into draining ditches along the streets of Paramaribo. After a water supply system was installed in Suriname (1933), the privies in the backyard were gradually replaced by flush toilets (water closets) in the homes and the excreta were disposed in septic tanks with drainage beds. The effluent of the septic tanks flows to the sewerage systems. This effluent can contain pathogenic micro-organisms and therefore, besides biological purification, it is sometimes necessary to treat the sewerage before it is disposed into the streams.

In 1945 a great deal of the draining ditches in Paramaribo was replaced by underground sewer-pipes, so that the open sewerage system became a closed system. In 1917 the "Ancylostoma Act" was proclaimed and according to this law every house in Paramaribo was compelled to have a good excreta disposal. Before rendering a permit to build a house, the Ministry of Public Works, in cooperation with the Bureau of Public Health, makes sure that there will be a sanitary excreta disposal. There are still a few privies in the backyards of old premises in Paramaribo, but these privies are gradually disappearing.

In the rural areas in Suriname excreta disposal was usually in pit latrines. These are simple holes, dug in the ground with a superstructure on top. When the hole is almost full of excreta, it is covered with soil and another spot is chosen to dig another hole. Nowadays these pit-latrines in the districts are also being replaced by flush toilets in the homes. Besides excreta disposal in a septic tank with drainage bed, there are four small community sewerage collection and treatment systems, two in Paramaribo and two in the districts.

3. The Health Dimension of Water Supply and Sanitation in Suriname

Many pathogens live, either facultatively or of necessity, in water or are otherwise dependent on it for their survival. Feachem *et al.* (1977) proposed a classification of water-associated diseases by using their relationship to water as the main distinctive criterion. Later Bradley (1991) added an additional group to this list.

- *Water-borne diseases*
- *Water-shortage diseases or water-washed diseases*
- *Water-based diseases or water-impounding diseases*
- *Water-associated vector-borne diseases or water-arthropod diseases*
- *Chemical constituents, either excess or shortage* (e.g. fluoride)
- *Water-dispersed diseases* (usually non-pathogenic amoeba may enter through the nose and cause damage, bacteria like *Legionella* in air-conditioning systems).

Here we will only discuss the first four categories.

3.1. Water-borne Diseases

These diseases are caused by organisms that can survive in water and are ingested when contaminated water is drunk. Before the water supply system was installed in Suriname (1933), water-borne diseases were widespread.

3.1.1. Dysentery

Bacillary dysentery (shigellosis), caused by bacteria, was endemic in Suriname and at the end of the dry season there were epidemics, for instance, from February to July 1929. After 63 years of absence an epidemic of shigellosis started in August 1992 in the neighborhood of the village of Mungo along the Cottica River. This epidemic extended to other areas with a peak in January 1993 (23 cases). The youngest patient was only one month old. Since March no cases have been reported. The epidemic lasted 7 months and altogether there were 7 deaths.

Amoebic dysentery, caused by Protozoa, also occurred frequently in periods when no potable water was available.

3.1.2. Typhoid Fever

Typhoid fever is caused by the bacteria *Salmonella typhi*. It is endemic in Suriname and formerly many epidemics occurred. Two epidemics among the bushnegro population are well described (Oostburg 1982). The first was in Auka villages along the Cottica and Patamakka Rivers, near the bauxite village of Mungo, and lasted from 18 May till 8 July 1961. A total of 102 cases of typhoid fever were diagnosed. The second epidemic was among the Saramacca tribe along the Suriname River in the neighbourhood of the village of Lombé. This epidemic was caused by the inundation of the area after a dam for hydroelectric energy was built across the Suriname River at Afobakka (see chapter 13). This epidemic lasted from 18 March till 30 May 1964 and altogether 48 cases were diagnosed.

3.1.3. Cholera

Cholera caused by the bacteria *Vibrio cholerae*, occurred in Suriname in the 19th century. In 1854 there was an epidemic in Nickerie, during which 29 of the 70 diagnosed cases died. In 1863 there were some imported cholera cases in Suriname. After this, until 1990, cholera was never again diagnosed in Suriname.

In 1990 there was a serious cholera epidemic in Peru, which passed into a serious pandemic in the western hemisphere. The health authorities in Suriname were alerted and on 1 March 1992 a case of cholera was diagnosed in the village of Cottica, along the Lawa River.

3.1.4. Acute Diarrhoea

The drinking of unsafe water may cause acute diarrhoea. Diarrhoea is a clinical syndrome of diverse etiology, associated with loose or watery stools and often vomiting and fever. It can be caused by bacterial, viral and parasitic agents, such as *Escherichia coli, Campylobacter, Shigella, Salmonella, Rotavirus, Giardia lamblia*, etc. In Suriname, water-borne acute diarrhoea is rare and epidemics have never occurred.

3.1.5. Leptospirosis or Weil's Disease

This disease is caused by a bacteria of the order Spirochaetalis. In the period 1982–1984, 78 suspected leptospirosis cases were admitted to hospitals in Paramaribo. Of these cases 54 were confirmed by laboratory tests. In the period 1985–1988, 95 suspected cases were hospitalized of which 50 were confirmed by laboratory tests. The water is usually polluted by the urine of rats, which can harbour the bacteria. Most leptospirosis cases occur in the major rainy season (May–July). The incidence in males was higher than in females. In 1987, 14 of the 16 confirmed leptospirosis cases (88%) were males.

3.2. Water-shortage Diseases or Water-washed Diseases

It has been noticed that all kinds of infections, some diarrhoeal diseases and contagious skin and eye infections are prevalent when the scarcity of accessible water supplies makes washing and personal cleanliness difficult and infrequent.

3.3. Water-based Diseases or Water-impounding Diseases

Some parasitic helminths spend part of their lifecycle in intermediate host organisms that live in freshwater. If man makes contact with this water, he can be infected.

3.3.1. Schistosomiasis or Bilharzia

Schistosoma mansoni is the only bilharzia species that occurs in Suriname. It is restricted to the coastal region, since the vector snail *Biomphalaria glabrata* only occurs in the region of coastal shell-ridges. It is presumed that the snail needs the calcium from the shells (Reyenga 1971).

3.3.2. Dracunculosis (or Infection with Guinea-worm)

This is caused by the filaria-worm *Dracunculus medinensis*. The female lives in subcutaneous tissues and can reach a length of 120 cm. The vector is a Copepod of the genus *Cyclops*. According to Stedman (1799) Guinea worm occurred in Suriname in the 18th century. It has now been eradicated.

3.4. Water-associated Vector-borne Diseases or Water-arthropod Diseases

These diseases occur when water provides a habitat for Arthropod vectors.

3.4.1. Malaria

Malaria is endemic only in the Interior of Suriname. The capital of Paramaribo and the coastal and savanna regions are almost free from malaria, since the main vector, *Anopheles darlingi*, occurs only in the Interior. Malaria is a disease characterized by headache, fever, chills and sweating. Depending on the species, there is a typical fever pattern (Table 1). In Suriname there are 3 species of malaria, but the majority of the cases is caused by *Plasmodium falciparum* (80%), followed by *Plasmodium vivax* (15%), whereas *Plasmodium malariae* was very rare in Suriname (Fig. 1). However, in the past 3 years more cases of *P. malariae* have been diagnosed. The interior of Suriname is mostly inhabited by bushnegroes, descendants from negro slaves who had escaped. They have a natural immunity against *P. vivax*, because of the absence of the so-called Duffy-factor in the blood. The occurrence of sickle cell haemoglobin (HbS) in the negro race appears to be a natural immunity against malaria as well. Beside this natural immunity, the bushnegroes living in the Interior for years also develop an acquired immunity. They harbour the malaria parasites in their blood, but are not sick. However, these asymptomatic positive cases infect the vector mosquito and cause malaria epidemics. It is assumed that the acquired immunity is completed by age 5. Therefore, children under five in the Interior and non-immune people from the capital Paramaribo or other non-malarious areas, are very susceptible to malaria tropica with complications (brain malaria, kidney insufficiency, etc.), which can lead to death. If prior to a visit to the Interior the health authorities of Suriname are consulted, malaria can be prevented, or if contracted, properly treated.

Anopheles darlingi breeds in the Interior. Most authors define the Interior as the area south of the Savanna Belt. According to Rozendaal (1990) a better

Table 1. Diseases in Suriname that are transmitted by water or of which the vector develops in water.

Pathogen	Disease	Most important symptoms	Vector	Remarks
Bacteria				
Salmonella typhi	Typhoid fever	fever, diarrhoea		
Vibrio cholerae	Cholera	serious diarrhoea		
Shigella sp.	Shigellosis (bacillary dysentery)	dysentery (diarrhoea with blood and mucus)		
Campylobacter sp.	acute diarrhoea	diarrhoea		
Escherichia coli	acute diarrhoea	diarrhoea		
Salmonella sp.	acute diarrhoea	diarrhoea		
Leptospira sp. (Spirochaetalis)	Leptospirosis (Weil's disease)	fever, jaundice		
Viruses				
Rota virus	Rotaviral enteritis	acute diarrhoea, fever		
Yellow fever virus	Yellow fever	fever, jaundice	*Aedus aegypti* *Haemagogus* sp.	
Dengue virus	Dengue	pian in joints and behind the eyeballs	*Aedes aegypti*	
Arbo viruses	Arboviral diseases	acute neurologic symptoms	mosquito spp.	
Protozoa				
Giardia lambia	Giardiasis	diarrhoea		
Entamoeba histolytica	Amoebic dysentery (Amoebiasis)	dysentery		
Plasmodium falciparum	Malaria tropica	fever and chills	*Anopheles darlingi*	
Plasmodium vivax	Malaria tertiana	fever and chills every third day	*Anopheles darlingi*	
Plasmodium malariae	Malaria quartana	fever and chills every fourth day	*Anopheles darlingi*	
Helminths				
Schistosoma mansoni	Bilharzia (Schistosomiasis)	diarrhoea, ascites, oesophagus varices	*Biomphalaria glabrata*	
Wuchereria bancrofti	Filaria (Wuchereriasis)	lymphangitis, chyluria, elephantiasis	*Culex quinquefasciatus*	
Mansonella ozzardi & *Mansonella perstans*	Savanna-filaria	usually asymptomatic	*Culiocoides* sp.	almost eridicated in Suriname
Dracunculus medinenses (Guinea-worm)	Dracunculosis Infection with Guinea worm	blister on the skin	*Cyclops.* sp.	eradicated in Suriname

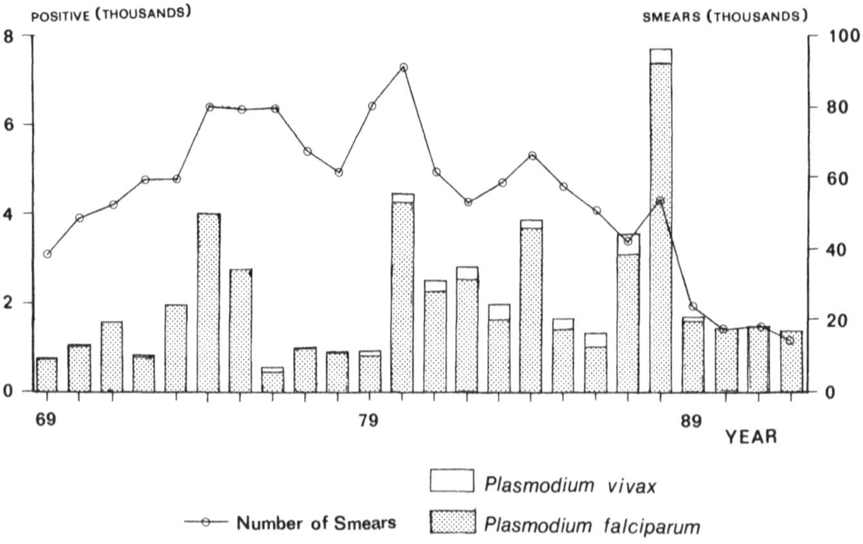

Figure 1. General malaria situation in Suriname for the period 1969–1992.

definition of the Interior is the area South of the limit of the tidal influence on the rivers. It is a fact that *A. darlingi* only breeds in areas of still water, where there are no low and high tides, hence the Interior. In most rivers this limit of tidal influence is just South of the first rapids (all the large rivers in Suriname run from South to North).

A. darlingi breeds in creeks and at partly sunlit spots in small forest streams, protected from the current by obstacles such as roots and fallen trunks. It also breeds near the river shores, among floating debris and water plants or protrusions from the river bank, which provide protection from the current, in habitats that are partly shaded by the nearby forest. Another breeding place of *A. darlingi* is the flooded forest, where it breeds in open, partly sunlit places, among emergent plants or floating debris in swamps. These habitats are created during the rainy season, when the river overflows its banks. Finally, this mosquito breeds in partly shaded pools, formed during the dry season in or near the river bed, following a drop in the water level of the river.

A. darlingi has occasionally been caught in the Savanna Belt. Bruijning (1952) believes that *A. darlingi* can be introduced in the savanna region by downstream flotation of larvae among patches of the water hyacinth (*Eichhornia crassipes*) in the large rivers which originate South of the savanna region.

Anopheles aquasalis. In the Coastal Plain *A. aquasalis* was indicated as the malaria vector by Bonne and Bonne-Wepster (1925). A correlation between an increase of the number of *A. aquasalis* and malaria epidemics in the Coastal Plain was found for the lower courses of the Saramacca, Suriname and Commewijne Rivers (Van der Kuyp 1951), Coronie (Stage and Geyskes 1946;

Van der Kuyp 1950), Mungo (Geijkes 1946; Panday 1974) and Galibi (Oostburg 1973). Nevertheless, *A. aquasalis* is not considered a proven malaria vector in Suriname. Elsewhere it appeared to be a better vector for *Plasmodium vivax*, which occurs among the Amerindians in South Suriname. The mere finding of increased numbers of *A. aquasalis* during malaria outbreaks is not enough to conclude that it is the vector. If *A. aquasalis* was as good a vector for *P. falciparum* as *A. darlingi*, autochtonous malaria should have occurred more in the coastal region, where many people with malaria parasites in the blood reside. Malaria in Suriname is, therefore, only considered to be endemic in the Interior, whereas incidentally there are indications that *A. aquasalis* might transmit the disease in the northern part of the country.

Anopheles nuneztovari. Since 1949 (Van der Kuyp) *A. nuneztovari* was reported in Suriname, but it was not considered to be a malaria vector. However, chromosome studies (Kitzmiller *et al.* 1973) have taught us that there are two siblings of *A. nuneztovari*. The sibling that occurs in northern Colombia and western Venezuela appears to be a vector of malaria, whereas the sibling occurring in Brazil, Ecuador, Bolivia, Peru and Suriname is not a malaria vector.

After the construction of the Afobakka dam across the Suriname river in 1960, the Brokopondo lake was formed. After this lake was completely filled by July 1965, the number of *A. nuneztovari* increased. Panday (1977) collected enormous numbers of female *A. nuneztovari* at Brownsberg and he observed extensive breeding in the Brokopondo Lake during certain periods of the year (1977). The only record of *A. nuneztovari* from the coastal region is by Van der Kuyp at Leiding XIa in June 1946. *A. nuneztovari* has been collected periodically in different locations in the Interior by Bruijning, Van Thiel and Panday.

3.4.2. Filaria

Wuchereria bancrofti is a filaria worm that lives in the lymphatic vessels and the lymphnodes. It may cause severe lymphangitis which is known as a "filaria – attack" and this disease may lead to elephantiasis (enlargement of legs, arms, breasts or scrotum). The vector is the mosquito *Culex quinquefasciatus*.

Culex quinquefasciatus breeds preferably in foul water, e.g. pools in the backyards of houses, where water from kitchens is collected, pails with dirty water and all kinds of refuse, privies, sewers, cesspools, ditches, drains, etc. In other words, *Culex quinquefasciatus* breeds in man-made breeding places, hence the density is higher in urban than in rural areas.

3.4.3. Savanna Filaria

Mansonella ozzardi and *Mansonella perstans* are savanna filarias, which occur mostly among the Amerindian population of the savanna region. The vector is a *Culicoides* species (midges or mampira), belonging to the family Ceratopogonidae.

3.4.4. Infection with Arthropod-borne Viruses (Arbo-viruses)

Arboviruses are animal viruses transmitted naturally between vertebrates and hematophagous Arthropods (Casals, 1967). Three arboviruses were discovered for the first time in Suriname. The *Paramaribo virus* was described in 1964 (Metselaar *et al.*). In the same period the *Restan virus* was described for Suriname and Trinidad (Jonkers *et al.* 1967). Finally, in 1964, de Haas, Jonkers and Heinemann discovered a new arbovirus in Suriname, the *Kwatta virus*. There is a variety of mosquito species which are vectors of arboviruses.

3.4.4.1. Yellow Fever. This is a serious disease with fever and jaundice leading to death within a few days. Formerly, epidemics of yellow fever occurred in Suriname, but the last epidemic of urban yellow fever occurred between December 1908 and January 1909 (32 cases with 8 deaths). However, it is known that the yellow fever virus circulates among monkeys in the Interior. People going to the Interior may run the risk of being infected and it has happened three times in this century, the last time on July 26, 1972 in an Amerindian woman. Fortunately, the infection with yellow fever can be prevented, because there is an effective immunization against this disease.

The vector of urban yellow fever is the mosquito *Aedes aegypti* and the vector of jungle yellow fever is a mosquito of the genus *Haemagogus*, which lives in the canopy of trees. When these trees are cut down, the *Haemagogus* mosquito gets the chance to bite man. Man can also be bitten by walking on a forest-road on a hill that is on the same level as the canopy of trees that grow in the valley. The breeding places of *Haemagogus* are water in treeholes, bamboo joints and similar locations.

3.4.4.2. Dengue. Dengue is a viral disease occurring in countries with a warm, moist climate, where the vector mosquito, *Aedes aegypti*, occurs. The disease is characterized by an abrupt onset with pain in the joints, the epiqigastric region and just behind the eyeballs. Also redness and swelling of the face and fever. There are two serious complications known, which may lead to death, i.e. Dengue hemorrhagic fever (DHF) and Dengue shock syndrome (DSS), mostly occurring during the second of two successive dengue infections. In some countries *Aedes albopictus* is a vector of dengue as well, but so far this mosquito has never been collected in Suriname.

Aedes aegypti is a day-biter and breeds in man-made breeding sites. The eggs are not laid in rafts, but single, very often not on the surface of the water, but along the sides of the container, near the surface. *Aedes aegypti* prefers clean water for breeding. The favorite breeding places are rainwater tanks, barrels and eaves-gutters of houses. Furthermore, any artificial container of water, flower vases, old car tyres, tin cans in the yard, and sometimes treeholes near the houses.

4. Annoyance by Mosquitoes

There are some mosquitoes, which do not transmit diseases, but are an annoyance for man because of their annoying buzzes and their painful bites. Such a mosquito in Suriname is *Aedes taeniorhynchus.*

Aedes taeniorhynchus breeds in saltwater swamps (Nickerie) and in very dirty puddles (urban areas). It does not usually occur in Paramaribo but is abundant in Nickerie and Coronie. In March and April 1983, this annoying mosquito all of a sudden appeared in Paramaribo, causing a real mosquito plague (Oostburg 1983). Bonne and Bonne-Wepster described a similar mosquito plague in Paramaribo in March and April 1919.

Summary

The significance of water for our health is discussed. Different water supply systems meant to provide potable water for the Surinamese population, based on WHO standards, are mentioned. Furthermore, the history of water supply and sewerage in Suriname is reviewed. Finally, the health dimension of water supply and sanitation in Suriname is discussed, whereby water-borne diseases and vectors that need water for their development are listed.

References

Bonne, C. and Bonne-Wepster, J., 1925. Mosquitoes of Surinam. Royal Colonial Institute, Amsterdam, 558 pp.

Bruijning, C.F.A., 1952. Verslag van de medische expeditie Tapanahony-Paloemeu. Mimeogr., Bureau of Public Health, Paramaribo, 18 pp.

Bradley, D.J., 1991. Communicable diseases related to water and water resources development. Unpubl. report, World Health Organization, Geneva.

Feachem, R., D.J. Bradley, H. Garelick, and D.D. Mara, 1980. Health aspects of excreta and sullage management. Appropriate technology for watersupply and sanitation. World Bank, Washington D.C.

Geijskes, D.C., 1946. Notes on the neotropical *Anopheles* in Moengo, Suriname. Mosq. News 6(3): 113–118.

Haas, R.A. de, A.H. Jonkers, and D.W. Heinemann, 1946. Kwatta-virus, a new agent isolated from *Culex* mosquitoes in Suriname. Amer. J. Trop. Med. Hyg. 15: 954.

Jonkers, A.H., D. Metselaar, A.H. Paes de Andrade, and E.S. Tikasingh, 1967. Restanvirus, a new group C arbovirus from Trinidad and Surinam. Amer. J. Trop. Med. Hyg. 16: 74.

Kitzmiller, J.B., R.D. Kreutzer, and E. Tallaferro, 1973. Chromosomal differences in a population of *Anopheles nuneztovari.* Bull. Wld. Hlth. Org. 48: 435–455.

Kuyp, E. van der, 1950. Contribution to the study of malarial epidemiology in Suriname. Meded. K. Vereen. Indisch Inst. Amsterdam 89, Trop. Hyg. 18.

Metselaar, D., J.D. Verlinde, and J. Versteeg, 1964. Paramaribo-virus. Properties of a group A arbovirus isolated from human blood in Suriname. Arch. ges. Virusforsch. 14: 336.

Reyenga, T.W., 1971. Verspreidingsecologie van *Biomphalaria glabrata* (Say 1880) in relatie tot Bilharziasis in Suriname. PhD thesis, Leyden.

Rozendaal, J.A., 1990. Observations on the distribution of Anophelines in Suriname, with particular reference to the malaria vector *Anopheles darlingi*. Mem. Inst. Oswaldo Cruz Rio de Janeiro 85: 221–234.
Stage, H.H. and D.C. Geijskes, 1946. Observation on mosquito and malaria control in the Caribbean area Post I - Suriname. Mosq. News 6: 177–180.
Stedman, J.G., 1799. Reize naar Surinamen en door de binnenste gedeelten van Guiana. Deel I t/m IV. S. Emmering, Amsterdam.

16. Conservation of freshwater ecosystems in Suriname

K. MOHADIN

1. Introduction

The conservation of freshwater ecosystems cannot be viewed separately from other components of nature conservation in Suriname. The Nature Preservation Law (Government Bulletin 1954, No. 26) provides the Government with the means to identify wildlands of special importance by designating them as Nature Reserves and, at the same time giving them full protection. Article 5 of this law states that it is forbidden to undertake any kind of activity in a nature reserve which will affect the integrity of the reserves in a detrimental way. It is also prohibited to camp, make fire, cut wood or make charcoal, and to hunt or fish, unless written permission for this has been obtained from the Head of the Suriname Forest Service, and then only on the conditions stipulated in the permit. On the basis of this law, thirteen nature reserves have already been established (Fig. 1; Table 1). Nine of these are in the freshwater region of the country, which is roughly located south of the road connecting Albina at the east border, with Nieuw-Nickerie at the west border (the so called Oost-West Verbinding), and which extends all the way to the south border with Brazil. It covers part of the Young Coastal Plain and the three other geomorphological regions, the Old Coastal Plain, the Savanna Belt, and the Interior, the latter being the hilly and mountainous region of the country. One nature reserve, the Galibi Nature Reserve, comprises both brackish-water and freshwater ecosystems. Two other areas, also situated in the freshwater region in the west of the country, await Government approval to be designated as nature reserves. Wildlife outside the nature reserves is protected by the Game Act (Government Bulletin 1954, No. 25), and fish by the Fish Protection Act (Government Bulletin 1961 No. 44), as amended in 1963, (Government Bulletin 1963, No. 153). The Game Act fully protects all mammals, birds, sea turtles, and other species mentioned, except those that are designated as game animals, cage birds, and predominantly harmful species. Of the reptiles, two caiman species, *Palaeosuchus trigonatus* and *P. palpebrosus*, and the iguana, *Iguana iguana*, are also protected. The Fish Protection Act protects eight species of fish, by setting limits on the size to be captured. The law also prohibits collecting or destruction of eggs or nests.

Paul E. Ouboter (ed.), Freshwater Ecosystems of Suriname, 275–284.

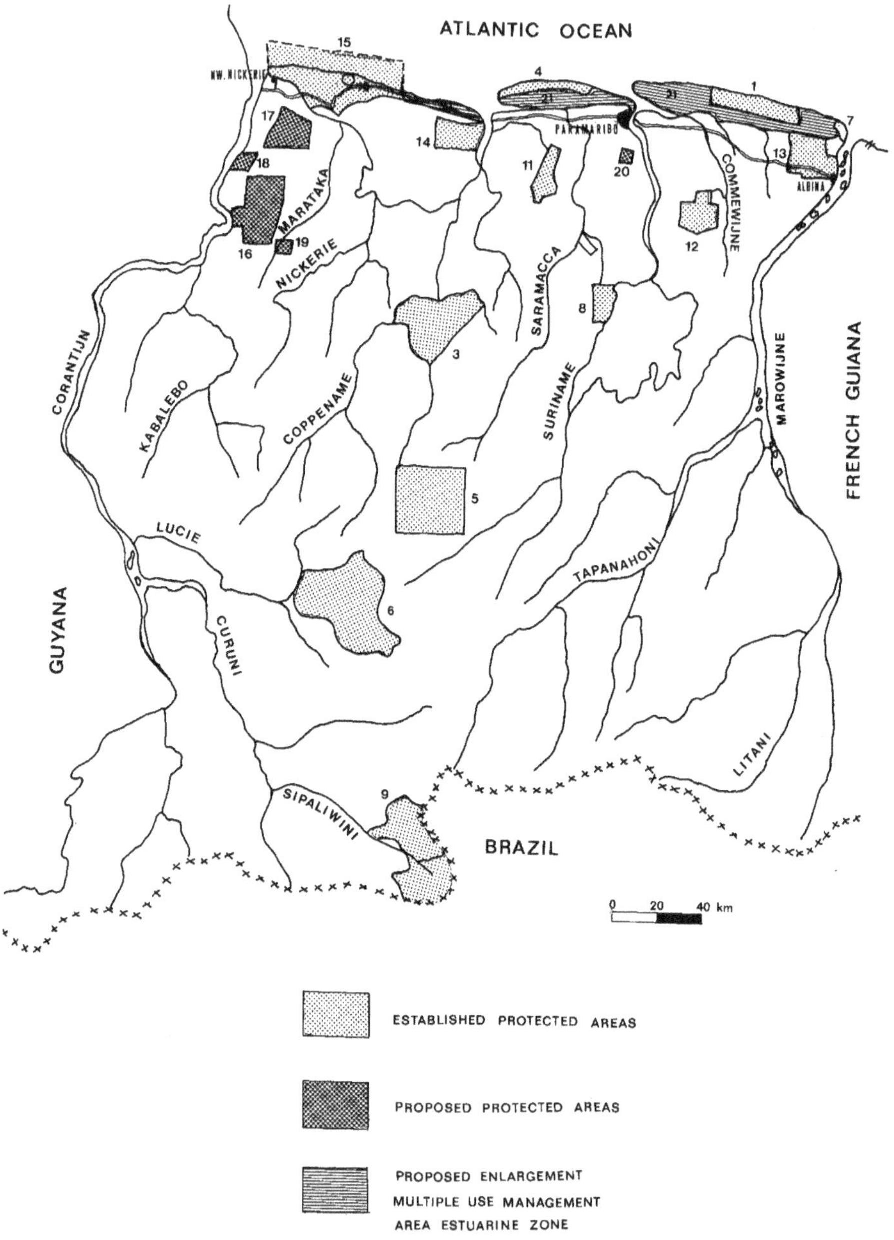

Figure 1. Existing and proposed protected areas in Suriname (for names see Table 1)

At present plants outside the nature reserves are also protected, but in a very limited way. The Timber Act (Government Bulletin 1947, no 42), as amended by Government Bulletin 1952 no 103, regulates logging and prohibits the logging

Table 1. Established and Proposed Protected Areas.

Established protected areas		Area
1. Wia Wia Nature Reserve		36,000 ha
2. Brinckheuvel Nature Reserve		6,000 ha
3. Raleighvallen/Voltz Berg Nature Reserve		78,170 ha
4. Coppename Monding Nature Reserve		12,000 ha
5. Tafelberg Nature Reserve		140,000 ha
6. Eilerts de Haan Gebergte Nature Reserve		220,000 ha
7. Galibi Nature Reserve		4,000 ha
8. Brownsberg Nature Park		8,400 ha
9. Sipaliwini Nature Reserve		100,000 ha
10. Hertenrits Nature Reserve		100 ha
11. Boven Cusewijne Nature Reserve		27,000 ha
12. Copi Nature Reserve		18,000 ha
13. Wane Kreek Reserve		45,000 ha
14. Peruvia Nature Reserve		31,000 ha
15. Bigi Pan Multiple-Use Management Area		68,300 ha
	subtotal	803,970 ha
Proposed protected areas		Area
16. Kaburikreek Nature Reserve		68,000 ha
17. Nani Nature Reserve		54,000 ha
18. Mac Clemen Forest Reserve		12,000 ha
19. Snake Kreek Forest Reserve		3,000 ha
20. Lelydorp Forest Reserve		500 ha
21. Multiple-Use Management Area Estuarine Zone		121,000 ha
	subtotal	259,000 ha
	total	1,063,670 ha

of certain tree species. Furthermore a permit is required for collecting forest by-products. The Forest Management Act, however, which will replace the Timber Act of 1949, will provide the possibility for maintaining forests that could have special scientific, educational, cultural, or recreational value because of their location, composition of flora and/or fauna, their aesthetic value, or their stabilizing effect on the natural environment – especially on soil and watershed.

The Acts on the Issuance of State-owned Lands ("Agricultural Act" of 1937 and Decree L-2 of 15 June 1982) also have mechanisms to protect certain wildlands. Based on these laws, part of the Brownsberg region (8,600 ha) in the Interior, was given to the Foundation for Nature Preservation in Suriname (STINASU) on long-term lease. STINASU manages this area as a National Park. Furthermore, the Planning Act (Government Bulletin 1973, No. 89) also provides for the creation of special management areas. The establishment of the Bigi Pan Multiple-Use Management Area in the estuarine zone is an example.

2. Nature Preservation in Suriname

Protection of nature in Suriname started in 1915 already, with the provision in article 44 of the Police Penal Code of 1915 (as ammended in 1942), which requires a permit from the Director of the Ministry of Natural Resources in order to hunt or to capture wildlife in certain areas of public land. Based on this article, the first sanctuary was established by Government Resolution (Government Bulletin 1953 No. 12) on 15 February 1953 in the estuary of the Coppename River to protect breeding colonies of the scarlet ibis (*Eudocimus ruber*). Later, as a result of the Nature Preservation Resolution of 22 April 1966, this sanctuary became the Coppename Monding Nature Reserve (on the basis of the Nature Preservation Act). The Law knows only one type of protected area: nature reserves.

With the same resolution an additional five nature reserves were created: the Wia Wia Nature Reserve (expanded in 1969), approx. 36,000 ha in the Young Coastal Plain, the Brinckheuvel Nature Reserve, approx. 6,000 ha, the Raleighvallen-Voltzberg Nature Reserve, (enlarged in 1986), approx. 78,170 ha, the Tafelberg Nature Reserve, approx. 140,000 ha and the Eilerts de Haan Gebergte Nature Reserve, approx. 220,000 ha, all situated in the Interior of Suriname.

In 1969 the Galibi Nature Reserve, approx. 4,000 ha in the Young Coastal Plain, was created by Government Resolution of 23 May 1969, and in 1972, also in the Young Coastal Plain, the Hertenrits Nature Reserve, approx. 100 ha. The Sipaliwini Nature Reserve, approx. 100,000 ha in the Interior at the border with Brazil, was established by Government Resolution of 31 January 1972.

Until then, no nature reserves had been established in the freshwater region of the Young Coastal Plain, the Old Coastal Plain, or the Savanna Belt, although it was known that unique ecosystems are located in these areas. The establishment of nature reserves here might be a problem because of human activities such as urban development, agriculture, mining and industry. The Government knew that caution should be exercised in conserving those areas, because of the rights and interests of local people, planned development strategy, and future urban development (Teunissen 1979). The need for conserving these unique ecosystems, however, led to a survey and mapping program of the ecosystems in the lowland area. This was initiated by the Foundation for Nature Preservation in Suriname (STINASU) and the Surinam Forest Service (LBB) and was conducted by P.A. Teunissen between 1976 and 1978. The result was the publication of "Recommendations for enlarging the system of nature reserves and forest reserves in the Suriname Lowlands" and of a "Reconnaissance Map of Suriname Lowland Ecosystems: scale 1: 200,000" (Teunissen 1977). Ten areas were selected for consideration for protection: six areas for the status of nature reserve and four areas for the status of forest reserves. In 1986, because of the Government Resolution of 26 August 1986, four of the six proposed nature reserves received legal status: the Peruvia Nature Reserve, approx. 31,000 ha, in the Young Coastal Plain; the Wanekreek Nature

Reserve, approx. 45,000 ha, in the Old Coastal Plain; the Copi Nature Reserve, approx. 28,000 ha, and the Boven-Cusewijne Nature Reserve approx. 27,000 ha, both in the Savanna Belt. The two other proposed nature reserves, the Nani Nature Reserve, approx. 54,000 ha, in the Young Coastal Plain, and the Kaburikreek Nature Reserve, approx. 68,000 ha, in the Old Coastal Plain, still await Government approval.

Until 1976, the establishment of nature reserves was based mainly on the presence of important flora, fauna, geological and cultural features, rather than on a systematic and scientific inventory of the area. This was indeed entirely in accordance with art. 2 of the Nature Preservation Act 1954, which states that for an area to be designated as a nature reserve, it should meet certain requirements, namely that it deserves protection by the Government because of its varied nature and landscape beauty and/or because of the presence, from a scientific or cultural point of view, of important flora, fauna, and geological objects.

Although several expeditions to explore the interior took place before 1976, none of them was specifically conservation-oriented. The main purpose of these expeditions was to map the areas visited. As a matter of fact, usually the physician who accompanyied these expeditions was incidentally entrusted with the collecting of zoological and botanical specimens.

After 1976 the selection for the establishment of nature reserves took place on the basis of a sound inventory and mapping of ecosystems, and previously established guidelines. One of the criteria was that, with the selection of these areas, the conservation aspects of the sites selected had to be taken into account. During the selection process, the development plans in adjacent areas were taken into account as much as possible to ensure that the activities, which are taking place (or will take place there) do not have a detrimental effect of the areas.

The wildlands have been selected as much as possible on larger tracts of public lands, including the upper reaches of creeks or rivers, in order to safeguard them from pollution caused by development activities in surrounding areas.

The presence of important populations of animal and plant species (although not necessarily endangered species), has also contributed to the selection of protected areas, as is the case with the Copi Nature Reserve, the Boven-Cusewijne Nature Reserve and the proposed Kaburikreek Nature Reserve. For the Peruvia Nature Reserve, which is located in a brackish-water region, the presence of large populations of possentri (*Hura crepitans*) was the main reason for establishing this reserve. Large population of the spectacled caiman (*Caiman crocodilus crocodilus*) are found in the Casewinica Creek in the Copi Nature Reserve. There is also a small population of the giant otter (*Pteronura brasiliensis*) in this creek. Unlike the caiman, this endangered species is fully protected by the Game Act.

The spectacled caiman is common and abundant in the entire coastal region of the country, and is only protected within a nature reserve on the basis of the Nature Protection Act. A large population of the spectacled caiman is found in

Figure 2. A group of giant otters (*Pteronura brasiliensis*) at Nani Creek (Photo P.A. Teunissen)

the Cusewijne River in the Boven-Cusewijne Nature Reserve (Ouboter and Nanhoe 1983). The river also has populations of manatee (*Trichechus manatus*) and giant otter. Another caiman species, *P. palpebrosus* is also present in this river (Ouboter and Nanhoe 1983). The manatee is fully protected by the Game Act.

The population of giant otters in the Kaburi Creek (Fig. 2) has contributed considerably to the selection of this area as a nature reserve. The establishment of this reserve has not yet been realized; nor has the proposed Nani Nature Reserve, which harbors large populations of manatee and spectacled caiman.

This area is predominantly a swamp, the water of which is used for the irrigation of the rice fields in the District of Nickerie. It is also for this reason that the area has been proposed as a nature reserve.

The Coppename River in the Raleighvallen-Voltzberg Nature Reserve consists of a system of rapids and waterfalls. It contains a number of plant and fish species adapted to the fast running water habitat. A number of species of the family of the Podostemaceae are common here, of which the species *Mourera fluviatilis* is very abundant and striking. In addition, the endangered species *Apinagia marowijnensis* of the same family has been recorded here (see also Werkhoven and Peeters, Chapter 5). Well-known fish species are the electric eel (*Electrophorus electricus*), several species of piranhas of the family Serrasalmidae, and members of the families Erythrinidae and Characidae.

3. Protection of Freshwater Ecosystems in the Interior

A detailed inventory of the ecosystems in the Interior has not yet taken place, due to the lack of qualified personnel and necessary funds. Biological surveys of the mountainous areas of the south have just started, and are likely to yield some interesting additions to the flora and fauna. It might be necessary to change or extend the system of protected areas on the basis of these future results. In addition, the development planned for the Interior, in particular logging, gold mining and bauxite mining, makes a detailed survey of the ecosystems here a high priority.

Important ecosystems must be protected because some of the planned activities could be detrimental to the integrity of these areas, and have a negative impact on their natural functions.

Table 2. Freshwater ecosystems in existing and proposed protected areas.

Name of protected area.	Ecosystem	Exceptional flora and fauna elements
2. Brinckheuvel NR	Subgraywacke landscape and conglomerate ridges and hills, with flat colluvial areas between, with one-layered sedge grass, xerophytic woods, and mesophytic forest.	
3. Raleighvallen-Voltzberg NR	Tropical lowland rain forest, Upper Coppename River with rapids and waterfalls.	*Mourera fluviatilis, Apinagia marowijnensis, Electrophorus electricus,* several Characidae, Erythrinidae, Serrasalmidae, *Paleosuchus trigonatus*
5. Tafelberg NR.	Scrub savanna of the Tafelberg type. Lowland and montane rainforest, and marsh forest.	

Table 2. (continued)

Name of protected area.	Ecosystem	Exceptional flora and fauna elements
6. Eilerts de Haan NR	Lowland and lower montane rainforest.	
7. Galibi NR	Grass swamps with *Typha angustifolia* and *Cyperus giganteus* Mixed mesophytic dryland and marsh forest.	Leatherback (*Dermochelys coriacea*), green turtle (*Chelonia mydas*), olive ridley (*Lepidochelys olivacea*).
8. Brownsberg Nature Park	Mesophytic tropical rainforest.	
9. Sipaliwini NR	Savanna of the Paru type. Gallery forests, freshwater swamps.	
11. Boven Cusewijne NR	Marshy clay savanna of Wanekreek type. Mixed mesophytic dryland- and marsh forest. Grass swamps, dominated by *Lagenocarpus guianensis*. Xerophytic dryland- and marshland woods. Dry white sand savannas of Cassipora type: short grass savannas with scattered shrubs. Marshy white sand savannas of Zanderij type: short grass savannas with scattered shrub and *Mauritia flexuosa* palm galleries. Cusewijne River.	Fish species: *Hoplias malabaricus, Hoplerythrinus unitaeniatus, Serrasalmus spilopleura.* Spectacled caiman *(Caiman crocodilus),* smooth-fronted caiman (*Palaeosuchus palpebrosus*). Several species of waterfowl, manatee (*Trichechus manatus*).
12. Copie NR	Dry savannas of Welgelegen type (clay savannas): short grass savannas with scattered gnarled treelets and bushes. Scattered *Pterocarpus-* bushes in grass and fern swamps. Mixed mesophytic dryland- and marsh forest.	Giant otter *(Pteronura brasiliensis*). Spectacled caiman (*Caiman crocodilus*). smooth-fronted caiman (*Paleosuchus palpebrosus*).
13. Wane Kreek	Ombrogenous peat swamps (incl. swaying swamps with floating peat layer, dominated by *Lagenocarpus guianensis.* Mixed mesophytic dryland- and marsh forest. Marshy clay-savanna of Wanekreek and Welgelegen type.	

Table 2. (continued)

Name of protected area.	Ecosystem	Exceptional flora and fauna elements
14. Peruvia NR	Mixed mesophytic dryland- and marsh forest. Hydrophytic swamp forest with possentri (*Hura crepitans*), mierenhout (*Triplaris surinamensis*), watrabebe (*Pterocarpus officinalis*), panta (*Tabebuia insignis*), pruimen (*Chrysobalanus icaco*), zuurzak (*Annona glabra*), and maurisi ((*Mauritia flexuosa*). Tall grass swamp dominated by langa-grasi (*Typha angustifolia*), prasoro-grasi (*Cyperus giganteus*), babun (*Virola surinamensis*). Grass and fern swamps. Ombrogenous peat swamps, also swaying swamps.	Possentri (*Hura crepitans*).
16. Kaburikreek NR	Scattered *Pterocarpus*-bushed in grass and fern swamps. Xerophytic swamp forest and woods with *Crudia glaberrima* and *Macrolobium acaciifolium*.	Giant otter (*Pteronura brasilensis*), Guiana otter (*Lutra enudris*). Manatee (*Trichechus manatus*).
17. Nani NR	Mixed mesophytic dryland- and marsh forest. Scattered *Mauritia flexuosa* palms in tall grass and fern swamps. Closed xerophytic palm swamp forest, dominated by *Mauritia flexuosa*. Floating aquatic vegetation.	Spectacled caiman (*Caiman crocodilus*). Manatee (*Trichechus manatus*)

NR = Nature Reserve

Summary

There is a solid legal basis for the protection of ecosystems and wildlife in Suriname. Thirteen nature reserves, one nature park and one multiple-use management area, have already been established, totaling an area of approx. 803,970 ha. Ten nature reserves are situated in the freshwater region of the country. Contrary to what was done prior to 1976, the selection of the new protected areas took place after a sound inventory and mapping of ecosystems, and on the basis of previously established guidelines. The presence of the and on the basis of previously established guidelines. The presence of the spectacled caiman (*Caiman crocodilus*), the giant otter (*Pteronura brasiliensis*) and the manatee (*Trichechus manatus*) has contributed considerably to the selection for protection of wildlands.

With the anticipated development, there is a great need for a further inventory of ecosystems in the Interior, which will form the basis for setting aside protected areas in order to mitigate the detrimental effects of the human activities.

Acknowledgements

I want to thank Mr. Henri A. Reichart for reviewing this paper. Many thanks also to Mrs. Ch. Kalika for the wordprocessing.

References

Baal, F.L.J., 1991. Legal aspects of protected areas in Suriname. Report prepared for the IV World Congress on National Parks and other protected areas, Caracas, Venezuela, February 1992.

Held, M.M. and H.A. Reichart, 1991. Managing Protected Areas in Suriname. Report prepared for the IV World Congress on National Parks and other protected areas, Caracas, Venezuela, February 1992.

Ouboter, P.E. and L.M.R. Nanhoe, 1983. An Ecological study of *Caiman crocodilus crocodilus* in Northern Surinam. Department of Animal Ecology, Catholic University of Nijmegen, The Netherlands.

Schulz, J.P., R.A. Mittermeier, and H.A. Reichart, 1977. Wildlife in Suriname. Journal of the Fauna Preservation Society 14 (2).

Teunissen, P.A., 1977. Reconnaissance Map-Surinam Lowland Ecosystems (Coastal Plain and Savanna Belt). De Walburg Pers Zutphen, Netherlands.

Teunissen, P.A., 1979. Aanbevelingen tot uitbreiding van het systeem van natuurreservaten en bosreserves in het Surinaamse Laagland. (2e druk)

Werkhoven, M.C.M. and M.M. Held, 1989. Conservation of Natural ecosystems versus Land-Use in Suriname.

Subject index

Index of scientific names

MONOGRAPHIAE BIOLOGICAE

1. *Physiologia Comparata et Oecologia*, Vol. I. 1948 out of print
2. *Physiologia Comparata et Oecologia*, Vol. II. 1952 out of print
3. *Physiologia Comparata et Oecologia*, Vol. III. 1954 out of print
4. *Physiologia Comparata et Oecologia*, Vol. IV. 1957 out of print
5. (1) R. A. Kennedy: *The Metzner Theory of Urine Formation.* 1957
   (2) M. D. Grmek: *On Ageing and Old Age.* Basic Problems and Historic Aspects of Gerontology and Geriatrics. 1958
   (3–4) O. Bassir: *Biochemical Aspects of Human Malnutrition in the Tropics.* 1962 out of print
6. F. S. Bodenheimer: *Animal Ecology Today.* 1958 out of print
7. R. R. Gates: *Taxonomy and Genetics of Oenothera.* 1958 ISBN 90-6193-061-8
8. A. Keast, R. L. Crocker and C. S. Christian (eds.): *Biogeography and Ecology in Australia.* 1959 out of print
9. S. Stanković: *The Balkan Lake Ohrid and Its Living World.* 1960 out of print
10. E. Rivnay: *Field Crop Pests in the Near East.* 1962 ISBN 90-6193-063-4
11. M. Kozhov: *Lake Baikal and Its Life.* 1962 ISBN 90-6193-064-2
12. M. S. Mani: *The Ecology of Plant Galls.* 1964 out of print
13. E. van Oye (ed.): *The World Problem of Salmonellosis.* 1964 ISBN 90-6193-066-9
14. D. H. S. Davis (ed.): *Ecological Studies in Southern Africa.* 1964 out of print
15. J. van Mieghem and P. van Oye (eds.): *Biogeography and Ecology in Antarctica.* 1965 ISBN 90-6193-067-7
16. H. Boyko (ed.): *Salinity and Aridity.* New Approaches to Old Problems. 1966 out of print
17. M. J. Coe: *The Ecology of the Alpine Zone of Mount Kenya.* 1967 ISBN 90-6193-068-5
18. E. J. Fittkau, J. Illies, H. Klinge, G. H. Schwabe and H. Sioli (eds.): *Biogeography and Ecology in South America*, Vol. I. 1968 out of print
19. E. J. Fittkau, J. Illies, H. Klinge, G. H. Schwabe and H. Stioli (eds.): *Biogeography and Ecology in South America*, Vol. II. 1969 ISBN 90-6193-071-5
20. R. N. Kaul (ed.): *Afforestation in Arid Zones.* 1970 out of print
21. R. Battistini and G. Richard-Vindard (eds.): *Biogeogaphy and Ecology of Madagascar.* 1972 ISBN 90-6193-073-1
22. J. E. Bishop: *Limnology of a Small Malayan River Sungai Gombak.* 1973 ISBN 90-6193-074-X
23. M. S. Mani (ed.): *Ecology and Biogeography in India.* 1974 ISBN 90-6193-075-8
24. E. K. Balon and A. G. Coche (eds.): *Lake Kariba.* A Man-made Tropical Ecosystem in Central Africa. 1974 ISBN 90-6193-076-6
25. W. D. Williams (ed.): *Biogeography and Ecology in Tasmania.* 1974 ISBN 90-6193-077-4
26. A. de Vos: *Africa, the Devastated Continent?* Man's Impact on the Ecology of Africa. 1975 ISBN 90-6193-078-2
27. G. Kuschel (ed.): *Biogeography and Ecology in New Zealand.* 1975 ISBN 90-6193-079-0
28. I. Prakash and P. K. Ghosh (eds.): *Rodents in Desert Environments.* 1975 ISBN 90-6193-080-4
29. J. Rzóska (ed.): *The Nile.* Biology of an Ancient River. 1976 ISBN 90-6193-081-2
30. G. Kunkel (ed.): *Biogeography and Ecology in the Canary Islands.* 1976 ISBN 90-6193-082-0

31. M. J. A. Werger (ed.): *Biogeography and Ecology of Southern Africa*, 2 vols. 1978 ISBN 90-6193-083-9
32. C. Serruya (ed.): *Lake Kinneret* [Lake of Tiberias / Sea of Galilee]. 1978 ISBN 90-6193-085-5
33. Ph. D. Mordukhaí-Boltovskoi (ed.): *The River Volga and Its Life.* 1979 ISBN 90-6193-084-7
34. Ch. W. Heckman: *Rice Field Ecology in Northeastern Thailand.* The Effect of Wet and Dry Seasons on a Cultivated Aquatic Ecosystem. 1979 ISBN 90-6193-086-3
35. M. Kalk, A. J. McLachlan and C. Howard-Williams (eds.): *Lake Chilwa.* Studies of Change in a Tropical Ecosystem. 1979 ISBN 90-6193-087-1
36. B. R. Allanson (ed.): *Lake Sibaya* [South Africa]. 1979 ISBN 90-6193-088-X
37. H. Löffler (ed.): *Neusiedlersee.* The Limnology of a Shallow Lake in Central Europe. 1979 ISBN 90-6193-089-8
38. J. Rzóska: *Euphrates and Tigris.* Mesopotamian Ecology and Destiny. With Contributions by J. F. Talling and K. E. Banister. 1980 ISBN 90-6193-090-1
39. K. A. Brodsky: *Mountain Torrent of the Tien Shan.* A Faunistic-Ecology Essay. Translated from Russian by V. V. Golosov. 1980 ISBN 90-6193-091-X
40. M. S. Mani and L. E. Giddings (eds.): *Ecology of Highlands.* 1980 ISBN 90-6193-093-6
41. A. Keast (ed.): *Ecological Biogeography of Australia*, 3 vols. 1981 ISBN 90-6193-092-8
42. J. L. Gressit (ed.): *Biogeography and Ecology of New Guinea*, 2 vols. 1982 ISBN 90-6193-094-4
43. H. Heathwole, T. Done and E. Cameron: *Community Ecology of a Coral Cay.* A Study of One-Tree Island, Great Barrier Reef, Australia. 1981 ISBN 90-6193-096-0
44. P. S. Maitland (ed.): *The Ecology of Scotland's Largest Lochs.* Lomond, Awe, Ness, Morar and Shiel. 1981 ISBN 90-6193-097-9
45. K. Müller (ed.): *Coastal Research in the Gulf of Bothnia.* 1982 ISBN 90-6193-098-7
46. G. K. Rutherford (ed.): *The Physical Environment of the Faeroe Islands.* 1982 ISBN 90-6193-099-5
47. J. I. Furtado and S. Mori (eds.): *Tasek Bera* [Malaysia]. The Ecology of a Feshwater Swamp. 1982 ISBN 90-6193-100-2
48. G. R. South (ed.): *Biogeography and Ecology of the Island of Newfoundland.* 1983 ISBN 90-6193-101-0
49. J. A. Thornton (ed.): *Lake McIlwaine.* The Eutrophication and Recovery of a Tropical African Man-made Lake. 1982 ISBN 90-6193-102-9
50. R. W. Edwards and M. P. Brooker: *The Ecology of the Wye* [U.K.]. 1982 ISBN 90-6193-103-7
51. T. Petr (ed.): *The Purari* [Papua New Guinea]. Tropical Environment of a High Rainfall River Basin. 1983 ISBN 90-6193-104-5
52. H. Kuhbier, J. A. Alcover and C. Guerau d'Arellano Tur (eds.): *Biogeography and Ecology of the Pityusic Islands.* 1984 ISBN 90-6193-105-3
53. J.-P. Carmouze, J.-R. Durand and C. Lévêque (eds.): *Lake Chad.* Ecology and Productivity of a Shallow Tropical Ecosystem. 1983 ISBN 90-6193-106-1
54. S. Horie (ed.): *Lake Biwa* [Japan]. 1984 ISBN 90-6193-095-2
55. D. R. Stoddart (ed.): *Biogeography and Ecology of the Seychelles Islands.* 1984 ISBN 90-6193-107-X
56. H. Sioli (ed.): *The Amazon.* Limnology and Landscape Ecology of a Mighty Tropical River and Its Basin. 1984 ISBN 90-6193-108-8

MONOGRAPHIAE BIOLOGICAE

57. C. H. Fernando (ed.): *Ecology and Biogeography in Sri Lanka.* 1984 ISBN 90-6193-109-6
58. S. J. Casper (ed.): *Lake Stechlin* [Germany]. A Temperate Oligotrophic Lake. 1985 ISBN 90-6193-512-1
59. U. Th. Hammer: *Saline Lake Ecosystems of the World.* 1985 ISBN 90-6193-535-0
60. B. R. Davies and K. F. Walker (eds.): *The Ecology of River Systems.* 1986 out of print
61. P. de Deckker and W. D. Williams (eds.): *Limnology in Australia.* 1986 ISBN 90-6193-578-4
62. Y. Yom-Tov and E. Tchernov (eds.): *The Zoogeography of Israel.* The Distribution and Abundance at a Zoogeographical Crossroad. 1988 ISBN 90-6193-650-0
63. F. D. Por: *The Legacy of Tethys.* An Aquatic Biogeography of the Levant. 1989 ISBN 0-7923-0189-7
64. B. R. Allanson, R. C. Hart, J. H. O'Keeffe and R. D. Robarts: *Inland Waters of Southern Africa.* An Ecological Perspective. 1990 ISBN 0-7923-0266-4
65. F. di Castri, A. J. Hansen and M. Debussche (eds.): *Biological Invasions in Europe and the Mediterranean Basin.* 1990 ISBN 0-7923-0411-X
66. R. W. Edwards, A. S. Gee and J. H. Stoner (eds.): *Acid Waters in Wales.* 1990 ISBN 0-7923-0493-4
67. S. K. Jain and L. W. Botsford (eds.): *Applied Populations Biology.* 1992 ISBN 0-7923-1425-5
68. C. Dejoux and A. Iltis (eds.): *Lake Titicaca* [Peru/Bolivia]. A Synthesis of Limnological Knowledge. 1992 ISBN 0-7923-1663-0
69. R.B. Wood and R.V. Smith (eds.): *Lough Neagh.* The Ecology of a Multipurpose Water Resource. 1993 ISBN 0-7923-2112-X
70. P.E. Ouboter (ed.): *The Freshwater Ecosystems of Suriname.* 1993 ISBN 0-7923-2408-0

KLUWER ACADEMIC PUBLISHERS – DORDRECHT / BOSTON / LONDON

Batch number: 10761697

Printed by Printforce, the Netherlands